BIOCHEMISTRY RESEARCH TRENDS

HYDROGEN PEROXIDE

DETECTION, APPLICATIONS AND HEALTH IMPLICATIONS

Biochemistry Research Trends

Additional books in this series can be found on Nova's website
under the Series tab.

Additional E-books in this series can be found on Nova's website
under the E-book tab.

Chemical Engineering Methods and Technology

Additional books in this series can be found on Nova's website
under the Series tab.

Additional E-books in this series can be found on Nova's website
under the E-book tab.

BIOCHEMISTRY RESEARCH TRENDS

HYDROGEN PEROXIDE

DETECTION, APPLICATIONS AND HEALTH IMPLICATIONS

GILBERTO AGUILAR
AND
RAPHAEL A. GUZMAN
EDITORS

nova publishers
New York

For permission to use material from this book please contact us:
Telephone 631-231-7269; Fax 631-231-8175
Web Site: http://www.novapublishers.com

LIBRARY OF CONGRESS CATALOGING-IN-PUBLICATION DATA

ISBN: 978-1-62257-414-8

Library of Congress Control Number: 2012942449

Published by Nova Science Publishers, Inc. † New York

CONTENTS

PREFACE

Hydrogen peroxide (H2O2) is a highly reactive compound produced by cells after secondary reactions as a result of the leakage of electrons from the electron transport chain in mitochondria or as a consequence of other enzymatic reactions. This book examines the detection, applications and health implications of hydrogen peroxide. Topics discussed include the ambient conditions of hydrogen peroxide advanced oxidation processes for wastewater treatment; the electrochemical approach to quantify cellular hydrogen peroxide and monitor its release process from living cells; H2O2 signalling and role in inflammation; H2O2 as a useful oxidative stressor to investigate various intracellular events under pathological conditions; uses of hydrogen peroxide in aquaculture; H2O2 in regulating fungal viability and pathogenicity; massive production of H2O2 upon ATP synthase activiation; and H2O2 based green oxidation reactions.

Chapter 1 - Hydrogen peroxide together to chlorine, chlorine dioxide, permanganate and ozone are one of the important oxidants used in water treatment. Although kinetics of hydrogen peroxide reactions with water pollutants, especially organic pollutants, is relatively low, hydrogen peroxide presents a high capacity to be combined with other oxidants and agents (catalysts or radiation) to initiate radical chain mechanisms that lead to the formation of hydroxyl radicals. These processes, named as advanced oxidation processes, are the main route to total elimination of pollutants that otherwise will remain in water or in another phase. Advanced oxidation processes involving hydrogen peroxide can be classified as ambient conditions (pressure and temperature) or wet oxidation conditions (high pressure and temperature) processes. This chapter will deal with the ambient conditions hydrogen peroxide advanced oxidation processes (H_2O_2AOP) which are suitable for both surface water or low concentrated wastewater treatment. H_2O_2AOP can also be classified as those where the oxidant is generated in situ (single ozonation, catalytic oxidation, etc) and other where hydrogen peroxide is added to the system (Fenton, ozone-peroxide, hydrogen peroxide photolysis, etc). Main mechanisms of reactions, kinetics and review of recent literature on these H_2O_2AOP for the treatment of water contaminants will be presented in the chapter.

Chapter 2 - Hydrogen peroxide (H_2O_2) is one significant representative of reactive oxygen species (ROS), it can express diverse physiological and pathological consequences within living systems in a concentration-dependent manner. At its low concentration, H_2O_2 functions as the second messenger for several growth factors, cytokines, and signal transduction. Whereas high concentration of H_2O_2 is a source of oxidative damage to cellular equilibrium leading to aging, Alzheimer's and related neurodegenerative diseases, cardiovascular disorders, even cancer. This chapter introduces the methods to detect cellular H_2O_2, including colorimetry, electrochemical methods, chemiluminescence, fluorescence, and electron spin resonance, etc., and mainly describes the electrochemical approach to quantify the level of H_2O_2 in various cell lines and monitor the release process of H_2O_2 from the living cells. Electrochemical detection approaches depicted in this chapter are based on electrochemical reduction of H_2O_2 catalyzed by enzymes (including proteins) and nanomaterials. Horseradish peroxidase (HRP), and hemoglobin (Hb) were selected as model enzyme and protein, nitrogen-doped graphene (N-graphene) was chosen as a representative of nanomaterials to catalyze H_2O_2. The preparation, immoblization, characterization, and application of the above mentioned catalysts to detect cellular H_2O_2, to study kinetics on the exocytotic process, and to evaluate the whole efficiency of the process in real-time are introduced. These results presented in this chapter open a new way for quantification cellular H_2O_2 level and also develop a new platform for a reliable collection of kinetic information on cellular H_2O_2 release, which could be potentially useful in study of downstream biological effects in physiology and pathology.

Chapter 3 - Hydrogen peroxide (H_2O_2) is a highly reactive compound produced by cells after secondary reactions as result of the leakage of electrons from the electron transport chain in mitochondria or as a consequence of other enzymatic reactions.

H_2O_2 is able to oxidize and thereby to induce damage in numerous cellular constituents. Over the last decade it has become accepted that H_2O_2 serves important biological roles outside of its capacity to cause oxidative damage. The main factor which determines H_2O_2 effects is its own level associated to the activation of different redox-sensitive transcription factors and to distinct biological responses. Once accumulates in cells, H_2O_2 can, for example, modify proteins and the underlining mechanisms involved attracting great interest at the moment. One of the main proteins target for H_2O_2 is the thiol group of the amino acid cysteine in protein regulating many cell responses (e.g. phosphatases and transcription factors).

Inflammation is a crucial function of the innate immune system that protects host tissues against dangerous insults that are detrimental to tissue homeostasis, including wound damage and pathogen invasion. During the course of inflammatory processes, H_2O_2 may be generated in large amounts. This occurrence may be due to the activation of mast cells, macrophages, eosinophils, and neutrophils, which generate superoxide radical ($O_2\cdot^-$), predominantly via NADPH oxidase (NOX). Superoxide dismutase (SOD) rapidly converts $O_2\cdot^-$ into H_2O_2 which can easily penetrate the membranes of surrounding cells. There is no known receptor for H_2O_2, but potentially H_2O_2 gradients might be "read" via intracellular sensors such as phosphatases that subsequently regulate the chemotactic response. Alternatively, there may be other extracellular (or intracellular) substrates that H_2O_2 modifies in some way to convert them into chemotactic factors or other modulators of cell motility. Several observations suggest that H_2O_2 acts as a common second messenger following cellular exposure to agents that induce inflammatory response. Interactions between H_2O_2 and inflammatory mediator (e.g. NF-κB) might be a component of the intracellular signalling process that leads to inflammation. Treatment with H_2O_2 directly activates NF-κB in some cells, and over expression of SOD, enhances the tumor necrosis factor-α induced activation of NF-κB. In this context H_2O_2 represent a messenger able to alter redox homeostasis contributing, at several levels, to inflammatory-related disease. Despite H_2O_2 is not an inherently reactive compound it can be also converted into highly reactive and deleterious products killing cells. This is a non-specific process direct to pathogens but it can induce damage also to the human body.

Here the authors describe the role of H_2O_2 in cellular signalling focusing the authors attention on H_2O_2 role in inflammation. Moreover the authors also examine the H_2O_2 signalling in some cardiovascular, renal and pulmonary inflammatory-based disease.

Chapter 4 - Hydrogen peroxide (H_2O_2) has gained much attention not only as a stressor that oxidizes components of the cell but also as signaling molecule in physiological processes. On the minus side, elevation of reactive oxygen species (ROS) including H_2O_2 causes oxidative stress, and is reported to be involved in many diseases, such as diabetes, atherosclerosis, Parkinson's disease, and Alzheimer's disease. On the plus side, H_2O_2 is produced when cells respond to various stimuli, and mediates downstream signals. Recently, the authors found that H_2O_2 treatment depleted HeLa cell endosomes of phosphatidylinositol-3-phosphate (PI3P). To quantify and visualize the intracellular localization of PI3P they have developed a novel analytical method that employs a semi-intact cell system, a permeabilized cell that enables the addition of various exogenous molecules into the cells, and protein probe for PI3P, GST-2xFYVE protein. Using this method, the authors found that H_2O_2 decreased the amount of PI3P in endosomes in a manner dependent upon p38 MAPK, a kinase that is activated during oxidative stress or pathogenic conditions. In addition, H_2O_2 treatment delayed vesicular transport in a broad range of endocytic pathways: the lysosomal degradation of EGFR, the retrograde transport of cholera toxin from the plasma membrane to the *trans*-Golgi network (TGN), and the uptake and recycling of transferrin.

The physiological role of PI3P in endosomes has received a lot of attention with respect to endocytosis and signal transduction via endosomes. For example, PI3P in the endosomal membrane appears to act as a microdomain for the recruitment of various proteins that contain a PI3P-binding (FYVE) domain. Interestingly, FYVE-containing proteins are involved not only in the regulation of endosomal processing but also in signal transduction via endosomes.

In this chapter, the authors demonstrate the effect of H_2O_2 on early endocytic pathways and discuss the utility of H_2O_2 as a useful oxidative stressor to investigate various intracellular events under pathological conditions.

Chapter 5 - In this chapter the authors revise the uses of hydrogen peroxide in aquaculture. At the beginning, hydrogen peroxide was used mainly as a treatment against ectoparasites of adult fish, such as sea lice (*Lepeophtherius salmonis*), or against the microbiota of fish eggs, mainly fungi in freshwater species.

Later, it became of interest for microbial control purposes rather than chlorine or antibiotics due to its relatively lower toxicity. Although it is one of the most powerful known oxidizers, it is easily decomposed in a water solution, particularly in the presence of organic matter, light and turbulence (all conditions easily met in aquaculture systems), leaving no toxic products. The spectrum of species and live stages has also increased; hydrogen peroxide has been tested on rotifers, crustacea, bivalves and fish, and on eggs, larvae, juveniles and adults.

Recently, issues concerning its impact on the aquaculture system (i.e. biofilters of the recirculation system), on the environment and on animal welfare of cultured species have risen.

Chapter 6 - As known as well, the active oxygen species hydrogen peroxide (H_2O_2) has been proven to play critical roles in various cellular activities, including the direct cause of membrane damage, the component of structural defense, the signal molecule operating within the challenged cell and the antimicrobial compound because of H_2O_2 acting as a signal molecule in the induction of systemic acquired resistance.

In addition, H_2O_2 is relatively stable, can diffuse freely throughout the cell, and has a similarly low reactivity toward biological molecules. In this review, the authors mainly summarize the functions of hydrogen peroxide in regulating the viability of antagonistic yeasts and the pathogenicity of fungal pathogens, as well as the molecular targets of hydrogen peroxide in fungal cells.

Chapter 7 - The mitochondrial reactive oxygen species (ROS) production is a crucial issue that has been involved in a variety of physio-pathological processes and aging. However, at the present measurements of mitochondrial ROS production is controversial. Indeed, the available tools are poorly specific and not direct.

Now taking advantage of the progress in microelectrode production it has been possible to approach the specific and direct measurement of O_2 species in a suspension of isolated mitochondria.

Such electrochemical determinations confirm active H_2O_2 release by mitochondria above the scavenging activity and reveal a significant role of ATP synthase activation in the control of mitochondrial ROS production.

These findings offer a new explanation for the effects of mitochondrial respiration on oxidative stress. Regardless the damaging effect of ROS that has been already associated to intense mitochondrial work during exercise in muscle for example, H_2O_2 spikes may represent specific cellular signal. The crosstalk between the mitochondrion and the cell is indeed an emergent issue. About this argument the authors' observation discloses a novel basic mechanism of mitochondrial ROS production and highlights the involvement of intracellular redox balance on health.

Chapter 8 - Hydrogen peroxide is one of the most important reactive oxygen species and it has gained an enormous attention due to its high permeability within cell and across cellular membranes, as an in vivo precursor of strong oxidants (namely hydroxyl radical and hypochlorous acid) and due to its implication in the generation of superoxide radical anion. Therefore, the development of reliable analytical methods to determine the scavenging capacity against hydrogen peroxide has been a topic of increased interest within pharmaceutical and biomedical area. Currently, several methods are available for performing the determination of scavenging capacity against hydrogen peroxide using UV–vis spectrophotometry, fluorescence and chemiluminescence as detection systems. In this chapter, a critical review of these methods will be undertaken, by comparing the figures of merit and limitations provided by some of the most applied methods (measurement of intrinsic absorption of H_2O_2 in UV region, fluorimetric assays based in scopoletin and homovanillic acid, chemiluminescence quenching of luminol and lucigenin, among others). Moreover, special emphasis will be given to methods developed in the last three years, based on automatic systems and nanotechnology breakthroughs.

Chapter 9 - This chapter summarizes the authors' recent developments for H_2O_2 based green oxidation reactions, using two different transition metal catalysts: palladium and cobalt. Results obtained from the oxidation of the natural olefins using as oxidant the hydrogen peroxide and as catalysts palladium and cobalt chlorides are presented and discussed in mechanistic terms. Monoterpenes are abundant and renewable raw materials and when converted into oxygenates products are valuable ingredient for fragrance, agrochemicals and pharmaceutical industries. Camphene, β-pinene and 3-carene are monoterpenes of high natural occurrence, which oxygenated derivates, are highly attractive for fine chemical and fragrances industry and for these reasons were the selected substrates. The different aspects involved in the oxidation of these substrates in presence of two different catalysts were assessed. Palladium(II)-catalyzed oxidation reactions proceeds via two electrons intermediates (heterolytic oxidation); conversely, Cobalt(II)-catalyzed oxidation reactions occurs via one electrons intermediates (homolytic oxidation). In all most of the catalytic reactions, high conversion substrates were achieved. Moreover, the selectivity of the oxidation products can be controlled by molar ratio substrate:oxidant and by temperature of reaction. Hydrogen peroxide, an environmentally benign reactant, regardless catalyst employed was an efficient oxidant in the conversion of monoterpenes into valuable oxygenates derivates.

In: Hydrogen Peroxide
Editors: Gilberto Aguilar and Raphael A. Guzman
ISBN: 978-1-62257-414-8

Chapter 1

THE ROLE OF HYDROGEN PEROXIDE ON ADVANCED OXIDATION PROCESSES OF WATER CONTAMINANTS AT AMBIENT CONDITIONS

Fernando J. Beltrán, Almudena Aguinaco and Ana Rey
Departamento de Ingeniería Química y Química Física,
Universidad de Extremadura, Badajoz, Spain

ABSTRACT

Hydrogen peroxide together to chlorine, chlorine dioxide, permanganate and ozone are one of the important oxidants used in water treatment. Although kinetics of hydrogen peroxide reactions with water pollutants, especially organic pollutants, is relatively low, hydrogen peroxide presents a high capacity to be combined with other oxidants and agents (catalysts or radiation) to initiate radical chain mechanisms that lead to the formation of hydroxyl radicals. These processes, named as advanced oxidation processes, are the main route to total elimination of pollutants that otherwise will remain in water or in another phase. Advanced oxidation processes involving hydrogen peroxide can be classified as ambient conditions (pressure and temperature) or wet oxidation conditions (high pressure and temperature) processes. This chapter will deal with the ambient conditions hydrogen peroxide advanced oxidation processes (H_2O_2AOP) which are suitable for both surface water or low concentrated wastewater treatment. H_2O_2AOP can also be classified as those where the oxidant is generated in situ (single ozonation, catalytic oxidation, etc) and other where hydrogen peroxide is added to the system (Fenton, ozone-peroxide, hydrogen peroxide photolysis, etc). Main mechanisms of reactions, kinetics and review of recent literature on these H_2O_2AOP for the treatment of water contaminants will be presented in the chapter.

1. INTRODUCTION

Hydrogen peroxide was discovered in 1818 by Jacques Thenard, a French scientist. It has a weak acid character according to its equilibrium in water:

$$H_2O_2 \xrightleftharpoons{pK=11.8} HO_2^- + H^+ \tag{1}$$

Then, hydrogen peroxide is in the molecular form when water is at neutral pH. Also, hydrogen peroxide presents an amphoteric character since it can act as oxidant and reductor in water processes according to the reactions:

$$H_2O_2 + 2H^+ + 2e \rightleftharpoons 2H_2O \quad E^o = 1{,}78\ V \tag{2}$$

and

$$H_2O_2 \rightleftharpoons O_2 + 2H^+ + 2e \quad E^o = 0{,}7\ V \tag{3}$$

Its redox potential according to Nerst relation is dependent on pH:

$$E = E^o - 0.06pH + 0.03\log C_{H2O2} \tag{4}$$

This means that hydrogen peroxide oxidant character increases with the decrease of pH. However, in spite of these figures hydrogen peroxide direct action on contaminants is rather limited because of the low reaction rates. Thus, only with certain inorganic compounds such as sulfides, it presents relatively high reaction rates (Doré, 1989). Most of possible hydrogen peroxide applications in water treatment rely on its high capacity of producing hydroxyl radicals when combined or produced in different oxidation processes as shown in this chapter. These processes, named as advanced oxidation processes, are the main route to total elimination of pollutants that otherwise will remain in water or in another phase.

Advanced oxidation processes involving hydrogen peroxide can be classified as ambient conditions (pressure and temperature) or wet oxidation conditions (high pressure and temperature) processes. This chapter will mostly deal with the ambient conditions hydrogen peroxide advanced oxidation processes which are suitable for both surface water and low concentrated wastewater treatment. H_2O_2AOP can also be classified as those where the oxidant is generated in situ (single ozonation, catalytic oxidation, etc) and others where hydrogen peroxide is added to the system (Fenton, ozone-peroxide, hydrogen peroxide photolysis, etc). Main mechanisms of reactions, kinetics and review of recent literature on these H_2O_2AOP for the treatment of water contaminants are presented in the chapter.

2. *In Situ* Generated Hydrogen Peroxide Oxidation Processes

Hydrogen peroxide is usually one of the products generated in water oxidation processes where it can act as initiator or promoter of hydroxyl free radicals. In this sense, the generated hydrogen peroxide will power the development of an advanced oxidation process (Glaze et al., 1987). This is the case when ozone or photocatalytic oxidation is used to give an example. However, in other cases, hydrogen peroxide can be formed from the catalytic decomposition of some organic substance such as formic acid (Yalfani et al., 2009; 2010) and then combined with some other agent to produce free radicals. Below, main features of some of these processes are presented.

2.1. *In Situ* Generated Hydrogen Peroxide in the Ozone Process

Ozone presents two reaction mechanism against the matter dissolved in water. (Beltrán, 2004; Langlais et al., 1991). On one hand, it directly reacts with compounds presenting certain functional groups in their molecules (unsaturated compounds or groups that activate electrophilic substitution reactions) (Rakowski and Cherneva, 1990; Mvula and von Sonntag, 2003; Leitzke and von Sonntag, 2009), and, on the other hand, and according to experimental conditions, ozone decomposes in hydroxyl free radicals which subsequently react with most of compounds in a non selective way (see Figure 1).

These mechanisms are the direct and indirect ways of ozone reactions in water. Since numerous contaminants do not directly react with ozone or they do with low reaction rates, the indirect way of ozone action develops and formation of hydroxyl radicals is a crucial step. These free radicals not only have a high redox potential (higher than ozone itself) but also the rate constants of their reactions with many contaminants are around 10^9-10^{10} $M^{-1}s^{-1}$ (Buxton et al., 1988). It is well known that when ozone is applied in water hydrogen peroxide is formed (Bühler et al., 1984; Staehelin et al., 1984). This reaction is particularly important when unsaturated compounds are present in water. Thus, as shown in Figure 2 ozone breaking of aromatic rings (for example, in phenols) or carbon double bond breaking through Criegge mechanism (Criegge, 1953; 1957) leads to saturated compounds and formation of hydrogen peroxide. The ionic form of hydrogen peroxide (the hydroperoxide ion, see reaction (1)) reacts fast with ozone to initiate a free radical mechanism eventually leading to the appearance of hydroxyl radicals (Staehelin y Hoigné, 1982):

$$O_3 + HO_2^- \xrightarrow{k\,=2.8x10^6\,M^{-1}s^{-1}} O_3^{\bullet-} + HO_2^{\bullet} \qquad (5)$$

In Figure 3, as example, the time evolution of the concentration of hydrogen peroxide formed when an aqueous solution of two pharmaceuticals (sulfamethoxazole and acetaminophen) is ozonated is shown.

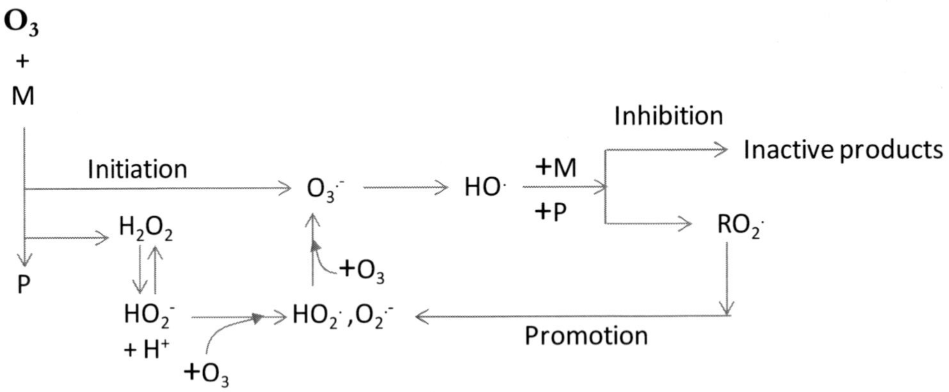

Figure 1. Mechanisms of ozone action on contaminants (M) of water.

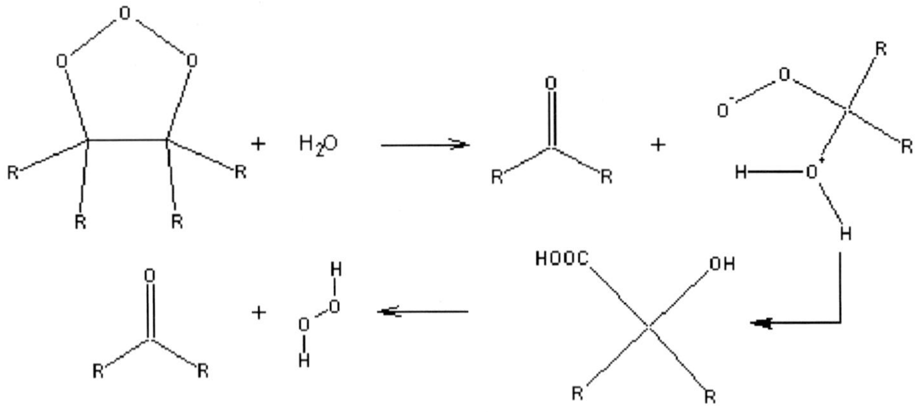

Figure 2. Formation of hydrogen peroxide from direct ozone reactions.

Also, there are other studies (Beltrán et al., 2008; García-Araya et al., 2010; Beltrán et al., 2009a; Beltrán et al., 2010; Beltrán et al., 2012) where the trend shown in Figure 3 is kept when different compounds are ozonated. The generated hydrogen peroxide allows the formation of hydroxyl free radicals from reaction (5).

It should be noted that, according to equilibrium (2), at neutral pH, very low concentrations of the hydroperoxide ion will be present. For example, at pH 7 and $5x10^{-5}$ M of total hydrogen peroxide concentration, the ionic form is about $5x10^{-8}$ M but given the high rate constant of reaction (5) the formation rate of free radicals will be about $1.5x10^{-6}$ Ms^{-1} (for an ozone concentration of 10^{-5} M), which is a relatively high reaction rate.

Obviously, some higher reaction rates are needed that is why the addition of hydrogen peroxide to an ozone system constitutes an important advanced oxidation system (see later).

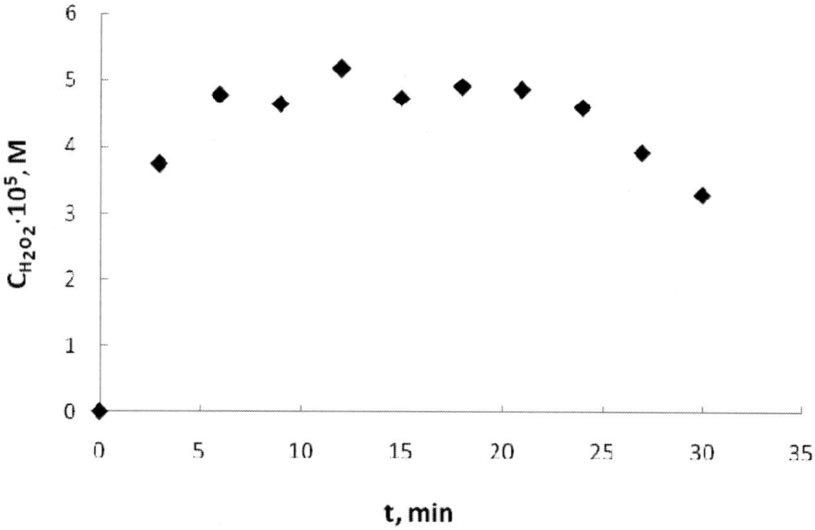

Figure 3. Time evolution of hydrogen peroxide concentration formed during the ozonation of two pharmaceutical compounds and their first intermediates in ultrapure water. Conditions: temperature: 20ºC, pH 7, gas flow rate: 20 $L \cdot h^{-1}$, inlet ozone gas concentration: 10 $mg \cdot L^{-1}$, initial sulfamethoxazole concentration: 13.5 $mg \cdot L^{-1}$, initial acetaminophen concentration 16.5 $mg \cdot L^{-1}$.

2.2. *In Situ* Generated Hydrogen Peroxide in an Activated Carbon Catalytic Ozonation Process

The ozone-activated carbon system is another example of ozonation process, this time a catalytic ozonation one, where hydrogen peroxide is formed in a greater concentration than in the ozone alone process. Thus, it has also been reported that ozone decomposes in water faster when activated carbon is present. In this reaction, hydroxyl free radicals are formed (Dusenbury y Cannon, 1996). Jans y Hoigné (1998) were among the first who reported an extensive study of the interactions between ozone and activated carbon.

These authors attributed to activated carbon a role of chain mechanism initiator but they did not clearly explain the nature of this reaction. Álvarez et al. (2005), later related some of these initiating reactions with the presence of hydrogen peroxide formed when ozone reacts with some surface oxygen groups of basic nature of the activated carbon surface such as pyrone or chromene groups (Álvarez et al., 2006). Thus, functional groups on the activated carbon surface fix the adsorbent properties regarding possible adsorbate compounds (Daifullah and Girgis, 1998).

Among surface groups, the so called oxygen complexes, are of acidic or basic nature (Vidic et al., 1997). Among the acidic ones, carboxylic, lactonic, phenolic or carbonylic groups can be highlighted, while among those of basic nature there are structures of pyrone, benzopyrone (chromene) and quinone types (Contescu et al., 1998; Menéndez et al., 1999). With the exception of carboxylic and lactonic groups, the rest of them are reported to react with ozone in water or in protic solvents (Bailey, 1978).

Another important fact to highlight is the formation of hydrogen peroxide in these ozone surface reactions. Then, the ozone-activated carbon is another system that generates hydrogen peroxide and initiates another more powerful advanced oxidation process: the ozone-peroxide process.

Besides, hydrogen peroxide formed not only reacts with ozone but also it decomposes on the carbon surface (Aguinaco et al., 2011, Lucking et al., 1998; Huang et al., 2003; Khalil et al., 2001; Rey et al., 2008, 2011). Thus, in addition of being a good support phase activated carbon has been applied as catalyst of hydrogen peroxide decomposition (Bansal et al., 1988).

In order to increase the efficiency of hydrogen peroxide decomposition on activated carbon, a series of factors such as concentration, temperature, carbon surface nature, carbon surface pH, water pH and surface treatment, have been studied (Bansal et al., 1988).

In these works, decomposition of hydrogen peroxide was assumed to be due to a primary interchange between an oxygen surface group and the hydroperoxide ion itself (Bansal et al., 1988). The absorbed peroxide undergoes a potential oxidation, decomposing in another oxygen molecule and regenerating the carbon surface.

Hydrogen peroxide decomposition is thought to be developed through its own dissociation as weak acid and because the anion HO_2^- is less stable than H_2O_2. Thus, in basic water decomposition is favored.

Other recent works have provided some evidences that activated carbon can promote hydrogen peroxide decomposition through the formation of hydroxyl radicals (Oliveira et al., 2004; Georgi and Kopinke, 2005; Santos et al., 2009).

These considerations have to be dealt with both for the used carbon surface and water conditions. The presence of basic and acid groups on the carbon surface plays a fundamental role on hydrogen peroxide decomposition. Thus, with carbons of high basic surface groups, conveniently distributed, the yield of decomposition increases. However, when a dense distribution is taking place an inhibiting effect could develop (Bansal et al., 1988). When the catalytic reaction is developed in buffered water, catalytic activity depends on the specific surface area (Bansal et al., 1988; Hassler et al., 1967).

Table 1 gives some recent examples of ozone-activated carbon systems to generate hydrogen peroxide to improve the removal of some compounds from water.

2.3. *In Situ* Generated Hydrogen Peroxide in the Ozone Photolysis Process

The combination of ultraviolet (UV) radiation with ozone may be a more effective advanced oxidation technique than using ozone alone for certain target materials due to the formation of additional H_2O_2 and $HO^\bullet$ radical generation via photolysis (Peyton and Glaze, 1988).

$$O_3 + H_2O \xrightarrow{h\nu} H_2O_2 + O_2 \tag{6}$$

$$H_2O_2 \xrightarrow{h\nu} 2HO^\bullet \tag{7}$$

However, the photolysis of hydrogen peroxide to produce two hydroxyl radicals is rather slow because the molar extinction coefficient of hydrogen peroxide is much lower than the one of ozone (see Figure 4). For example, at 254 nm it is 19.6 $M^{-1}cm^{-1}$ and 3300 $M^{-1}cm^{-1}$, respectively (Glaze et al., 1987).

Besides, a fraction of hydrogen peroxide is dissociated into HO_2^- ($pK_a = 11.8$) (see equation (1)) and the ionic form of hydrogen peroxide reacts with further ozone by producing the ozonide ion radical, $O_3^{\bullet -}$, and it therefore acts as a further chain carrier (Staehelin y Hoigné, 1982).

The photolytic ozonation (UV/O_3) process is considered as an advanced oxidation process that applies ozonation, UV irradiation and action of HO· radicals simultaneously to the target compounds, and converts these compounds completely to simple inorganic species, such as carbonate and water, if the reaction is taking long enough for completion (See Table 2 for examples of this system). Photolytic ozonation thus has the potential for being an environmentally clean process for treating organic contaminants and has attracted attention at present for a wide range of environmental applications. However, the initial and operating costs of photolytic ozonation are still relatively expensive comparing to traditional treatment technologies. Therefore, the determination of optimum operational conditions is one of the primary concerns for the future development and potential application of photolytic ozonation process (Ku and Lin, 2002).

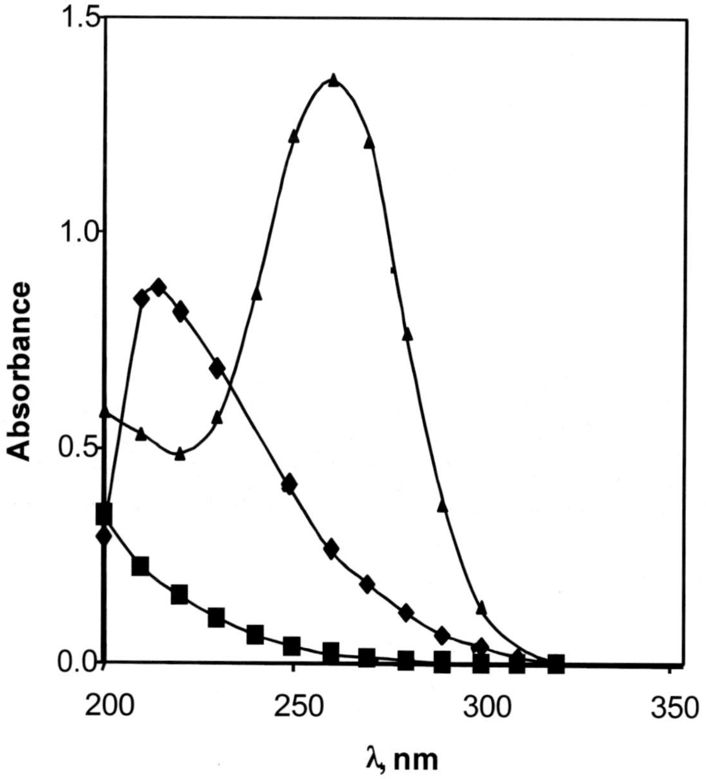

Figure 4. Absorption spectrum of H_2O_2 (■), O_3 (▲) and HO_2^- (♦). Conditions: H_2O_2 concentration: $2.2x10^{-3}$ M, O_3 concentration: $4.4x10^{-4}$ M and HO_2^- concentration: $2.2x10^{-3}$ M.

Table 1. Recent examples of works on ozone-activated carbon system to remove compounds from water

Compound and Process	Reactor, conditions	Main features	Reference
Polyphenol compounds (gallic acid, tyrosol, and syringic acid). Ozone and Ozone/Activated Carbon	750 cm^3 semibatch reactor. pH: 2.5 and 5. Carbon concentration: 10 g/L. Ozone gas concentration: 40 mg L^{-1}	AC ozonation processes significantly improve both polyphenol conversion and mineralization. Hydrogen peroxide formed during the process seems to play an important role in accelerating the oxidation rate.	Giráldez et al. (2007)
Real wastewater from chemical industry. Ozone and Ozone/Activated Carbon	Column reactor (65 mm in i.d. and 500 mm in length). (Batch mode). Carbon dosage: 30 g. Continuous test runs: column (i.d.: 65 mm; *h*: 500 mm) packed with about 280 g of the GAC adsorbent (15 cm). pH: 4.4, 6, 8 9, 10 and 12.	The integrated process enhanced COD removal, increased ozone utilization and avoided frequent GAC regeneration. Two mechanisms: adsorption–ozonation and catalytic oxidation. The pH value affects the reaction mechanism.	Lei et al. (2007)
Nitroimidazoles Ozone and Ozone/Activated Carbon	2-L reactor (batch mode). pH: 2, 4, 6, 7, 9 and 10. Carbons: Two commercial carbons from Sorbo (S) and Merck (M) and a third carbon prepared by chemical activation of petroleum coke. Carbon dose: 0.25–0.50 g L^{-1}	kO_3 values <350 M^{-1} s^{-1} k_{HO} values of around 10^{10} M^{-1} s^{-1} Presence of activated carbon produces an increase in the removal rate, a reduction in the toxicity of oxidation by-products, and a reduction in the concentration of dissolved organic matter. The system O_3/activated carbon also shows a great efficiency when using real waters (surface, subterranean and wastewater)	Sánchez-Polo et al. (2008)
Diclofenac Ozone and Ozone/Activated Carbon	Tubular glass reactor (300 mm-long and 50 mm-diameter) pH 7 Carbons: P110 Hydraffin, Darco 12–20, and GMI2000. 5 g of activated carbón.Ozone concentration of 20 mg L^{-1}	Hydrogen peroxide has been detected. TOC removal is mainly due to the catalytic or promoting action of activated carbon on the ozone decomposition to yield hydrogen peroxide and then hydroxyl radicals.	Beltrán et al. (2009b)
17α-ethynylestradiol. Ozone and Ozone/Activated Carbon	Tubular glass reactor (300 mm long, 50 mm diameter) pH: 2, 5, 7 and 9. Carbons: GMI2000, P110 Hydraffin, 12–20 Darco. Carbon concentration: 20 g/L.	The ozone/AC system led not only to higher TOC removal but also to lower ozone consumption. AC ozonation of 17α-ethynylestradiol presents the lowest ecotoxicity level among the different systems applied.	Beltrán et al. (2009c)

Compound and Process	Reactor, conditions	Main features	Reference
Tinidazole (TNZ) Ozone and Ozone/Activated Carbon	2-L reactor (batch mode). Carbon dose range: 0.25–0.5 g L^{-1}pH 7 Dynamic ozonation experiments: 7-cm height column of 2 g granulated activated carbon. $(O_3)_o = 1.04 \times 10^{-4}$ M A commercial carbon from Merck	Presence of activated carbon increases TNZ removal rate, reduces oxidation by-product toxicity, and reduces the concentration of dissolved organic matter.	Rivera-Utrilla et al. (2010)
Diethyl phthalate (DEP) Ozone and Ozone/Activated Carbon	1 L batch reactor pH 2.5, 3.5, 5.6, 6.2 and 7.2. Carbon: L27, X17, F22 and S21 (from Pica) Carbon dose: 2 g. Ozone concentration 50 g Nm^{-3}	Presence of AC enhance DEP degradation kinetics High AC external and microporous surface favour the O_3/AC coupling efficiency. Pseudo-first order kinetic model.	Ferreira de Oliveira et al. (2011)
Atrazine Ozone and Ozone/Activated Carbon	800 mL laboratory-scale slurry reactor (batch mode). pH: 3, 5 and 7. Carbon: Norit. Catalysts dose range of 0–1 g L^{-1} Ozone concentration: 5 mg L^{-1}	AC accelerates ozone decomposition leading to the production of hydroxyl radicals. Variables: presence of radical scavenger, activated carbon dose, and solution pH. Reaction rate constant increases linearly with mass of catalyst.	Guzman-Perez et al. (2011)

Table 2. Recent examples of works on ozone-radiation system to remove compounds from water

Compound and Process	Reactor, conditions	Main features	Reference
Butylated hydroxyanisole (BHA). O_3, UV and O_3/UV	1 L (95 mm ID×191 mm H) quartz beaker. pH: 3, 5, 7, 9 and 11. Low-pressure mercury lamps (253.7 nm). Light intensity: 1.5×10^{-6} Einstein L^{-1} s^{-1}	UV/O_3 process has exhibited the best overall performance. It shown advantages not only to the elimination of the separable solids formed but also achieving 90% mineralization at the end of the process (180 min). Intermediates identified. Decay pathways proposed.	Lau et al. (2007)
Dissolved organic matter in the biotreated effluent of a municipal wastewater treatment plant. O_3 and O_3/UV	Glass reactor with a height of 300 mm and an inner diameter of 74 mm. pH: 7.06-7.80. 25 W low-pressure mercury UV lamp (254 nm). Light intensity of 4700 μW/cm^2	UV/O_3 was more effective for demineralization, achieving a DOC reduction of 90% vs. 36% by ozonation alone. The UV/O_3 process was less selective than O_3 and therefore effective for removing DOM. The effectiveness of oxidation was dependent on polarity of the specific organic constituents.	Gong et al. (2008)

Table 2. (Continued)

Compound and Process	Reactor, conditions	Main features	Reference
Linuron O_3, UV and O_3/UV	800 mL (97.8 mm ID × 125 mm H) quartz beaker. pH: 3.3, 6.1, 9.1, 10, 10.7 and 11. Low-pressure mercury lamps (253.7 nm. Intensity of 1.5×10^{-6} Einstein L^{-1} s^{-1}	The decay rate of linuron by UV/O_3 process was found to be around 3.5 times and 2.5 times faster than sole-UV and ozone-alone, respectively. The overall rate constants increase exponentially with pH above 9.0. Predominating parameters involved in the three processes were determined.	Rao and Chu (2009)
4-chloro-3,5-dimethylphenol (PCMX). O_3 and O_3/UV.	A cylindrical Pyrex glass reactor (diameter 80 mm, height 260 mm). pH: 2, 4, 6, 8 and 11. 175 W high-pressure mercury lamp (250-450 nm) with a peak at 365 nm. Irradiation intensity (at 365 nm): 24 mW cm^{-2}	The UV/O_3 experimental results showed a pronounced synergistic effect, leading to the nearly complete elimination of total organic carbon (TOC) within 75 min. The efficiency of PCMX mineralization increased with an increase in ozone dose up to 3.1 g h^{-1}. The optimal pH value was 4.0, and the preferred bulk temperature was 20 °C . Intemediates identified. Degradation mechanism proposed.	Song et al. (2009)
Linuron O_3, UV and O_3/UV	800 mL (97.8 mm ID × 125 mm H) quartz beaker. pH: 3 and 7. Low-pressure mercury lamps (253.7 nm. Intensity of 1.5×10^{-6} Einstein L^{-1} s^{-1}	UV/O_3 process has exhibited the best overall performance achieving 86% mineralization in 120 min. New intermediates identified. The effect of various anions on the performance of ozonation was examined.	Rao and Chu (2010)
N-Nitrosodiethylamine (NDEA). UV and O_3/UV	A cylindrical quartz photoreactor (700 mL) pH: 6. 8 W low-pressure Hg lamp (253.7 nm). light quantum flow rate: 3.3×10^{-6} Einstein L^{-1} s^{-1}	The results showed that UV/O_3 process had many advantages on NDEA removal in water compared with sole UV irradiation: (1) it was a fast, effective and feasible method for water treatment; (2) it would not be influenced by solution pH; (3) it was suitable for reducing the yields of diethylamine and NO_2^- and inhibiting the NDEA regeneration. Products degradation identified.	Xu et al. (2010)
Mefenamic acid (MEF) O_3 and O_3/UV	3.5 L (50 mm ID × 250 mm H) quartz beaker. pH: 4, 7 and 9. Low-pressure mercury lamp (254 nm).	A decrease of pH could enhance MEF removal by both the ozonation and O_3/UV processes. The O_3/UV process could obviously enhance the formation of nitrate and the reduction of TOC. The degradation of MEF can be successfully illustrated through the parallel rapid and slow reactions model. The rapid reaction can enhance the efficiency and effectiveness of degradation in the O_3/UV process.	Chang et al. (2012)
Polycyclic musk HHCB. O_3, UV and O_3/UV	A photochemical chamber and a reaction UV transparent quartz reaction vessel (3L) (13 cm IDx 30 cm H). 16 installed positions for UV lamps. pH: 5, 7 and 9. Low-pressure mercury lamps. UV intensity of 0.42, 0.88, and 1.71 W L^{-1}	The UV/O_3 process displayed the best performance for TOC removal in the presence of ozone residual. The TOC removal rates after reaction time of 5 min by O_3 and UV/O_3 process are 5.9% and 12.1%, respectively. The effect of ozone dosage and UV intensity on the removal of the target HHCB was investigated. Kinetic study.	Liu et al. (2012)

2.4. *In Situ* Generated Hydrogen Peroxide in Photocatalytic Oxidation

In general, photocatalytic oxidation is also considered as another example of the innovative technologies collectively known as "Advanced Oxidation Processes" that rely on the generation of very reactive hydroxyl radicals (*e.g.*, $^{\bullet}OH$). Those reactive species are subsequently used to degrade the organic pollutants.

The principles of photocatalytic oxidation and its environmental applications have been reviewed extensively since 1985 (*e.g.* Ollis, 1985; Serpone and Pelizzetti, 1989; Legrini et al., 1993; Fox and Dulay, 1993; Herrmann, 1999; Herrmann, 2010; Mills and Lee, 2002; Bahnemann, 2004; Uyguner and Bekbolet, 2007; Fujishima et al., 2008; Choi et al., 2010 and Chong et al., 2010).

Basically, the heterogeneous photocatalytic process relies on the utilization of the near UV radiation to photoexcite the semiconductor catalyst in the presence of oxygen. Under these circumstances oxidizing species either as free or surface bound hydroxyl radical or mobile holes are generated. Many other semiconducting oxides have been tested as catalysts, although TiO_2 in the anatase form performed the most interesting and efficient features such as high stability, good performance, low toxicity and low cost.

The optical absorption of TiO_2 in the near UV region is the major advantage of the photocatalytic method over UV-C driven AOPs (O_3/UVC, H_2O_2/UVC) requiring light of shorter wavelengths (λ<300nm).

As bibliography reports (Turchi and Ollis, 1990) main steps of this AOP are also well known:

*Absorption of light by the semiconductor leads to oxidant positive hole formation on the valence band and electron migration to the conduction band:

$$TiO_2 \underset{k_r}{\overset{\lambda<365nm, k_{TiO2}}{\rightleftarrows}} e^- + h^+ \tag{8}$$

* Adsorption of water on activated points of the semiconductor:

$$TiO_2^* + H_2O \overset{K_2}{\rightleftarrows} TiO_2^* - H_2O \tag{9}$$

* Formation of hydroxyl radicals from positive hole oxidation:

$$TiO_2 - H_2O + h^+ \xrightarrow{k_{16}} TiO_2 - OH^{\bullet} + H^+ \tag{10}$$

* Oxygen electron trapping in the conduction band to avoid electron-hole recombination and process inhibition:

$$TiO_2 + e^- \longrightarrow Ti(III) \tag{11}$$

$$O_2 + Ti(III) \xrightarrow{k_{18}} TiO_2 - O_2^{\bullet -} \tag{12}$$

Furthermore, the superoxide ion radical is a moderate oxidant with reduction potential $\left[E_0(O_2^{\bullet-}/HO_2^{\bullet})=1.7\,V; E_0(O_2^{\bullet-}/H_2O_2)=1.7\,V\right]$, which suffices to oxidize organic compounds (Song et al., 2007).

$$TiO_2-O_2^{\bullet-}+B\xrightarrow{k_{19}}Interm \tag{13}$$

or be recombined to yield hydrogen peroxide:

$$2TiO_2-O_2^{\bullet-}\xrightarrow{k_{20}=5x10^7\,M^{-1}s^{-1}}H_2O_2+2TiO_2 \tag{14}$$

where hydrogen peroxide can also be adsorbed or in solution. As a result, reactions (15), (17) and (18) develop here.

$$HO_2^{\bullet}\xrightleftharpoons{pK=4.8}O_2^{\bullet-}+H^+ \tag{15}$$

$$H_2O_2+HO^{\bullet}\xrightarrow{k_{11}=2.7x10^7\,M^{-1}s^{-1}}HO_2^{\bullet}+H_2O \tag{16}$$

$$HO_2^{-}+HO^{\bullet}\xrightarrow{k_{12}=7.5x10^9\,M^{-1}s^{-1}}HO_2^{\bullet}+OH^- \tag{17}$$

Besides, hydroxyl radicals can be formed from reactions in bulk water, from hydrogen peroxide photolysis following reaction (7) (Baxendale and Wilson, 1957) and also from hydrogen peroxide oxidation of Ti(III) reduced sites:

$$H_2O_2+Ti(III)\xrightarrow{k_{22}}TiO_2-HO^{\bullet}+OH^- \tag{18}$$

As far as the reactions between hydroxyl radical and organic compounds and intermediates are concerned, there are four possible situations or reactions with adsorbed and non-adsorbed hydroxyl radicals (Turchi and Ollis, 1990). Also, direct photolysis of organics has to be considered.

Literature shows some examples of works on photocatalytic oxidation where hydrogen peroxide is formed and followed. Some of these recent works are related below.

Chen et al. (1999) made a detailed description of various mechanisms for the photocatalytic oxidation of alcohols and organochlorides in an aerated or a deaerated system. The description includes mechanisms of surface reactions and radical reactions based on the experimental data presented. They deduced that although Pt enhances the total rate of methanol and ethanol oxidation, the rapid reduction of oxygen on Pt/TiO_2 does not increase the hydrogen peroxide concentration in solution, and reactions (19) and (20) may not occur easily:

$$H_2O_{2S}\longrightarrow H_2O_2+s \tag{19}$$

$$O_2 + 2H^+ + 2e^- \longrightarrow H_2O_2 \tag{20}$$

Reactive radicals such as hydroxyl radical (·OH), hydrogen peroxide (H_2O_2), and superoxide anion radical ($O_2^{\bullet -}$) are the main power driving titanium dioxide (TiO_2) photocatalytic reactions. Cai et al. (2003) investigated the effect of copper ions on the formation of H_2O_2 over photocatalytic TiO_2. Under an oxygen-purged solution, formation of H_2O_2 was increased dramatically up to 20 times by the addition of a small amount of copper ions.

Mrowetz and Selli (2006) studied the photocatalytic degradation of formic acid (FA) and benzoic acid (BA), chosen as model organic molecules with acidic properties, in TiO_2 and ZnO water suspensions under different experimental conditions. Hydrogen peroxide evolution, formed through a reductive pathway started by conduction band electrons, was also simultaneously monitored during the degradation runs. A much higher amount of hydrogen peroxide was detected in ZnO irradiated suspensions, both in the presence and in the absence of the substrates, mainly because of its lower photocatalytic decomposition rate on such oxide. The rate of hydrogen peroxide evolution during the photocatalytic oxidation of BA on TiO_2 could be related to the rate of the oxidation process, while H_2O_2 could not be detected during the photocatalytic degradation of FA on this oxide, mainly because of the reduced shielding ability of this substrate.

2.4.1. The Photocatalytic Ozonation Process

Photocatalytic ozonation is the combination of photocatalytic oxidation (with oxygen or air) and ozonation. Radiation can be of ultraviolet or even visible wavelength (Agustina et al., 2005). The yield of the photocatalytic process increases even for treating contaminants in wastewater (Suri et al., 1993) but it has the drawback of a slow electron transfer from TiO_2 to oxygen and the need of three electrons to generate one hydroxyl free radical when using oxygen, instead of just one single electron required in the case of using ozone. Thus, in most of photocatalytic processes, oxygen is used due to its availability and low cost although low mineralization is achieved.

In this sense, there are studies that report the application of ozone to significantly improve the elimination and mineralization of water contaminants (Sánchez et al., 1998; Gimeno et al., 2005; Addamo et al., 2005; Zou y Zhu, 2008), suggesting a synergism between photocatalysis and ozonation (Wang et al., 2002). So, *a priori*, for the whole process, both ozone and photocatalytic oxidation reactions shown above have to be considered. This mechanism of reactions, however, can be simplified to some extent, depending on the experimental conditions and nature of compounds treated. For example, it is assumed that ozone, because of its higher oxidation character, will preferentially react with reduced TiO_2 sites, Ti(III), to yield adsorbed ozonide ion radicals that eventually lead to more hydroxyl radicals:

$$O_3 + Ti(III) \xrightarrow{k = 1.1x10^{10} M^{-1} s^{-1}} TiO_2 - O_3^{\bullet -} \tag{21}$$

It will also be assumed that ozone adsorption steps are very fast so that in reaction (21) ozone can be on an adsorbed phase. Then, reactions (12) and (18) are considered negligible against reaction (21). Also, reaction (14) is neglected.

For reaction (21) due to the lack of data, the rate constant of the reaction between aqueous electron and ozone (Buxton et al., 1988) has been considered. Also, ozone photolysis has to be included (reaction (6))

In photocatalytic ozonation, organic degradation can be due to the ozone direct reaction, by hydroxyl radical oxidation, by positive hole oxidation in the valence band, reaction (8) or by direct photolysis. On the other hand, formation of hydroxyl radicals develops through four posible ways:

1. Direct reaction ozone-contaminants (Beltrán et al., 2010)
2. Positive hole oxidation (reaction (10)).
3. Ozone reduction in the conduction band of the semiconductor, reaction (21), which is the synergism of this system.
4. Also, reaction of ozone and the hysdroperoxide ion has to be considered.

As an example, Beltrán et al. (2010) reported the degradation of sulphamethosazole, with different advanced oxidation systems involving ozone, radiation and TiO_2. They concluded that photocatalytic ozonation allows the highest mineralization degree (TOC elimination) while of direct ozone and photolysis were negligible (lower than 10%). TOC elimination is due to hydroxyl radicals from photocatalytic reactions or the ozone-hydrogen peroxide reaction. From results of mass balances of species acting in the system, the authors concluded that hydroxyl radicals formed in the ozone-hydroperoxide ion reaction is becoming more important as long as the reaction time increases. For example, after 5 and 15 minute, contribution of this reaction goes from only 1.8 to 100%.

2.5. *In Situ* Electro-Generated H_2O_2 in Fenton Related Processes

Consumption of hydrogen peroxide, main reagent in Fenton processes (see later), is intimately related to the final cost of treatment. Then, dose optimization and the way it is fed to the reactor are key factors to be considered. In situ hydrogen peroxide generation is then an attractive idea to minimize the operating costs. In this sense, the electro-Fenton process has been extensively studied because it yields hydrogen peroxide in the cathode of the electrochemical reactor (Brillas et al., 1997, 2004, 2009; Panizza and Cerisola, 2001; Rosales et al., 2009). Electrochemical oxidation is based on the application of electric power (2 to 20 A) between two electrodes in water, which generates hydroxyl free radicals and hydrogen peroxide. Oxidation of water molecules in the anode (reaction (22)) yields free radicals while oxygen reduction in the cathode leads to hydrogen peroxide (reaction (23)) (Brillas et al., 2009):

$$H_2O \longrightarrow {}^{\bullet}OH + H^+ + e^- \tag{22}$$

$$O_2 + 2H^+ + 2e^- \longrightarrow H_2O_2 \tag{23}$$

When iron is added to this system an electro-Fenton oxidation takes place (see later Fenton processes). This increases the formation of hydroxyl radicals. Boron doped diamonds (BDD) are the most used anodes in this process since the high hydroxyl radical concentration they produce. Other alternative electrodes present some problems: amorphous carbon deactivates fast, IrO_2 shows a selective oxidation and others such as PbO_2 introduce toxic elements in the water. If an iron electrode is used additional iron stoichiometric amounts are supplied and the process is called peroxy-coagulation. Main drawback of BDD anodes is the fragility and cost. However, if the system is optimized, their associated costs could be assumed by eliminating that due to hydrogen peroxide consumption. On the other hand, oxygen diffusion cathode is the most used. It consists in a conductive surface of active carbon through which oxygen circulates. Fundamentals, application and results of these processes have been recently reviewed by Brillas et al. (2009). Operating variables affecting the process are oxygen (or air) flow rate, speed of agitation or water flow, temperature, pH, electrolyte composition, voltage applied and iron and contaminant concentrations. The electro-Fenton process has been used to remove some organic contaminants from water or wastewater. Data from some relevant published works on this process are given in Table 3.

2.6. Direct Reaction of Hydrogen and Oxygen to Produce Hydrogen Peroxide

The direct reaction between hydrogen and oxygen in the presence of an appropriated catalyst (i.e. paladium) yields hydrogen peroxide according to:

$$H_2 + O_2 \xrightarrow{Pd-CAT} H_2O_2 \tag{24}$$

Some authors have focused their attention on the improvement of Pd catalyst (influence of support phase, bimetallic catalysts, particle size, etc) to increase selectivity and conversion (Abate et al., 2005, 2011).

Other authors (Yalfani et al., 2009) have suggested the use of this process for in situ generation of hydrogen peroxide and their subsequent application to oxidize water contaminants in the presence of another catalyst to yield the following mechanism:

$$R + H_2 + O_2 \xrightarrow{CAT1(Pd)/CAT2(Fe)} H_2O + RO \tag{25}$$

$$R + H_2 + O_2 \xrightarrow{CAT1(Pd)} H_2O_2 + R \xrightarrow{CAT2(Fe)} H_2O + RO$$

Thus, the use of solid catalyst of Pd and Fe in water or bimetallic catalysts of Pd and Fe that allow the oxidation process have been proposed (Yalfani et al., 2009, 2010, 2011a, 2011b)

Table 3. Recent works on in-situ generated H_2O_2 in electro-Fenton process

Target compound Wastewater	Operating conditions	Main features	Reference
Triclosan Triclocarban C_0=5 mgL^{-1} in water C_0=50 mgL^{-1} in AcN:H_2O (20:80)	200 mL undivided batch cell T=20±1 °C pH=3 Current=60 mA $C_{Fe(III)}$=0.20 mM 4 electro-Fenton systems (anode/cathode) 1) Pt/carbon felt 2) BDD/carbon felt 3) Pt/O_2 diffusion 4) BDD/O_2 diffusion Q_{air}=1 $Lmin^{-1}$ (1 and 2) Q_{O2}=12 $mLmin^{-1}$ (3 and 4)	Complete degradation of triclosan and triclocarban by different systems following the rate decreasing sequence: Pt/carbon felt>BDD/carbon felt>Pt/O_2 diffusion>BDD/O_2 diffusion according to the generation rate of hydroxyl radicals. Role of the Fe-oxalate complexes.	Sirés et al. (2007)
Phenol COD_0=670 mgL^{-1} 4-chlorophenol COD_0=290 mgL^{-1} Aniline COD_0=2400 mgL^{-1}	1L recirculated solution pH=3 T=20°C Carbon gas diffusion cathode (GDE) Ti-IrO_2-SnO_2 mesh anode Current=5A Q_{O2}=2 $Lmin^{-1}$ Three experimental modes: 1) Electro-Fenton undivided batch cell $C_{Fe(II)}$= 1 mM 2) In cell electro-Fenton, membrane divided cell (MK-40 PEM) $C_{Fe(II)}$= 1 mM 3) Ex cell Fenton with electrogenerated H_2O_2 (another reactor) Fe(II)/H_2O_2 =1	Almost complete destruction (90-98%) of aromatic organic pollutants achieved in all cases. The ex-cell process with the membrane reactor showed the best performance with lower energy consumption.	Agladze et al. (2007)
Phenol C_0=0.33 and 1.05 mM	400 mL undivided cylindrical cell T=25±1 °C pH=3 Current=100 mA Pt grid anode Carbon felt cathode Different catalysts: $C_{Co(II)}$=0.05-1 mM $C_{Cu(II)}$=1-10 mM $C_{Fe(II)}$=0.05-1 mM $C_{Mn(II)}$=0.1-1 mM	Fe(II) was the most effective with an optimum concentration of 0.1 mM, allowing total phenol mineralization. Pseudo first order kinetics for phenol elimination (k_{app}=0.037 min^{-1}). Two stages in mineralization according to the different reaction rate of aromatics or organic acids with hydroxyl radicals.	Pimentel et al. (2008)

Target compound Wastewater	Operating conditions	Main features	Reference
Triphenilmethane dyes (TMPs): malachite green (MG), crystal violet (CV), methyl green (MeG), fast green (FCF). MG as model for some experimental conditions $TMPs_0$=0.05-0.5 mM	250 mL undivided batch cell Pt anode Carbon felt cathode T=23±2 °C pH=3 Current=20-400 mA Q_{O2}=1 $Lmin^{-1}$ Some tests: MG/Fe(III)=0.1-2.5	Experiments with MG to analyze the effect of operating conditions. Individual experiments, reactivity order: MeG<FCF<CV<MG Mixture experiments, COD_0=1000 mgL^{-1}, around 100% efficiency removal at 400 mA and $C_{Fe(III)}$=0.2 mM after 25 h	Sirés et al. (2008)
Wastewater from toluente nitration (di- and tri-nitrotoluenes) TOC_0=450 mgL^{-1}	300 mL batch cell Pt electrodes T=30-70°C pH=0.2-2 Potential=4-6 V Q_{O2}=0-200 $mLmin^{-1}$ $C_{Fe(II)}$=5-50 mgL^{-1}	Optimum conditions: 6V, 30°C, 100 $mLmin^{-1}$ oxygen, 15 mgL^{-1} Fe(II), pH=0.2, H_2O_2 production around 140 mgL^{-1} and completely TOC removal.	Chen and Lin (2009)
Alizarin red (dye) C_0=120 mgL^{-1}	250 mL undivided batch cell Pt grid anode Gas diffusion cathode (GDE) (carbon-PTFE layer in a nickel mesh) T=25-45 °C pH=3-6 Current=100-200 mA Q_{air}=20 mLs^{-1} $C_{Fe(II)}$=0-2.0 mM	Influence of variables, optimum conditions: pH=3, $C_{Fe(II)}$= 1.0 mM, I=200 mA, T=35° C, total alizarin degradation and almost complete COD removal (higher than 95%) were reached in 4 h.	Panizza and Cerisola (2009)
Basic Yellow 2 (BY2 dye) C_0=5-25 mgL^{-1}	600 mL undivided cylindrical cell Fe sheet anode Gas diffusion cathode (GDE) (own-made carbon nanotube-PTFE) T=25°C pH=2-6 Current=100-300 mA Q_{O2}=140 $mLmin^{-1}$ Electrolysis time=2-30 min	Response surface methodology (RSM) to study the influence of variables. Optimum conditions: t=16 min, pH=3, I=200 mA, C_0=15 mgL^{-1}, decolorization higher than 95% was achieved.	Zarei et al. (2010)

On the other hand, given the limitations in the use of hydrogen, hydrazine, hydroxylamine or formic acid can be used as reductors in the production of hydrogen peroxide. It can be highlighted that these processes are in a first stage of research and literature on this subject is very limited so far. Some of the reported works on this system are given in Table 4.

3. Added Hydrogen Peroxide Oxidation Processes

Since hydrogen peroxide constitutes in oxidation processes a key substance to generate hydroxyl free radicals, in many of them it is directly added to increase the oxidation rate. This happens, for example, with the ozone process.

Thus, in this section attention will focus on processes some of them already shown in the preceding section where, in addition, hydrogen peroxide is added to the system to increase the formation of free radicals.

3.1. The Ozone-Hydrogen Peroxide Process

In ozone processes, the amount of hydrogen peroxide formed to activate the ozone decomposition (see reaction (5)) is not enough to produce a high concentration of hydroxyl radicals. Thus, it is well known that ozonation alone is not able to significantly reduce the organic carbon of water (TOC) (see Figure 5).

To solve this problem, the addition of hydrogen peroxide has been proposed (see Table 5). The concentration of added hydrogen peroxide, however, should keep the balance between the formation of free radicals through reaction (5) and the scavenging character of hydrogen peroxide itself through reactions (26) and (27) (Tomiyasu et al, 1985):

$$HO^{\bullet} + H_2O_2 \xrightarrow{k=2.7x10^7\,M^{-1}s^{-1}} H_2O + HO_2^{\bullet} \tag{26}$$

$$HO^{\bullet} + HO_2^{-} \xrightarrow{k=7.5x10^9\,M^{-1}s^{-1}} OH^{-} + HO_2^{\bullet} \tag{27}$$

3.2. Hydrogen Peroxide Photocatalytic Oxidation Processes

Because of its oxidizing character, hydrogen peroxide has been previously used for enhancing the rate of photocatalytic reactions. Hydrogen peroxide is a better electron acceptor than molecular oxygen (Ollis et al., 1991). The minimum energy requirement of oxygen reduction to produce hydroxyl radical is 3.0 eV while that of hydrogen peroxide is 2.2 eV (Anheden et al., 1996).

Table 4. Recent works on direct in-situ generated H_2O_2 applied to AOPs

Target compound Wastewater	Operating conditions	Main features	Reference
Phenol C_0=100 mgL^{-1}	50 mL batch reactor Solid catalyst: Pd/Al_2O_3 (0.1-5 wt.% Pd) C_{CAT}=2 gL^{-1} Reducing agent: formic acid (C=40-500 mM) Q_{O2}=20 $mLmin^{-1}$ $C_{Fe(II)}$=10 mgL^{-1} T=25-45°C pH=2.2-2.5	Total phenol removal and higher mineralization (60%) with the "in-situ" H_2O_2 generated process using Pd/Al_2O_3 catalyst with 5 wt.% Pd. Enhanced productivity of H_2O_2 compared to batch and semicontinuous addition. High stability of the catalyst without Pd leaching.	Yalfani et al. (2009)
Phenol 2,4-dichlorophenol Clofibric acid C_0=100 mgL^{-1}	50 mL batch reactor Solid catalysts supported on Al_2O_3 : Pd5Al (5 wt.% Pd) Pd5FeCI (5 wt.% Pd, 1 wt.% Fe, co-impregnation) Pd5Fe (5 wt.% Pd, 1 wt.% Fe, successive) FePd5 ((1 wt.% Fe, 5 wt.% Pd, successive) C_{CAT}=2 gL^{-1} Reducing agent: formic acid (C=40+20 (4h) mM) Q_{O2}=20 $mLmin^{-1}$ T=25°C pH=2.5	Phenol conversion 92% and higher mineralization (78%) with the "in-situ" H_2O_2 generated process using FePd5 catalyst after 6 h and a second addition of formic acid. Enhanced productivity of H_2O_2 and mineralization compared to conventional Fenton process and the combination of Pd5Al and ferrous ion. The best catalyst did not show Fe-Pd alloy formation. High stability of the best catalyst without Pd and Fe leaching.	Yalfani et al. (2010)
Phenol C_0=100 mgL^{-1}	50 mL batch reactor Catalytic systems: Pd/Al_2O_3 (5 wt.% Pd) + Fe(II) (10 mgL^{-1}) $FePd/Al_2O_3$ (1 wt.% Fe, 5 wt.% Pd) C_{CAT}=2 gL^{-1} Reducing agents: Formic acid (40 mM) Hydrazine (20 mM) Hydroxylamine (40 mM) Q_{O2}=20 $mLmin^{-1}$ T=25°C pH=3-9	Formic acid was the most effective at acid pH 3-4 with both catalytic systems (semi-heterogeneous and full heterogeneous) reaching 60-70% mineralization whereas hydroxylamine shown the best performance at pH 7-9 in the full heterogeneous system reaching 60% mineralization. High stability of the catalysts without Pd and Fe leaching.	Yalfani et al. (2011a)
2,4-dichlorophenol C_0=306 µM Pentachlorophenol C_0=27 µM	100 mL batch reactor Solid catalysts supported on Al_2O_3 : $PdFe/Al_2O_3$ (5 wt.% Pd, 1 wt.% Fe) C_{CAT}=1 gL^{-1} Reducing agent: formic acid (C=20 mM) T=25°C, pH=2.5 3 protocols: CRO-He: consecutive reduction-oxidation (with He purging) CRO: without He purging SRO: simultaneous reduction-oxidation Q_{O2}=20 $mLmin^{-1}$	DCP results: 70% mineralization in CRO-He and SRO, this last better detoxification. PCP results: self-inhibition in the SRO process, better results with CRO-He process where phenol is formed during hydrodechlorination stage and subsequently oxidized. High stability of the best catalyst without Pd and Fe leaching.	Yalfani et al. (2011b)

Table 5. Some recent works on ozone-hydrogen peroxide system

Compound and Process	Reactor, conditions	Main features	Reference
2,4-dichlorophenoxyacetic acid, O_3; O_3/H_2O_2	Bubble contactor column (50 X 1850 mm), pH: 3.65; 6.68; 8.02; 9.87; 11.83. H202 applied: 0, 1,2,4mg/L	H_2O_2 highly enhanced the oxidation of 2,4-D by ozone. Proper H202/03molar Ratio around 0.6. pH value varied from 3.65 to 9.87, the removal efficiency increased from 22.3% to 74.01%. Identification of degradation products and intermediate. Mechanism proposed.	Yu (2006)
N-nitrosodimethylamine (NDMA) O_3; O_3/H_2O_2	500 ml glass bottle equipped with a dispenser. pH 7 and 8	Second-order rate constants for reactions of NDMA with ozone and $^{\bullet}OH$ were determined. $(O_3)_0:(H_2O_2)_0=2$. The effects of solution pH, H_2O_2 and HCO_3^- and ozone concentration on NDMA oxidation were investigated.	Lee et al. (2007)
Trichloroethylene O_3; O_3/H_2O_2	Diffuser type reactor Ozone dose: 32 and 28 mg/L. pH 6–8	Efficiency of O_3/H_2O_2 was 40% better than that of ozone treatment. H_2O_2/O_3 ratio was maintained at 0.07–0.18 g-H_2O_2/g-O_3. Ozone dose was reduced from 32 mg/L to 28 mg/L by the 2-port injection method.	Yasunaga and Hirotsuji (2008)
Dissolved organic chemicals. O_3; O_3/H_2O_2 and alkaline ozonation	5-L glass jacketed reactor (batch mode) Injections of 0.15 mL of H_2O_2 (30%, w/v). pH: 8.5	The purpose was to increase the wastewater's reuse potential by improving water quality. Periodic pulses of hydrogen peroxide. H_2O_2/O_3 molar ratio lower than 0.5. O3/H2O2: TOC decay ranging from 75 to 100%. O3: mineralization was always below 35%. Mineralization model	Rosal et al. (2009)
Dibutylsulfide (DBS) O_3; O_3/H_2O_2	Batch reactor (200 mL). pH values: 2, 5, 7, 9 and 12 Concentrations of hydrogen peroxide: 0, 10^{-4}, 10^{-3}, 10^{-2}, 10^{-1}, 1 mol/L,	Increasing H_2O_2 concentration from 0.1 to 1mol/L causes significant increase in DBS oxidation rate (regardlessthe pH). At pH 9, and at high concentration of hydrogen peroxide (from 0.1 to 1 mol/L), a clear increase in DBS decomposition rate. Depending on pH value there are two mechanisms of DBS oxidation present: ionic and radical.	Stanisław et al. (2009)
Effluent from a Wastewater Treatment Plant O_3; O_3/H_2O_2	20 L semi-batch reactor. Ozone dosage of 15 mg/L pH 4; 7 and 12.	O_3 provided superior disinfection, however, lower reductions in DOC, COD and endocrine disrupting compounds were exhibited compared to O_3/H_2O_2. DOC, COD, UV_{254}, total coliform, dissolved ozone and some endocrine disrupting compounds were monitored. 0.5 molar ratio of O_3/H_2O_2. The benefit of H_2O_2 reduced when the mole ratio was increased to 2.	Can and Çakır (2010)
Bromoxynil and trifluralin; O_3; O_3/H_2O_2	Batch reactor (400 mL) pH: 7 and 9. Hydrogen peroxide levels were between 10^{-5} M and $5 \cdot 10^{-4}$ M,	Enhancement of the level of degradation was observed using O_3/H_2O_2 process. Acute toxicity analysis performed. Quantification of oxidation by-products generated. The highest degradation was observed at pH 9 using O_3/H_2O_2 An optimum H_2O_2/O_3 ratio of 0.4-0.6 was found Kinetic study	Chelme-Ayala (2011)

As a result, addition of hydrogen peroxide (along with oxygen) in the suspension will result in decreasing the rates of the electron-hole recombination process, and thus, increasing the degradation rates. Moreover, hydrogen peroxide may produce hydroxyl radicals by photolytic splitting depending on the wavelength of the incident radiation and may avoid oxygen starvation of the solution due to inadequate oxygenation or mass transfer limitations (Ollis et al., 1991).

Hydrogen peroxide can react directly with the photogenerated holes (Ulmann et al., 1986) and like other molecules (i.e., alkaline formates, EDTA) can adsorb onto TiO_2 which may increase the charge transfer rates on the semiconductor surface (Ulmann et al., 1986). It is apparent, however, that the effect of hydrogen peroxide on the degradation rates is not always positive and it depends on the specific photocatalytic system. For example, it may depend on a number of factors including the type and concentration of the organic contaminants, the solution matrix, the pH, the wavelength of UV radiation and the type of photocatalyst.

Chiou et al. (2008) investigated the use of hydrogen peroxide (H_2O_2) for improving the photocatalytic degradation of phenol in aqueous suspension of commercial TiO_2 powders (Degussa P-25).

They examined the effects of operating parameters such as catalyst dosage, light intensity, pH of the solution, initial phenol, and H_2O_2 concentrations on photodegradation process. It was shown that photodegradation using $H_2O_2/TiO_2/UV$ process was much more effective than using either H_2O_2/UV or TiO_2/UV process.

Evgenidou et al. (2007) studied the heterogeneous photocatalytic degradation of prometryn using TiO_2 as photocatalyst. They examined the influence of various parameters such as initial concentration of pesticide or catalyst and presence of oxidants (H_2O_2 and $K_2S_2O_8$).

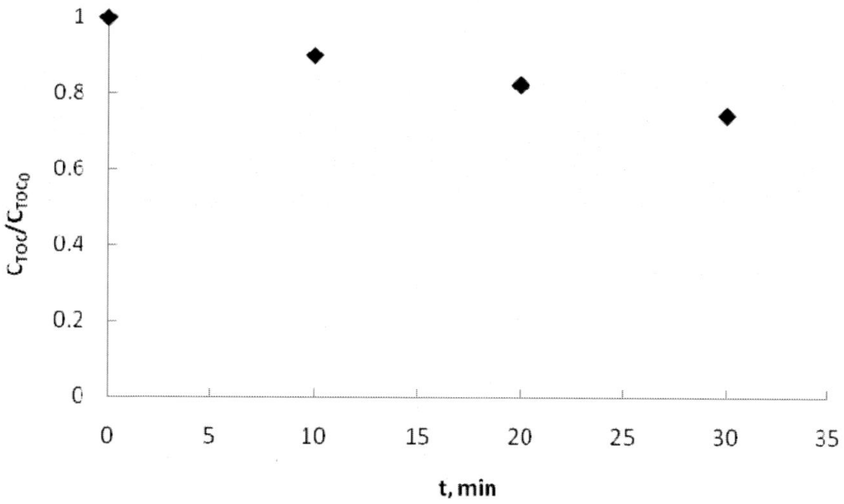

Figure 5. Time evolution of total organic carbon during the ozonation of pharmaceutical compounds in ultrapure water. Conditions as in Figure 3.

Authors observed that the addition of the oxidants although enhances the transformation of the parent compound is unable to achieve higher mineralization leading to 70% reduction of DOC. Jain and Shrivastava (2008) studied the photocatalytic degradation of a hazardous water soluble xanthene dye cyanosine in aqueous suspensions of titanium dioxide under a variety of conditions: catalyst concentration, substrate concentration, pH, temperature and electron acceptor hydrogen peroxide (H_2O_2). The degradation rate was increased significantly by increasing the amount of catalyst, while on increasing substrate concentration decolorisation decreases.

Alkaline pH conditions were found to significantly affect the dye degradation. The maximum degradation efficiency of dye was achieved with the combination of $UV+H_2O_2+TiO_2$. In the heterogeneous photocatalytic system, the characteristics of the solution matrix, such as the pH and the ionic strength, are critical in determining the speciation distribution of acidic and basic molecules and the magnitude of the solute-surface interactions in the solid oxide/solution interface, respectively.

Consequently, these parameters can significantly affect the extent of adsorption or chemisorption of the solute on the semiconductor surface as well as the concentration of the key oxidizing species formed during the photocatalytic process. Therefore, these factors will be crucial in controlling the rates and mechanism of the degradation process. However Dionysiou et al. (2000) demonstrated that the rates of the photocatalytic reactions were not significantly affected by an increase in the ionic strength by a factor of 50. On the other hand, the reaction rate was a strong function of catalyst crystals and loading. On the other hand, addition of hydrogen peroxide up to 248 mg/L resulted in an increase of the reaction rates with a corresponding increase in photonic efficiency by $\approx$20%. Above this concentration, hydrogen peroxide either did not enhance or caused a significant inhibition of the mineralization rates.

3.3. Fenton Processes

This section is focused on the well-known Fenton and related technologies using mainly iron or other transition metals (homogeneous or heterogeneous) to decompose hydrogen peroxide to produce hydroxyl radicals capable of oxidizing water contaminants at ambient conditions. The use of iron and H_2O_2 for wastewater treatment is attractive due to the facts: (a) iron is a highly abundant and non-toxic element, and (b) hydrogen peroxide is easy to handle and decomposes in environmentally benign products. The Fenton reaction dates to 1894, when H.J. Fenton reported the ferrous ion catalyzed tartaric acid oxidation in the presence of H_2O_2 (Fenton, 1894). However, the use of the Fenton reagent (i.e. $Fe(II)/H_2O_2$) to oxidize organic pollutants in water began only in the late sixties (Goi and Trapido, 2002). Since then, Fenton reaction has been the focus of numerous investigations in the last two decades as can be seen in Figure 6 where a search results with the keywords “Fenton” and “wastewater” show that about 1000 peer-review scientific publications have been published in this field in this period of time. Fenton and related reactions applied to organic pollutants removal from water, focused on the role of hydrogen peroxide in these treatments, are shortly revised in the next subsections.

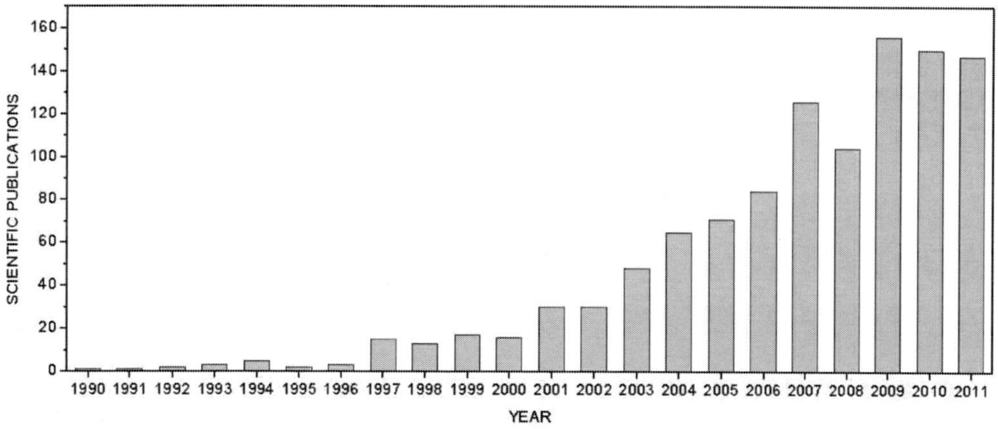

Figure 6. Number of scientific publications related to Fenton reaction for wastewater treatment in the last 20 years (Web of Science-WOK, 2012).

3.3.1. Fenton Reactions and Fundamentals

In the 1930s, Haber and Weiss suggested the formation of hydroxyl radicals in the system $Fe(II)/H_2O_2$ (Haber and Weiss, 1934), where hydrogen peroxide oxidizes Fe(II) to Fe(III). The "classical" commonly accepted reaction mechanism of the Fenton reactions is listed below (Pignatello et al., 2006):

$$Fe(II) + H_2O_2 \longrightarrow Fe(III) + {}^{\bullet}OH + OH^- \quad (28)$$

$$Fe(III) + H_2O_2 \longrightarrow Fe(II) + HO_2^{\bullet} + H^+ \quad (29)$$

$${}^{\bullet}OH + Fe(II) \longrightarrow Fe(III) + OH^- \quad (30)$$

$${}^{\bullet}OH + H_2O_2 \longrightarrow HO_2^{\bullet} + H_2O \quad (31)$$

$$Fe(III) + HO_2^{\bullet} \longrightarrow Fe(II) + O_2 + H^+ \quad (32)$$

$$Fe(II) + HO_2^{\bullet} + H^+ \longrightarrow Fe(III) + H_2O_2 \quad (33)$$

$$HO_2^{\bullet} + HO_2^{\bullet} \longrightarrow H_2O_2 + O_2 \quad (34)$$

In the presence of an organic substrate, hydroxyl radicals react with it mainly through hydrogen abstraction from aliphatic carbon atoms, electrophilic addition to double bonds or aromatic rings and electron transfer reactions (Legrini et al., 1993), forming organic radicals that continue reacting prolonging the chain reaction. In addition, "non-classical" mechanisms have also been reported (Bossmann et al., 1998; Gogate and Pandit, 2004) where ferryl species (higher valence iron) can also react through direct electron transfer with the organic substrate. Both mechanisms ("classical" and "non-classical") can coexist and predominate one or the other depending on the operating conditions (Pignatello et al., 2006; Magario et al., 2012).

Fenton's reagent can be employed to treat a variety of organic compounds or wastewater containing them, like phenols and chlorophenols, pesticides, dyes, pharmaceuticals, etc. (Pignatello et al., 2006; Zazo et al., 2007; Bautista et al., 2008; Homem and Santos, 2011; Magario et al., 2012). The mechanisms, kinetics, operating conditions and applications of the Fenton treatment have been recently reviewed (Neyens and Baeyens, 2003; Gogate and Pandit, 2004; Pignatello et al., 2006; Bautista et al.; 2008; Magario et al., 2012) and are not deeply commented in this chapter.

The main factors that influence the Fenton process are the concentration of iron species, the pH value of the medium, the organic contaminant nature and concentration, the hydrogen peroxide quantity required for oxidation, and the temperature (Pignatello et al., 2006). Fe(II) concentration in the reaction medium is a key factor that conditions the reaction rate and is mainly regenerated through reaction (29), although the rate of this reaction (0.002 $M^{-1}s^{-1}$, Pignatello, 1992) is much lower than that of reaction (28) (70 $M^{-1}s^{-1}$, Rigg et al., 1954). Therefore, Fe(III) reduction to Fe(II) is crucial to the global process and it can be favored by some organic molecules such as quinones or organic radicals formed as intermediates (Pignatello et al., 2006, Al-Hayek et al., 1985, Duesterberg and Waite, 2007).

On the other hand, some reaction intermediates such as oxalic acid form very stable complexes with Fe(III) which avoid the reduction process decreasing the oxidation capacity of the system (Duesterberg and Waite, 2007). Some of the inorganic ions commonly present in water also may cause the removal of iron from the reaction medium (De Laat et al., 2004).

The process shows a pH optimum value around 3 (Neyens and Baeyens, 2003; Rivas et al., 2004; Pignatello et al., 2006). At higher pH Fe(III) could precipitate (depending on concentration) as $Fe(OH)_3$ that leads to H_2O_2 decomposition to form water and oxygen. On the other hand, at lower pH, the catalyst regeneration is inhibited (Pignatello, 1992).

Dosage of Fe(II) determines the reaction rate, and H_2O_2 concentration is related to the reaction extent, playing both key roles in the kinetics of the process. Neither H_2O_2 nor Fe(II) must be overdosed, to ensure that the maximum amount of hydroxyl radicals are available for the oxidation of organic compounds (Tang and Huang, 1996). Thus, the H_2O_2 to Fe(II) molar ratio is an important operational parameter to control, together with the COD or TOC to H_2O_2 ratio, and to optimize the use of the H_2O_2.

Respect to Fe(II) concentration, high ratios of Fe(II) to H_2O_2 can decrease the efficiency of hydroxyl radicals for organics degradation due to reaction (30), acting Fe(II) as a radical scavenger. On the other hand, higher amounts of iron yields bigger amount of iron sludge that would need to be separated after the treatment (Yoon et al., 2001).

Regarding hydrogen peroxide, many times the amount used is higher than the theoretical stoichiometric quantity (respect to the organic matter loading) since the consumption of H_2O_2 is not equal to the formation rate of hydroxyl radicals and thus, a part of the oxidant decomposes into water and oxygen (Pignatello et al., 2006). However, there is a maximal concentration above which the process performance is not improved (Voelker and Sulzberger, 1996) due to hydroxyl radicals scavenging reaction (31), among others, which constitute a waste of bulk oxidant (Pignatello et al., 2006; De Laat and Le, 2006).

On the other hand, adding H_2O_2 in a single batch leaves much of it available to be attacked by hydroxyl radicals, whereas continuously feeding smaller quantities of the reagent to the system would allow the majority of these radicals to react with the organic contaminant, improving thus the use of the oxidant (Duesterberg and Waite, 2006).

Therefore, the doses and dosing regimes of Fenton reagent have to be carefully analyzed and varied according to the application and type of wastewater, the required disposal limits and economic factors since the overall cost of the treatment is closely related to the amount of hydrogen peroxide needed and the iron sludge produced that needs to be adequately managed.

In this line, some of the research works published from 2007 related to Fenton processes at near ambient conditions that present relevant results attending to hydrogen peroxide dosage or use optimization are listed in Table 6.

3.3.2. Fenton-Like Treatments

When ferrous ions, Fe(II), are substituted by ferric ions Fe(III) the process has been defined as Fenton-like reaction. However, many authors use this term for any Fenton-type reaction which does not use dissolved Fe(II) both, in homogeneous or heterogeneous systems. The particularities and some relevant aspects of these processes are shortly commented in the following sections.

a) Homogeneous Catalysts

Numerous works have been focused on using Fe(III) as catalyst in Fenton-like treatments according to reaction (29), that it has been accepted to take place in two steps:

$$Fe(III) + H_2O_2 \longleftrightarrow Fe - OOH^{2+} + H^+ \qquad (35)$$

$$Fe - OOH^{2+} \longrightarrow HO_2^{\bullet} + Fe(II) \qquad (36)$$

where ferric ions catalyzes hydrogen peroxide decomposition to produce ferrous ions that gives place the mechanism described in the previous section which leads to the generation of hydroxyl radicals.

In fact, when Fenton reaction ($Fe(II)/H_2O_2$) is carried out in large excess of hydrogen peroxide, Fe(II) is readily oxidized to Fe(III) and the system will behave independent of the initial oxidation state of iron since Fenton-like reaction becomes the rate limiting step after an initial period, according to the difference in their rate constants.

Fenton-like processes with different sources of Fe(III) have been studied (De Laat et al., 2004) to remove different organic compounds or to treat wastewater from different industrial sectors, e.g. landfill leachates (Rivas et al., 2003), dyes (Hrvoje et al., 2007; Wang, 2008), olive-oil mill wastewater (Martínez-Nieto et al., 2011), phenols and chlorophenols (Yingxun et al., 2006; Bach et al., 2010; Muñoz et al., 2011), pharmaceutical compounds (Benítez et al., 2011) among others, some of them comparing their results also with Fe(II) catalyst.

Commonly, the main difference when using Fe(III) salts is a slower initial reaction rate than that found with Fe (II) (Pignatello et al., 2006), although the extent of degradation is usually comparable since it is mainly related to the hydrogen peroxide concentration.

Table 6. Recent works on Fenton process highlighting hydrogen peroxide use

Target compound Wastewater	Operating conditions	Main features	Reference
Cosmetic industry wastewater COD_0=2395-4255 mgL^{-1} TOC_0=705-1090 mgL^{-1}	50 mL batch reactors T=25-50°C pH=2-4 C_{H2O2}=2500-12725 mgL^{-1} $C_{Fe(II)}$=40-1000 mgL^{-1}	Stoichiometric H_2O_2/COD ratio (2.12 w/w) leads to TOC conversion 45-60% with COD in required limits at pH=3. TOC kinetics (second order).	Bautista et al. (2007)
Acid orange 6 azo dye C_0=200 mgL^{-1} COD_0=169 mgL^{-1} TOC_0=54 mgL^{-1}	Batch experiments T=25°C pH=2-7 $C_{Fe(II)}$=8.93 mM H_2O_2/Fe(II)=0.82-8.25 (M/M)	Optimum H_2O_2/Fe(II)=6.58 Optimum pH=3 AO6 conversion higher than 90%, TOC removal higher than 40%. Comparison with different AOPs	Hsing et al. (2007)
Textile wastewater COD_0=8100 mgL^{-1} TOC_0=3100 mgL^{-1}	2L batch reactor T=ambient pH=3 C_{H2O2}=0.440-1.764 mM $C_{Fe(II)}$=0.358-0.895 mM H_2O_2/Fe(II)=0.492-4.927 (M/M)	Optimum ratios H_2O_2/COD=0.0038 (w/w) H_2O_2/Fe(II)=1.2 (M/M) lead to 40% TOC conversion and 48% COD removal.	Papadopoulos et al. (2007)
p-Nitroaniline C_0=0.072-0.217 mM	200 mL batch reactor T=10-50°C pH=2-6 C_{H2O2}=2.5-40 mM $C_{Fe(II)}$=0.025-0.1 mM	Optimum pH=3 At T=20°C Optimum C_{H2O2}=10 mM Optimum $C_{Fe(II)}$=0.05 Kinetic model for PNA depletion (first order), activation energy 53.96 $KJmol^{-1}$	Sun et al. (2007)
p-Nitrophenol C_0=0.1-0.4 mM	Batch experiments T=25°C pH=3 C_{H2O2}=2-20 mM $C_{Fe(II)}$=0.1-0.8 mM	No improvement observed at higher C_{H2O2}>10 mM. Kinetic model for PNPh initial rate degradation based on snowflak-type network.	Chang et al. (2008)
Livestock wastewater COD_0=5000-5700 mgL^{-1} (chromate) Color(287nm)=2.1	200 mL batch experiments T=22-25°C pH=3.5-7 C_{H2O2}=2993-17955 mgL^{-1} $C_{Fe(II)}$=1230-9486 mgL^{-1} H_2O_2/Fe(II)=1-8 (M/M)	Optimum pH=3.5-4 Optimum ratios H_2O_2/COD=1.05 (w/w) H_2O_2/Fe(II)=2 (w/w) lead to 88% COD removal and 95.4% color elimination. No improvement was observed by adding Fenton's reagent in several aliquots.	Lee and Shoda (2008)

Target compound Wastewater	Operating conditions	Main features	Reference
Olive mill wastewater (diluted 1:1) COD_0=61639 mgL^{-1} TOC_0= =23018 mgL^{-1} Total Ph=3182 mgL^{-1}	Batch experiments T=25°C pH=2-5 H_2O_2/COD=1.1-4.4 (w/w) H_2O_2/Fe(II)=10-50 (w/w)	Optimum pH=3 Optimum ratios H_2O_2/COD=3.3 (w/w) H_2O_2/Fe(II)=10 (w/w) lead to 96% COD removal, 89% TOC conversion and 84% phenols removal. H2O2 was dosed during the first 2h. Integrated with a biological post-treatment.	El-Gohary et al. (2009)
Antibiotics: amoxicillin, ampicillin and cloxacillin C_0=100 mgL-1 each COD_0=520 mgL-1	500 mL batch reactor T=25°C pH=2-4 C_{H2O2}=15-54 mM $C_{Fe(II)}$=0.32-24.3 mM H_2O_2/Fe(II)=2-150 (M/M) H_2O_2/COD=1-3.5	Optimum pH=3 Optimum ratios COD/H_2O_2/Fe(II)=1/3/0.3 lead to complete removal of antibiotics in 2 min and 81.4% COD removal in 1 h.	Elmolla and Chaudhuri (2009)
Simulated industrial wastewater Organics: 1,2-dichloroethane sodium formate TOC_0=370-5500 mgL^{-1}	100 mL batch experiments T=23°C pH=3.5 C_{H2O2}=0.11-2.20 M $C_{Fe(II)}$=0.01-0.2 M	Box-Behnken experimental design. Optimum Fe(II)/H_2O_2=0.05 (M/M) that lead to 94% TOC removal. Comparison with Fe(III)-Fenton like process.	Grcic et al. (2009)
Olive mill wastewater COD_0=2000 mgL^{-1}	500 mL batch reactor T=10-50°C pH=2-9 H_2O_2/Fe(II)=7.5-30 (M/M) H_2O_2/COD=0.35-1.75 (w/w)	Optimum pH=3.5 Optimum ratios H_2O_2/Fe(II)=15 (M/M) H_2O_2/COD=1.75 (w/w) At T=30°C lead to 70% COD. Modified pseudo-first-order kinetics, activation energy of 28.2 $KJmol^{-1}$	Lucas and Peres (2009)
Phenol C_0=100 mgL^{-1}	1 L stirred reactor (batch except H_2O_2) T=25°C pH=3 C_{H2O2}=500-5000 mgL^{-1} $C_{Fe(II)}$=1-100 mgL^{-1}	H_2O_2 continuously feeding compared with conventional batch operation leads to more efficient use of H_2O_2. Kinetic models for TOC and phenol oxidation. Complete phenol conversion and 40% TOC removal C_{H2O2}=5000 mgL^{-1} $C_{Fe(II)}$=100 mgL^{-1}	Zazo et al. (2009a)

Table 6. (Continued)

Target compound Wastewater	Operating conditions	Main features	Reference
Phenol C_0=500-1500 mgL^{-1}	Stirred batch reactor T=50ºC pH=3.5 C_{H2O2}=250-7500 mgL^{-1} (10-100% of stoichiometric amount) $C_{Fe(II)}$=10 mgL^{-1}	Kinetic model for mineralization considering H_2O_2 disappearance. The model includes an empirical coefficient that provides a value of the more convenient dosage of H_2O_2 to avoid its loss.	Santos et al. (2010)
Industrial's wastewater treatment plant effluent COD_0=11.62-85.30 gL^{-1}	Stirred 100 mL batch reactor T=30-35ºC pH=3 H_2O_2/Fe(II)=10-20 (M/M) COD/H_2O_2=0.23-3.40 (w/w)	Experimental factorial analysis to determine a model response function. Second-order model including COD to account the wide range of COD variation. 80% COD removal (COD_0=32 gL^{-1}) with COD/H_2O_2=0.58 and H_2O_2/Fe(II)=15	Bianco et al. (2011)
Phenol C_0=100 mgL^{-1}	1 L stirred reactor (batch except H_2O_2) T=25ºC pH=3 C_{H2O2}=500-5000 mgL^{-1} $C_{Fe(II)}$=1-100 mgL^{-1}	H_2O_2 continuously feeding compared with conventional batch operation leads to more efficient use of H_2O_2. Kinetic models for TOC and phenol oxidation. Complete phenol conversion and 40% TOC removal C_{H2O2}=5000 mgL^{-1} $C_{Fe(II)}$=100 mgL^{-1}	Zazo et al. (2009a)
Phenol C_0=500-1500 mgL^{-1}	Stirred batch reactor T=50ºC pH=3.5 C_{H2O2}=250-7500 mgL^{-1} (10-100% of stoichiometric amount) $C_{Fe(II)}$=10 mgL^{-1}	Kinetic model for mineralization considering H_2O_2 disappearance. The model includes an empirical coefficient that provides a value of the more convenient dosage of H_2O_2 to avoid its loss.	Santos et al. (2010)
Industrial's wastewater treatment plant effluent COD_0=11.62-85.30 gL^{-1}	Stirred 100 mL batch reactor T=30-35ºC pH=3 H_2O_2/Fe(II)=10-20 (M/M) COD/H_2O_2=0.23-3.40 (w/w)	Experimental factorial analysis to determine a model response function. Second-order model including COD to account the wide range of COD variation. 80% COD removal (COD_0=32 gL^{-1}) with COD/H_2O_2=0.58 and H_2O_2/Fe(II)=15	Bianco et al. (2011)

Table 7. Recent works on homogeneous Fenton-like processes

Target compound Wastewater	Operating conditions	Main features	Reference
Reactive blue 49 Reactive blue 137 (dyes) C_0=20 mgL^{-1}	250 mL batch reactor T=ambient pH=2,3,4 $C_{Fe(III)}$=0.5-1 mM Fe(III)/ H_2O_2=1/5-1/50	Optimum pH=3, 87-99% color removal although mineralization was around 40% at the best conditions: Fe(III)/ H_2O_2=1/10 for RB49 and 1/40 for RB137. Comparison with Fe(II) conventional Fenton, better for RB49 but not for RB137.	Kusic et al. (2007)
1-butyl 3-methylimidazolium chloride (ionic liquid aqueous solution) C_0=0.072-0.217 mM	300 mL batch reactor T=25°C pH=2.5-4.5 C_{H2O2}=10-400 mM $C_{Fe(III)}$=0.5-1.5 mM	Optimum pH=3, Fe(III)= 1.5 mM with H_2O_2=100 mM. Complex reaction pathway. H_2O_2 can provoke ionic liquid depletion. Higher reaction rate with higher H_2O_2 concentration.	Siedlecka et al. (2008)
Crystal violet dye (CV) C_0=0.15 mM	Batch experiments (100 mL flasks) T= not described pH=3-5 C_{H2O2}=0.5-500 mM $C_{Fe(III)}$=0.25-1 mM $C_{Fe(II)}$=0.25-1 mM	Comparison of Fenton and Fenton-like processes. With Fenton-like process lower CV reaction rate was observed. Optimum pH=3 and Fe(III)/H2O2 ratio 1/50 (twice the Fenton one). Identification of reaction intermediates.	Fan et al. (2009)
Benzoate C_0=0.5 mM Methanol C_0=200 mM	60 mL batch experiments T=20°C pH=6-9 C_{H2O2}=50 mM $C_{Fe(III)}$=1-4 mM Addition of phosphotungstate $PW_{12}O_{40}^{3-}$ (C_0=1 mM)	The addition of the polyoxometalate extends the working pH range up to pH 8.5. PT forms a soluble complex with Fe(III) that converts H_2O_2 into oxidants	Lee and Sedlak (2009)
Olive mill wastewater (2 different effluents) $COD1_{,0}$=1673 gL^{-1} $COD2_{,0}$=4137 gL^{-1}	1 L batch experiments T=5-40°C pH=3-6 C_{H2O2}=20-200 gL^{-1} $C_{Fe(III)}$=0.04-4 gL^{-1} ($FeCl_3$) Fe(III)/ H_2O_2=0.002-0.02 (COD1) Fe(III)/ H_2O_2=0.02-0.2 (COD2)	Optimum pH=3 $FeCl_3$ and H_2O_2 strong effect on the reaction rate. With Fe(III)/H_2O_2=0.02-0.04 higher of 90% organic matter removal. Kinetics described by two step first order model.	Nieto et al. (2011)
4-chlorophenol (4-CP) C_0=100 mgL^{-1}	500 mL batch reactor T=50°C pH=3 C_{H2O2}=17.2-344 mgL^{-1} $C_{Fe(III)}$=1-20 mgL^{-1} H_2O_2/4-CP=5-100% stoichiometric dose	H_2O_2 stoichiometric amount and Fe(III)=10-20 mgL^{-1} led to complete breakdown of 4-CP up to CO_2 and organic acids. Substoichiometric amounts or lower Fe(III) concentration led to the formation of condensation byproducts including chlorinated toxic compounds (PCBs, dioxins and dichlorodiphenyl ethers were detected).	Muñoz et al. (2011)

Besides iron salts, other metal ions have been also suggested as catalysts for Fenton-type reactions since exhibit at least two oxidation states such as copper, ruthenium, cerium, manganese (Mantzavinos, 2003; Anipsitakis and Dionysiou, 2004; Watts et al., 2005; Caudo et al., 2006; Heckert et al., 2008). This species acts as electron donors/acceptors similar to iron ions promoting the generation of hydroxyl radicals from H_2O_2 according to the general reactions:

$$M^{n+} + H_2O_2 \longrightarrow M^{(n+1)+} + {}^{\bullet}OH \quad (37)$$

$$M^{(n+1)+} + H_2O_2 \longrightarrow M^{n+} + HO_2^{\bullet} \quad (38)$$

However, due to economic and environmental concerns, iron is usually preferred for the process (Navalon et al., 2010). Also, these metals have usually shown lower catalytic activity than iron (Anipsitakis and Dionysiou, 2004). In general, using Fe(III) or other transition metals as catalysts, similar results can be expected regarding to the hydrogen peroxide to catalyst ratio (H_2O_2/M^{n+}) or hydrogen peroxide to organic matter ratio (H_2O_2/COD). However optimum pH of the treatment can change with different transition metals according to their chemistry in aqueous solution. Table 7 shows some of the recent works on homogeneous Fenton-like application.

b) Heterogeneous Fenton-Like or Catalytic Wet Peroxide Oxidation Treatments

The main drawbacks of the homogeneous Fenton processes are the need of removing iron before water discharge, mainly through precipitation which generates large amounts of iron sludge that must be adequately managed, and the pH control. These problems can be partially overcome when heterogeneous catalysts are used (Perathoner and Centi, 2005; Zazo et al., 2006; Caudo et al., 2006; Navalon et al., 2010).

Heterogeneous Fenton-type processes have also been named Catalytic Wet Peroxide Oxidation (CWPO) and Wet Hydrogen Peroxide Catalytic Oxidation (WHPCO). In these processes solid iron compounds or active metals supported on porous solids are used. These catalysts can be easily recovered after the treatment. In addition, the use of heterogeneous catalysts allow operating in continuous reactors (fixed bed or slurry reactors), offering new advantages to optimize the reaction system.

Some attempts have been carried out to develop active and stable heterogeneous catalysts for the CWPO process. The first near heterogeneous Fenton catalysts tested were metallic iron and iron oxides/hydroxides. Lücking et al. (1998) used Fe(0) powders to oxidize 4-chlorophenol and observed that when pH decreased, dissolved Fe was increased at the same time than oxidation rate, showing an important effect of homogeneous contribution. Similar results were found by Lu (2000) where the reaction evolution was closely related to the amount of iron leached from goethite mineral (α-FeOOH).

Chou et al. (2001) used γ-FeOOH to treat benzoic acid and observed that heterogeneous catalysis contribution was important at the beginning whereas homogeneous contribution was very important at pH lower than 4.4 due to iron dissolution.

However, other authors have found evidences of surface reactions over heterogeneous iron oxides catalysts. Thus, Chou and Huang (1999) observed that at pH higher than 6 most of benzoic acid depletion was attributable to heterogeneous reactions over goethite. Also,

Huang et al. (2001) found a great dependence between reaction rate with different iron oxides and their surface area corroborating the heterogeneous contribution.

Although the reaction mechanism of hydrogen peroxide decomposition over a heterogeneous catalyst has not been unambiguously established, the following simplified steps have been suggested in most of the works for the generation of hydroxyl radicals (Lin and Gurol, 1998):

$$\equiv Fe(III) + H_2O_2 \leftrightarrow \equiv Fe(III) - H_2O_2 \quad (39)$$

$$\equiv Fe(III) - H_2O_2 \longrightarrow \equiv Fe(II) + H_2O + HO_2^{\bullet} \quad (40)$$

$$\equiv Fe(II) + H_2O_2 \longrightarrow \equiv Fe(III) + {}^{\bullet}OH + OH^{-} \quad (41)$$

similar to the Fenton-like process with iron species located at the catalyst surface and considering the adsorption of hydrogen peroxide onto the Fe(III) sites.

Some authors also suggest the adsorption of the organics onto the catalyst surface (Feng et al., 2003a; González-Olmos et al., 2009). Thus, Andreozzi et al. (2002) have suggested a non-radical mechanism where the absorbed substrate and hydrogen peroxide react on the catalyst surface. The main point, however, is whether hydroxyl radical is formed and is free to diffuse or is coordinated to the iron on the surface (Magario et al., 2012).

Besides pure iron oxides as catalysts, some mixed oxides have been successfully applied to the CWPO of phenolic compounds or dyes (Ovejero et al., 2001; Sotelo et al., 2004; Massa et al., 2008; Oliveira et al.; 2007, 2008). However, most of the CWPO research efforts have been focused on the use of a porous support/constituent such as zeolites (Fajerwerg et al., 2000; Centi et al., 2000), mesoporous silica (Crowther and Larachi, 2003; Gokulakrishnan et al., 2009; Martínez et al., 2007a, 2007b, 2008; Melero et al., 2004, 2006, 2007), clays/pillared clays (Barrault et al., 2000; Guélou et al., 2003; Catrinescu et al., 2003; Carriazzo et al., 2007; Molina et al., 2006, 2010) or carbonaceous materials (Zazo et al., 2006, 2009b; Ramírez et al., 2007a; Duarte et al., 2009; Rey et al., 2009, 2010), to develop active and stable catalysts with higher surface areas and porosity. Heterogeneous Fenton research works using zeolites, mesoporous silica, clays and carbon materials as supports and/or catalysts have been recently reviewed where more details can be found (Navalon et al., 2010, 2011; Garrido-Ramírez, 2010). Other materials have been used as a support such as polymers and resins (Parra et al., 2004; Liou et al., 2005), alumina (Bautista et al., 2011) or silica (Huang et al., 2008) but at low extent.

In general, the main problem of these catalytic systems is the leaching of the iron ions to the reaction medium which limits the stability and reusability of the catalysts. Leaching is favored at acid pH and due to the presence of organic acids (like oxalic acid) that can form very stable complexes with iron. Zeolites and pillared clays have shown to be more stable to the leaching phenomena and the activity of the catalysts is less sensible to pH, allowing high TOC reduction even at pH value higher than 5 (Caudo et al., 2006).

Like in homogeneous Fenton reactions, by far, the most used active phase has been iron. However, some transition metals have been used to synthesize heterogeneous catalysts such as Cu, Co, Ni (Perathoner and Centi, 2005) or rare earth metals like Ce (Massa et al., 2008). Among them, generally Cu has been the most active.

Regarding hydrogen peroxide, catalysts should also be selective in the use of H_2O_2 to generate hydroxyl radicals and minimize the waste of the bulk oxidant. In addition, the ratios H_2O_2/catalyst and H_2O_2/COD should be optimized to ensure the best use of the hydrogen peroxide avoiding the loss of active species through reactions similar to (30) and (31).

Table 8 summarizes some of recent results of CWPO with different heterogeneous catalysts and reactors.

3.3.3. Photo-Fenton Reactions

The use of Fe(II)/Fe(III) in the presence of hydrogen peroxide and radiation has been named photo-Fenton. In the fifties, Bates and Uri (1953) suggested that the oxidation of some aromatic compounds in aqueous solution in the presence of iron species was due to the formation of hydroxyl radicals by photo-excited electron transfer in iron complexes.

Afterwards, first works of the application of the photo-Fenton process (or photoassisted/light enhanced Fenton process) for wastewater treatments were published by the groups of Pignatello, Lipcznska-Kochany, Kiwi, Pulgarín and Bauer (Pignatello et al., 2006; Malato et al., 2009). Since then increasingly research efforts have been focused on photo-Fenton processes to treat numerous organic compounds and different effluents. These works have been quite recently reviewed by Pignatello et al. (2006), Pupo-Nogueira et al., (2007) and Malato et al. (2009).

a) Homogeneous Photo-Fenton

To understand the improvement observed by combining Fenton's reagent and light it is necessary to take into account the speciation of Fe(III) in aqueous solution. According to hydrolysis equilibrium of Fe(III) given by reactions (42)-(45):

$$Fe(III) + H_2O \xrightleftharpoons{K=6.3x10^{-3}\,M} Fe(OH)^{2+} + H^+ \tag{42}$$

$$Fe(III) + 2H_2O \xrightleftharpoons{K=3.2x10^{-4}\,M} Fe(OH)_2^+ + 2H^+ \tag{43}$$

$$2Fe(III) + 2H_2O \xrightleftharpoons{K=1.3x10^{-3}\,M} Fe_2(OH)_2^{4+} + 2H^+ \tag{44}$$

$$Fe(III) + 3H_2O \xrightleftharpoons{K=10^{-6}\,M} Fe(OH)_3 + 3H^+ \tag{45}$$

where hydration water has been omitted to simplify the reaction scheme. With the values of the equilibrium constants (Faust and Hoigné, 1990; Benkelberg and Warneck, 1995) it can be deduced that at pH in the range 2.5-5, most of iron is in the form of $Fe(OH)^{2+}$. This complex is able to absorb light appreciably in the UV and visible region to undergo photolysis according to reaction (46):

$$Fe(OH)^{2+} \xrightarrow{h\nu} Fe(II) + {}^{\bullet}OH \tag{46}$$

that presents a quantum yield between 0.7-0.05 mol·einstein^{-1} at wavelengths between 254-380 nm. Thus, in presence of hydrogen peroxide the Fe(II) generated reacts following the Fenton reactions sequence (reactions (28)-(34)) where Fe(II) is easily regenerated through reaction (46). Thus, differences observed when iron is supplied as Fe(II) or Fe(III) are quite lower than in the dark-Fenton. This reaction also involves the production of an extra hydroxyl radical, enhancing the overall reaction rate.

In addition, under UV radiation hydrogen peroxide can undergo photolysis giving place to an additional $^{\bullet}OH$ radical production way (Baxendale and Wilson, 1957):

$$H_2O_2 \xrightarrow{h\nu \leq 320nm} 2^{\bullet}OH \qquad (47)$$

The total quantum yield of this reaction is 1 mol·einstein^{-1} (Goldstein et al., 2007) but it is of importance only when using radiation of wavelenght lower than 320 nm due to the low value of molar absorptivity at higher wavelengths (Chu and Anastasio, 2005) (see sections 2.3 and 4.1.1). These processes take place not only with UV radiation sources but also with solar light that is more favorable from an economic and environmental point of view (Malato et al., 2009). The group of Malato has been pioneer in the application of solar light at pilot plant scale to photo-Fenton and photocatalytic wastewater treatment. Main operating conditions that affect the process are similar to the dark-Fenton reaction, i.e. pH control (optimum around pH=3), iron concentration, temperature, substrate concentration, hydrogen peroxide concentration, and other light-affected conditions such as wavelength or light intensity used. Quite detailed comments about this can be also found in the review of Malato et al. (2009).

This process has been suggested to be feasible and promising to remove pollutants from natural and industrial wastewaters and increase their biodegradability, being used as a pre-treatment method (Miller et al., 1988; Fallman et al., 1998; Malato et al., 2009).

Regarding hydrogen peroxide, similar results to dark-Fenton can be expected related to the dosage procedure or the concentration used. In fact, hydrogen peroxide can be rate-limiting if applied in concentrations that are too low. On the contrary, at very high concentration, H_2O_2 can compete with contaminants for the hydroxyl radicals and also decompose into oxygen and water (Prieto-Rodríguez et al., 2011).

Several works have been focused on analyzing the influence of hydrogen peroxide concentration on the reaction rate or mineralization reached (Fallman et al., 1999; Pariente et al., 2008). Also, the effect of different dosing ways has been pointed out (Ince, 1999; Oller et al., 2007; Prieto-Rodríguez et al., 2011).

b) Use of Carboxylate Complexes

In presence of some carboxylic acids, Fe(III) ions may form ferricarboxylate complexes that photolyse to yield free radicals with a quantum yield higher than that of reaction (46). The efficacy of some carboxylic acids (oxalic, citric, tartaric, malic, etc.) has been demonstrated for the degradation of different organic compounds through the formation of Fe(III)-complexes followed by photoreduction (Faust and Zepp, 1993; Abrahamson et al., 1994; Zuo and Hoigné, 1994; Kuma et al., 1995; Deng et al., 1998; Li et al., 2006).

In a recent work of Rodríguez et al. (2009), the effect of the structure of different carboxylic acids on the ability to form photoactive ferricarboxylate complexes has been discussed. Thus, the presence of one HO group in the α position with respect to the carboxylic acid group gave place to photoactive ferricarboxylate complexes that absorb UVA radiation and photodecompose in Fe(II) and free radicals accelerating the oxidation rate of the target compound (muconic acid in this case).

Among the carboxylic acids able to form ferricarboxylate complexes, oxalic acid deserves special attention since the complex formed absorbs radiation from 250-450 nm with high absorptivity coefficients.

In fact, photolysis of trisoxalateferrate (III) ($(Fe(C_2O_4)_3)^{3-}$) is widely used in actinometry. When light is present, the ferrioxalate complex is photoreduced and decarboxilated giving place to Fe(II) with a quantum yield around 1-1.2 mol·einstein^{-1} between 250-450 nm (Balzani and Carassiti, 1970; Goldstein and Rabani, 2008) and free radicals (Jeong and Yoon, 2004) according to the reactions (48)-(51):

$$(Fe(C_2O_4)_3)^{3-} \xrightarrow{h\nu} Fe(II) + 2C_2O_4^{2-} + C_2O_4^{\bullet -} \quad (48)$$

$$C_2O_4^{\bullet -} + (Fe(C_2O_4)_3)^{3-} \longrightarrow Fe(II) + 3C_2O_4^{2-} + 2CO_2 \quad (49)$$

$$C_2O_4^{\bullet -} \longrightarrow CO_2^{\bullet -} + CO_2 \quad (50)$$

$$CO_2^{\bullet -} + O_2 \longrightarrow O_2^{\bullet -} + CO_2 \quad (51)$$

The superoxide ion radical ($O_2^{\bullet -}$) will react according to the known reactions mechanisms commented in the previous sections, mechanism through H_2O_2 can be eventually formed and undergo Fenton reactions (if not added).Some of the recent works on photo-Fenton and carboxylate assisted photo-Fenton are summarized in Table 9.

c) Heterogeneous Catalysts

Taking into account the main drawbacks associated to the homogeneous photo-Fenton process as commented in the previous section for dark-Fenton, namely the formation of high amounts of iron sludge that need to be adequately managed, the continuous loss of catalyst and the need of controlling pH in a narrow range, several efforts have been recently focused on the synthesis and application of heterogeneous catalysts for photo-Fenton reactions. However some disadvantages have been also observed with heterogeneous photo-Fenton catalysts such as diminished reaction rates compared to the homogeneous treatment due to mass transfer limitations and worsened light penetration into the photoreactor in the presence of solids (Pérez-Estrada et al., 2005; Malato et al., 2009).

Different catalysts have been synthesized and tested in photo-Fenton processes such as pure iron oxides (He et al., 2002a, 2002b, 2005; Du et al., 2008), doped iron oxides (Gajovik et al., 2011; Liang et al., 2012) or mixed iron oxides (Zhang et al., 2007), iron species supported or embedded in a porous matrix like zeolites (Noorjahan et al., 2005; Tekbas et al., 2008; Kasiri et al., 2008; Duarte and Madeira, 2010), mesoporous silica (Martínez et al., 2005, 2007c; Pariente et al., 2008, Lam and Hu, 2007), clays and pillared clays (Feng et al., 2003a, 2003b, 2004, 2005; Perathoner and Centi, 2005; De León et al., 2008; Iurascu et al., 2009; González-Bahamon et al., 2011) and polymeric materials (Fernández et al., 1998, 2000; Kiwi et al., 2002; Parra et al., 2003, 2004) among others. Most studied catalysts have been those based in clays and pillared clays which have been recently reviewed by Ramírez et al. (2010). As in other Fenton-related treatments, iron has been commonly used as active phase although other elements such as Cu or Ce have been also tested (Perathoner and Centi, 2005; Zhang et al., 2007).

The reaction mechanism of the heterogeneous photo-Fenton processes has not been unambiguously established yet. In general simple mechanisms equivalent to homogeneous photo-Fenton reactions have been proposed, where photo-reduction of Fe(III) surface species is the first step:

$$CAT \equiv Fe(III) \xrightarrow{h\nu} CAT \equiv Fe(II) \tag{52}$$

$$CAT \equiv Fe(II) + H_2O_2 \longrightarrow CAT \equiv Fe(III) + {}^{\bullet}OH + OH^- \tag{53}$$

$$CAT \equiv +S \longleftrightarrow CAT \equiv S \tag{54}$$

$$CAT \equiv S + {}^{\bullet}OH \longrightarrow CAT \equiv P \tag{55}$$

$$CAT \equiv P \longleftrightarrow CAT \equiv +P \tag{56}$$

Although the role of substrate adsorption is still not clear, the importance of this process onto the photocatalyst surface has been established in some works (De León et al., 2008; Zhang et al., 2010b). In addition, iron leaching phenomena has been also observed which contributes to homogeneous photo-Fenton reactions (Ramírez et al., 2010).

Main operating conditions affecting the reaction are similar to homogeneous photo-Fenton process, i.e. wavelength and light intensity, catalyst load, initial H_2O_2 concentration and dosage form, parent compound concentration, pH and temperature. Regarding catalyst load, an optimum can be reached since high concentration may provoke high turbidity of the suspension and the transmission of light to the photoreactor is inhibited. In general, an increase in light intensity exerts a positive effect in the oxidation rate and lower wavelength is also more effective.

The optimum pH in this process is also around 3 although at acidic pH the leaching of the active phase needs to be considered. However the range of pH that can be used is wider than in the homogeneous treatment. Also, an increase of the temperature exerts a positive effect in the reaction rate although most studies are focused on the application at ambient temperature.

Lastly, the role of hydrogen peroxide concentration and dosage form has been discussed in the review of Ramírez et al., (2010). Similar conclusions than in previous Fenton related processes can be addressed. Ratios of H_2O_2/COD and H_2O_2/Catalyst need to be optimized to ensure the best use of the oxidant avoiding self-scavenging reactions. For the same reason, usually the dosage of hydrogen peroxide in several steps or even continuously feed give place to a better use of the oxidant.

Some of the recent works on heterogeneous photo-Fenton process have been summarized in Table 10.

Table 8. Recent works on CWPO processes

Target compound Wastewater	Operating conditions	Main features	Reference
Azo dye Orange II C_0=0.1 mM	200 mL batch reactors Two carbon catalysts tested (7% Fe): Fe/Activated carbon Fe/Carbon aerogel With the best catalyst, operating conditions study: T=10-70°C pH=2-4 C_{CAT}=0.15-0.30 gL^{-1} C_{H2O2}=3-48 mM	High adsorption capacity of both catalysts. Best performance of Fe/Carbon aerogel due to higher iron dispersion. Compromise between reaction rate and low iron leaching. Conditions: pH=3, T=30°C, H_2O_2 6mM, complete decolorization is observed with 80% TOC removal.	Ramírez et al. (2007a)
Azo dye Orange II C_0=0.1 mM	1 L batch reactor Twelve catalysts (saponite clay loaded with Fe): Wt.% Fe=7.5, 13, 17 Iron precursor: Fe(II)-acetate, Fe(II)-oxalate, Fe(II)-acetylacetonate, Fe(III)-acetylacetonate. T=10-70°C pH=2-4 C_{CAT}=92 mgL^{-1} C_{H2O2}=3-24 mM	Influence of iron precursor and content. Best catalyst: Fe(II)-oxalate with 17 wt.% Fe. Best conditions: pH=3, T=70°C, C_{H2O2}=6 mM led to 99% dye conversion and 91% TOC removal. At T=30°C also 96% dye conversion and 82% TOC removal. Very low Fe leached, lower than 1 mgL^{-1} in the worst case.	Ramírez et al. (2007b)
Methylene blue (MB) C_0=100 mgL^{-1}	10 mL batch reactors 4 Catalysts (different T reduction): Fe_3O_4 Fe^0/Fe_3O_4-300 Fe^0/Fe_3O_4-400 Fe^0/Fe_3O_4-500 T=25°C pH=6 C_{CAT}=3 gL^{-1} C_{H2O2}=300 mM	High active systems can be obtained by controlled reduction of Fe_3O_4 with H_2. Complete characterization studies. Best performance with heat-treatment T=400°C (47% Fe^0), total discoloration and 75% TOC removal in 2 h. No Fe leached was detected.	Costa et al. (2008)

Target compound Wastewater	Operating conditions	Main features	Reference
MTBE C_0=100 mgL^{-1}	120 mL batch reactors Zeolite catalysts: Fe-ZSM-5 (2 wt.% Fe) Fe-BETA (1 wt.% Fe) C_{CAT}=5-25 gL^{-1} T=25°C pH=7 C_{H2O2}=1.0-6.3 gL^{-1}	Fe-ZSM-5 higher catalytic activity for H_2O_2 decomposition and MTBE depletion. Fe-BETA more efficient in the use of H_2O_2. Pseudo-first order kinetic model. Adsorption onto the zeolite surface plays a significant role. High consumption of H_2O_2. 96% MTBE conversion with 25 gL^{-1} Fe-ZSM-5 and 1 gL^{-1} H_2O_2.	González-Olmos et al. (2009)
Phenol C_0=10 mgL^{-1}	50 mL batch reactors Catalysts: Fe/AC (4 wt.% Fe) Different precursors: Iron nitrate and iron pentacarbonyl Different AC supports: CM, CN, CC C_{CAT}= 0.5 gL^{-1} C_{H2O2}=500 mgL^{-1} (stoichiometric) T=50°C pH=3	AC support plays a crucial role in the use of H_2O_2, CC carbon provokes inefficient H_2O_2 decomposition. With CM and CN catalysts 100% phenol conversion and 80% TOC removal were achieved in 4 h regardless iron precursor. Lower iron leaching when using iron nitrate precursor.	Rey et al. (2009)
Cosmetic industry wastewater COD_0=2300-4730 mgL^{-1} TOC_0=686-1220 mgL^{-1} BOD/COD=0.22	50 mL batch experiments Catalysts: Fe/Al_2O_3 (4 wt.% Fe) Fe/AC (4 wt.% Fe) C_{CAT}=5 gL^{-1} T=55-85°C pH=3 C_{H2O2} stoichiometric respect to COD_0	Better results with Fe/Al_2O_3 catalyst. BOD/COD=0.54 after 4h at 85°C with TOC conversion 55.9% and COD conversion 83.5% No leaching was observed.	Bautista et al. (2010)
Phenol C_0=200 mgL^{-1}	50 mL batch experiments Catalyst: Fe-Cu-Al-MCM-41 C_{CAT}=1.5 gL^{-1} T=25-60°C pH=4-7 C_{H2O2}=0.024-0.098 M	At pH=4, T=60°C and C_{H2O2}=0.049 M, 47% TOC removal. The incorporation of Al in the mesoporous silica MCM-41 improves H_2O_2 utilization.	Xia et al. (2011)

Table 9. Recent works on photo-Fenton and carboxylate assisted processes

Target compound Wastewater	Operating conditions	Main features	Reference
Bleaching paper mill wastewater COD_0=1286 mgL^{-1} TOC_0=332 mgL^{-1}	Dish reactor (22x15x3 cm, 330 cm^2, 500 mL used) Simulated solar-light radiation (Xe lamp) T=30-50°C pH=2-6 $C_{Fe(II)}$=31-310 mgL^{-1} C_{H2O2}=0.5-3 stoichiometric amount for TOC depletion (1883 mgL^{-1})	82% TOC removal at 120 min pH=3.5, H_2O_2/TOC=2, H_2O_2/Fe=10, T=30°C All of the treatments shown 60% TOC conversion, photo-Fenton has a potential application in TOC removal from bleaching paper mill wastewater.	Xu et al. (2007)
4-chlorophenol C_0=4-222 mgL^{-1}	500 mL batch reactor T=40°C pH=1.5-4 (not varied) C_{H2O2}=200-800 mgL^{-1} $C_{Fe(II)}$=10-50 mgL^{-1}	Design of Experiments Completely TOC removal at 60 min with adequate operating conditions (40 mgL^{-1} Fe and 800 mgL^{-1} H_2O_2 for 222 mgL^{-1} 4-CP) Empirical models applied.	Pérez-Moya et al. (2008)
Pharmaceuticals: amoxicillin, paracetamol. C_0=0.1 mM benzafibrate C_0=0.055 mM Effect of water matrix, sewage treatment plant effluent.	Black light experiments: 800 mL batch recirculation mode, 280 mL irradiated (anular photoreactor) Solar light experiments: 800 mL batch recirculation mode, 510 mL irradiated (tubular photoreactor) Room T pH=2.5 C_{H2O2}=1-10 mM C_{Fe}=0.02 mM (iron nitrate or potassium ferrioxalate)	Degradation rate is favored using potassium ferrioxalate (see later). H_2O_2 exerts a positive effect in the degradation rate of AMX and BZF but not PCT. TOC removal was around 85% at 2 mM H_2O_2 with no further improvement. Water matrix affects the degradation rate of PCT and BZF but not AMX.	Trovo et al. (2008)
Orange II dye C_0=20 mgL^{-1}	35 L CPC in batch recirculation mode, 16 L irradiated volume Solar light radiation T=20-30°C	Central composite experimental design-Naural Networks.	Monteagudo et al. (2009)

Target compound Wastewater	Operating conditions	Main features	Reference
	pH=2-6 $C_{Fe(II)}$=2.90-10 mgL^{-1} C_{Oxalic}=17-60 mgL^{-1} $C_{H2O2,0}$=75 mgL^{-1} Q_{H2O2}=0-1 $mLmin^{-1}$ (30%) Q_{air}=0.60-1.60 m^3h^{-1}	Continuous addition of H_2O_2 increased the overall efficiency (100% dye conversion and 95% TOC removal) of the oxalate assisted photo-Fenton. Oxalic acid can be used to pH adjustment.	
Pesticide mixture (oxamyl, methomyl, imidacloprid, dimethoate, pyrimethanil) DOC_0=100-500 mgL^{-1}	75 L CPC in batch recirculation mode, 44.6 L irradiated volume Solar light radiation T=25-50°C pH=2.7-2.9 $C_{Fe(II)}$=20 mgL^{-1} Maintaining H_2O_2 concentration in the range 100-300mgL^{-1}	Temperatures below 45°C are needed to avoid significant loss of iron due to precipitation during the process. H_2O_2 needs to be carefully dosed to avoid its inefficient use, more important at low DOC concentration. 45% of DOC removal is achieved in the best case.	Zapata et al. (2010)
Pesticide mixture (oxamyl, methomyl, imidacloprid, dimethoate, pyrimethanil) DOC_0=200 mgL^{-1}	75 L CPC in batch recirculation mode, 44.6 L irradiated volume Solar light radiation T=35°C pH=2.6-2.8 $C_{Fe(II)}$=20 mgL^{-1} 5 type of experiment: -Maintaining H_2O_2 concentration in the range 100-400 mgL^{-1} -Amount to 60% DOC elimination at the beginning (C_{H2O2}=1200 mgL^{-1}) -H_2O_2 in 80 mgL^{-1} doses. -DOC and H_2O_2 as 40 mgL^{-1} and 80 mgL^{-1} doses, respectively. -Automatically controlled H_2O_2 addition according to the dissolved oxygen	H_2O_2 and dissolved oxygen where found very close related. This relationship was used to manage automatic H_2O_2 dosing. On this form, photo-Fenton treatment can be optimized to H_2O_2 consumption.	Prieto-Rodríguez et al. (2011)

Table 10. Recent works on heterogeneous photo-Fenton processes

Target compound Wastewater	Operating conditions	Main features	Reference
Phenol C_0=500 mgL^{-1} TOC_0=380 mgL^{-1}	1L batch photoreactor UVA>313 nm radiation Fe supported on amorphous silica, zeolite material, mesoporous SBA-15 C_{cat}=0.5 gL^{-1} T=25°C pH=6 (not buffered) C_{H2O2}=2450 mgL^{-1} (95% stoichiometric amount)	Best results with Fe-SBA-15 catalyst with supported crystalline iron oxide particles. Fe leaching lower than 2% with the best catalyst. Complete phenol oxidation and 50% TOC removal after 240 min irradiation time.	Martínez et al. (2007)
Brilliant Red X-3B C_0=100 mgL^{-1}	1L immersion batch photoreactor UVC-254 nm radiation Fe-Ce oxide catalyst C_{cat}=0.5 gL^{-1} Room T pH=3 C_{H2O2}=34 mgL^{-1}	100 mgL^{-1} of X-3B were effectively removed (99%) at 30 min. Maximum Fe leached 1 mgL^{-1} Comparison of the process with homogeneous treatments	Zhang et al. (2007)
Benzoic acid C_0=25-450 mgL^{-1}	350 mL immersion batch photoreactor UVA-365 nm radiation Fe-SBA-15 catalyst (16-19 wt.% Fe onto mesoporous silica) C_{cat}=0.3-1.2 gL^{-1} T=27±3°C pH=3.1 C_{H2O2}=20-100% of the stoichiometric amount	60% of the stoichiometric amount of H_2O_2 enough to complete benzoic acid conversion with an acceptable catalytic performance. 0.6 gL^{-1} Fe-SBA-15 showed similar results than higher concentrations. Low Fe leaching (lower than 5%)	Pariente et al. (2008)
Acid blue 74 dye C_0=0.0214-0.168 mM	Cubic 2L batch photoreactor UV-C radiation Fe-ZSM5 catalyst (3 wt.% Fe onto zeolite) C_{cat}=0.125-1 gL^{-1} T=25°C (increased 7-8°C during the experiments) pH=3-7 C_{H2O2}=10.7-85.6 mM	At optimum conditions (21.4 mM H_2O_2, 0.5 gL catalyst, pH 5) a 57% of TOC removal is achieved in 120 min. The catalyst is stable with only 0.3 mgL^{-1} of Fe detected in the reaction medium. Wider range of pH can be used compared to homogeneous treatment.	Kasiri et al. (2008)

Target compound Wastewater	Operating conditions	Main features	Reference
Phenol C_0=0.5-1.5 mM	10 mL batch reactors (quartz cells) Different Fe-Laponite catalysts (pillared clay). Heat-treated (55% Fe2O3) Non heat-treated (33% Fe2O3) All the samples were calcined T=250-550°C C_{cat}=0-2 gL^{-1} T=30°C pH=2.5-3.5 C_{H2O2}=20-100 mM	At t=5 min, total conversion of phenol can be reached with UVC radiation, pH=3, 1gL^{-1} of catalyst, 50 mM H_2O_2 for 1 mM phenol. The catalyst prepared under thermal treatment and calcined at 350° C showed the best performance. Kinetic model proposed.	Iurascu et al. (2009)
Acid orange 7 C_0=35 mgL^{-1} Phenol C_0=40 mgL^{-1}	100 mL cylindrical batch photoreactor Visible and UVB/UVA radiation S doped α-Fe_2O_3 C_{cat}=0.1 gL^{-1} T=25°C pH=6.85 C_{H2O2}=1.94 mM	Complete characterization study. S-αFe_2O_3 showed little dark Fenton reactivity but higher reactivity in photo-Fenton under UV or visible radiation due to the effect of S. 0.12-0.14 % Fe leached that can contribute to the reaction rate.	Guo et al. (2010)

Table 11. Recent examples of works on hydrogen peroxide-radiation system to remove compounds from water

Compound and Process	Reactor, conditions	Main features	Reference
Rhodamine B UV/H_2O_2	50 ml beaker. pH: 1, 7 and 9. UV Lamp (254 nm). H_2O_2 concentration: 0.835, 1.67, 3.34 or 6.68 mM.	The optimum operating conditions of the method were found to be (Dye) = 10 μM; (H2O2) = 1.67 mM at pH 7. A maximum of 73% decoloration of the dye was achieved. Kinetic study. Degradation products identified.	Alhamedi et al. (2009)
Amoxicillin UV and UV/H_2O_2	A cylindrical reactor (Pyrex, 500 mL). pH: 7. Low pressure Hg arc-UV lamp (254 nm). Photon fluence: 8×10^{-7} Einstein L^{-1} s^{-1} Effective light path: 5.5 cm.	*Pseudo* first-order kinetics. Rates increased up to six fold with increasing H_2O_2 addition at 10 mM H_2O_2 compared to direct photolysis.	Jung et al. (2012)

Table 11. (Continued)

Compound and Process	Reactor, conditions	Main features	Reference
	H_2O_2 concentration: 0.4, 2, 3, 4, 5 and 10 mM.	Maximum of 50% TOC removal with UV/H_2O_2 after a reaction time of 80 min. Intermediates identified. UV/H_2O_2 process effectively eliminated any antibacterial activity	
17α-ethinylestradiol (EE_2) UV and UV/H_2O_2	0.5 L capacity thermostat batch glass reactor (ID 100 mm). pH: 6.8. Low-pressure UV lamp (254 nm). UV intensities: 2.456, 2.669, 4.299, 7.771 and 15.093 W/m^2. H_2O_2 concentration: 5, 10, 20, 30 and 50 ppm.	Pseudo-first-order kinetics. The EE_2 degradation rate enhanced from 0.0364 to 0.0684 min^{-1} when H_2O_2 was combined with the photolysis process. Regression model developed as a function of the UV intensity, H_2O_2 concentration and initial EE_2 concentration. The estrogenicity was effectively decreased by UV/H_2O_2 technology.	Zhang et al. (2010a)
Thiacloprid UV and UV/H_2O_2	pH: 2, 5, 7 and 9. 125 W high-pressure mercury lamp ($\lambda > 290$ nm) Emission bands in the UV region at 304, 314, 335 and 366 nm, with maximum emission at 366 nm. Photonic flux: 2.54×10^{-7} E s^{-1}. H_2O_2 concentration: 5, 14, 23, 45, 71, 162 mM.	UV/H_2O_2 system: maximum first-order rate constant 2.7×10^{-2} min^{-1} at the H_2O_2/thiacloprid molar ratio of 220 and pH 2.8. 97% of the thiacloprid was removed in about 120 min. Ionic byproducts identified. In natural water the rate of removal of thiacloprid showed that it is dominantly influenced by the presence of HCO_3^-.	Abramović et al. (2010)
n-butylparaben (BP) and 4-*tert*-octylphenol (OP). UV/H_2O_2	Merry-go-round device with quartz test tubes (volume: 10 cm^3, average optical path length: 0.85 cm) placed between two exposure panels consisted of five low-pressure mercury lamps (254 nm). pH: 5, 6, 7, 8, 11 and 12. Photon fluence rate: 0.4×10^{-5} einstein dm^{-3} s^{-1} to 1.5×10^{-5} einstein dm^{-3} s^{-1}. H_2O_2 concentration: 0.0001, 0.001, 0.005, 0.01, 0.05, 0.1 M.	For optimum hydrogen peroxide concentration the reactions in alkaline solutions were slower. Influence of humic acids and nitrate ions on degradation was not noticed. BP oxidation in the presence of hydroxyl radical scavengers did not inhibit the reaction course. Kinetic study.	BŁędzka et al. (2010)

Compound and Process	Reactor, conditions	Main features	Reference
Fenofibric acid UV and UV/H_2O_2	700 mL batch reactor. 15 W low-pressure mercury vapour lamp (254 and 185 nm). pH: 6.5. Fluence rate: 18.73 ± 0.23 mW cm^{-2}. Path of radiation through the reactor of 2.37 ± 0.07 cm. H_2O_2 concentration: 10, 50, 100, 200 ppm.	Maximum TOC removal observed for UV/H_2O_2 was slightly over 50% after 4 min of reaction. Maximum radical exposure per fluence ratio obtained was 1.4×10^{-10} M L^{-1} mW^{-1}. Low values for the toxicity of UV/H_2O_2 treated samples. Kinetic study. Intermediates identified. Reaction pathway proposed.	Santiago et al. (2011)
Meprobamate, carbamazepine, dilantin, atenolol, primidone and trimethoprim UV/H_2O_2	250 mL glass beaker. pH: 7.1, 7.6 and 8.2. A custom-made low pressure UV collimated beam system. Two G15T8 germicidal lamps. UV fluencies: 300, 500, and 700 mJ/cm^2 H_2O_2 concentration: 2, 5, 10, 15, 20 mg/L.	Three different wastewater effluents. Efficacy of UV/H_2O_2 treatment for the removal of pharmaceuticals from wastewater was a function of not only the concentration of effluent organic matter but also its inherent reactivity towards HO·. The overall removals were between 0 and >99%, and were dependent on the overall scavenging rate.	Rosario-Ortiz et al. (2010)
Ibuprofen (IBU), diphenhydramine (DP), phenazone (PZ), and phenytoin (PHT) UV and UV/H_2O_2	500 mL reactor. pH: 7. 11 W low-pressure Hg vapor lamp (254 nm). Photon fluence rate: 4.5×10^{-5} E m^{-2} s^{-1}. Path length: 1.9 cm. H_2O_2 concentration: 0.29 and 1 mM.	Considerably higher removal of IBU and DP was obtained with the addition of 0.29 mM H_2O_2 compared to no addition. Intermediates identified. Pseudo-first-order kinetics. Quantum yields determined.	Yuan et al. (2009)
Calcon UV and UV/H_2O_2	Multi lamp photoreactor pH: 1, 3, 5, 7, 9 and 12.6. 12 + 12 numbers of 8 W Mercury UV lamps of 254 nm and 365 nm. H_2O_2 concentration: 30 to 260 mM.	Higher degradation of Calcon is achieved in strong basic medium (99.4% at pH = 12.6 at 1260 μW cm^{-2}) than in strong acid medium (≈96% at pH 1.0 and 3.0 at 1260 μW cm^{-2}). The degree of mineralization is more in the presence of H_2O_2 than in its absence and is pH dependent (pH 1 > pH 12.6 > pH3).	Sahoo et al. (2011)
Sulfamethoxazole, sulfamethazine, sulfadiazine, trimethoprim, bisphenol A and diclofenac. UV and UV/H_2O_2	Bench scale quasi-collimated beam (QCB) apparatus. pH: 3.6 and 7.85. Four low-pressure UV lamps. UV fluences : 40 , 140 and 540 mJ cm^{-2} H_2O_2 concentration: 2, 6 and 10 ppm.	UV/H_2O_2 transformation efficiency was DCL > SMX > SMZ > SDZ> BPA ≈ TMP. Quantum yields and second order rate constants determined. At typical UV/H_2O_2 treatment conditions (fluence = 540 mJ cm^{-2}, H_2O_2 dose = 6 mg L^{-1}): (1) transformation percentages in lake water ranged from 43% for trimethoprim to 98% for diclofenac. (2) In wastewater treatment plant effluent, transformation percentages were lower (31–97%).	Baeza and Knappe (2011)

4. Other Combined Treatments

4.1. Combination of Ozone/Hydrogen Peroxide/Radiation

4.1.1. Hydrogen Peroxide Photolysis

When UV light is absorbed directly by hydrogen peroxide, $^{\bullet}OH$ radicals are generated by photolysis of the –O–O– peroxidic bond (7). Hydrogen peroxide absorbs light (depending on its concentration) in the range of the 185–300 nm (see Figure 4), the highest hydroxyl radical yields are obtained when short-wave ultraviolet radiations (200–280 nm) are used. Low pressure lamps having an irradiation at 253.7 nm have been found very effective for the degradation of recalcitrant organic pollutants in UV/H_2O_2 mediated processes (Peyton, 1990). The molar absorption of hydrogen peroxide at 253.7 nm is 19.6 $M^{-1}cm^{-1}$ and the quantum yield for the hydroxyl radical production approximately 1.0 (Baxendale and Wilson, 1957). Moreover, at 253.7 nm, the rate of H_2O_2 photolysis in aqueous solutions increases with pH because of the higher molar absorption coefficient of the peroxide anion (240 $M^{-1}cm^{-1}$ beginning at pH 11.63) compared with that of un-dissociated hydrogen peroxide (Guittonneau et al., 1988; Doré, 1989; Nicole et al., 1990).

Hydroxyl radicals can react to: oxidize organic compounds; recombine with other hydroxyl species to form hydrogen peroxide or initiate a radical chain degradation of hydrogen peroxide (Clark, G. Knowles, 1982; Glaze et al., 1987). Furthermore, hydroxyl radicals can attack organic molecules by: abstracting a hydrogen atom, adding hydroxyl groups, transferring electrons (Huang et al, 1993) and the optimum H_2O_2 dose in the UV/H_2O_2 process should be empirically determined because excess H_2O_2 can be a hydroxyl radical scavenger (Wang et al., 2000).

UV/H_2O_2 processes have been effective in the degradation of various water contaminants (see Table 11). Besides, this process appears to be cost effective and environmentally friendly. The use of hydrogen peroxide as an oxidant has a number of advantages: it is cheaper, readily available, easy to store, relatively safe to handle and environmentally friendly since it slowly decomposes into oxygen and water; very cost-effective source of hydroxyl radicals (two hydroxyl radicals are formed for each molecule of H_2O_2 photolyzed); peroxyl radicals are generated after $HO^{\bullet}$ attack of most organic substrates, thus facilitating oxidation reactions (Legrini et al., 1993; Pan et al., 1993)

The role of UV is to catalyze the oxidation process in two ways: one through excitation of the target molecule, thereby increasing their reactivity toward oxidants and the other by promoting homolysis of H_2O_2 to hydroxyl radicals, which are very strong oxidizing species (Sahoo et al., 2011).

Almost all the photo-oxidation studies using UV/H_2O_2 technique are carried out with 254 nm UV lamp. But 254 nm light is not environmentally relevant as light of wavelengths below 290 nm reaching the earth is minimal. Sun produces 0.2–0.3 mol photons $m^{-2}h^{-1}$ in the range of 300–400 nm near the earth's surface with a typical UV flux of 20–30 Wm^{-2}(Goslich et al.,1997).

Thus sunlight can be used as an economically and ecologically sensible light source in the wastewater treatment process. Therefore, it would be ideal to use UV lamps of wavelength 365 nm and verify the possibility of using sunlight in the photo-degradation process involving H_2O_2 (Abramović et al., 2010; Sahoo et al., 2011).

4.1.2. The O_3/UV/H_2O_2 System

O_3/UV/H_2O_2/O_3 system is a very powerful method which allows the fast and complete mineralization of pollutants. This process generally involves two oxidation/photolysis routes to remove organic compounds present in water. Thus, ozone, hydrogen peroxide and/or UV radiation can react individually or directly photolyze the organic pollutants in water. However, when they are used in combination, they can degrade pollutants by oxidation through hydroxyl free radicals generated in situ (Peternel et al., 2006). The reaction pathways leading to the generation of HO· radicals have been summarized by Legrini et al. (1983). The O_3/UV/H_2O_2 system is considered one of the most effective treatment process for high-polluted effluents (Kusic et al., 2006) and (Teresa et al., 2007). Thus, it is a promising technology for contaminant removal from water and for practical application, it is important to optimize the experimental conditions and reagent concentrations to develop a cost-effective O_3/UV/H_2O_2 process and to enhance removal efficiency (Im et al., 2011).

Peternel et al. (2006) studied the degradation of the organic dye C.I. Reactive Red 45 in water solution and they determined that O_3/UV/H_2O_2 process was the most efficient with 61.1% total organic carbon removal and 72.0% adsorbable organic halides removal, respectively, achieved after a 1-h treatment.

Kusic et al. (2006) compared the efficiency of several ozone- and/or UV-based processes: O_3, O_3/H_2O_2, UV/H_2O_2, UV/O_3 and O_3/UV/H_2O_2, for the minimization of phenol as a model hazardous pollutant in wastewater. They concluded that the highest overall mineralization extent, 58.0% total organic carbon removal, was achieved by O_3/UV/H_2O_2 process and that from the aspect of both eco-effectiveness and cost-effectiveness, this system was shown as the most suitable process.

Similarly, isopropyl alcohol (IPA) degradation and its major degradation intermediate, acetone formation and total organic carbon (TOC) removal efficiency were studied (Wu et al., 2008), using O_3, O_3/UV, H_2O_2/UV, H_2O_2/O_3, and O_3/UV/H_2O_2 processes at various physico-chemical conditions. Authors determined that O_3/UV/H_2O_2 was the most efficient process compared to any other selected AOPs and that the removal efficiency is almost identical at pH 7 and pH 10 and complete degradation of IPA was achieved in 30 min at both pHs, at the same time. 83% of degradation is noted at pH 3. However, the TOC removal efficiency is higher at pH 3 than at pH 10. About 58%, 79%, and 37% of TOC removal was observed at pH 3, pH 7, and pH 10, respectively. It should be noted that the increase of pH leads to an increase of the hydroxyl ion that catalyzes the decomposition of ozone into hydroxyl radicals and decreases the ozone concentration for direct photolysis. Therefore, in a basic medium the removal rate is mainly due to hydrogen peroxide reaction with ozone and in an acidic medium UV/O_3 process is mainly responsible for a higher degree of mineralization. However, under higher pH the formation of bicarbonate/carbonate ion would trap hydroxyl radicals generated in the process to prevent TOC from further mineralization when compared to other pHs.

Popiel et al. (2008) applied different methods to get the destruction of sulfur mustard. Among them, the O_3/UV/H_2O_2 system was tested changing hydrogen peroxide concentration and pH value of the solution. Hydrogen peroxide concentration of 5×10^{-3} $molL^{-1}$ proved to be optimal for the given oxidative system. Besides, authors determined that this system is largely dependent on pH and they determined that mustard oxidation rate is the highest in strong acidic solution (pH 2), moderate in neutral and alkaline solutions, and the rate was the lowest with the pH value close to 5.

Lucas et al. (2010) observed with the $O_3/UV/H_2O_2$ process, under optimum oxidant dose ($COD/H_2O_2=2$), the fastest rate constant (6.5×10^{-3} min^{-1}), at the natural pH of the wastewater. Besides, this process was shown to be highly dependent on the initial pH of the wastewater. On the other hand, an economic analysis of the operating costs of the AOPs processes investigated revealed the $O_3/UV/H_2O_2$ process to be the most economical one (1.31 € $m^{-3}g^{-1}$ of TOC mineralised under optimised conditions) to treat a winery wastewater.

Im et al. (2011) determined that when the $O_3/UV/H_2O_2$ system is applied to remove carbamazepine (CBZ), the ozone concentration significantly influenced CBZ and total organic carbon (TOC) removal as well as total inorganic nitrogen ion production. However, CBZ, TOC removal, and total inorganic nitrogen ion production (T-N) were enhanced with increasing ozone and hydrogen peroxide concentrations up to certain levels, and further increases in ozone and hydrogen peroxide resulted in adverse effects due to hydroxyl radical scavenging by higher oxidant and catalyst concentrations. Finally, UV intensity had the most significant effect on T-N production.

4.2. Combination of Ozone and Fenton Related Treatments

The simultaneous combination of ozone and Fenton related chemistry needs to consider several contributions to the oxidation of organic matter in water. On one hand, direct and indirect ozone reactions according to the mechanism presented in Figure 1, ozone-hydrogen peroxide reactions (reaction (5) and related), Fenton related reactions (28)-(34) and the role of ozone with iron (or other transition metals), this last contribution described through reactions:

$$Fe(II) + O_3 \longrightarrow FeO^{2+} + O_2 \qquad (57)$$

$$FeO^{2+} + H_2O_2 \longrightarrow Fe(III) + {}^{\bullet}OH + OH^- \qquad (58)$$

that gives place to extra-hydroxyl radical production. However, at higher concentrations of Fe(II), FeO^{2+} can be oxidized to Fe(III) without the formation of radical intermediate species (Sauleda and Brillas, 2001; Nawrocki and Kasprzyk-Hodern, 2010), being an important limitation of the process:

$$FeO^{2+} + Fe(II) + 2H^+ \longrightarrow 2Fe(III) + H_2O \qquad (59)$$

Nevertheless, in the presence of light (ozone-photo-Fenton process) photochemical reduction of Fe(III) plays a crucial role giving place to additional hydroxyl radicals according to reaction (46) (Piera et al., 2000; Rodríguez et al., 2012). In this process, when short wavelengths are used (lower than 320 nm) photolysis of hydrogen peroxide and ozone need also to be considered.

On the other hand, oxalic acid is a common by-product of organic matter oxidation or can be added to the photo-Fenton system to improve the efficiency by photo-reducing Fe(III) as it has been shown in reactions (48)-(51).

In addition to this, in presence of ozone, reactions (60)-(65) have been reported (Beltrán et al., 2005) which implies the formation of the ozonide ion radical:

$$Fe(III) + C_2O_4^{2-} \longrightarrow FeC_2O_4^{+} \quad (60)$$

$$FeC_2O_4^{+} + C_2O_4^{2-} \longrightarrow Fe(C_2O_4)_2^{-} \quad (61)$$

$$Fe(C_2O_4)^{-} + C_2O_4^{2-} \longrightarrow Fe(C_2O_4)_3^{3-} \quad (62)$$

$$FeC_2O_4^{+} + O_3 \longrightarrow 2CO_2 + Fe(III) + 2O_3^{\bullet-} \quad (63)$$

$$Fe(C_2O_4)_2^{-} + 2O_3 \longrightarrow 2CO_2 + FeC_2O_4^{+} + 2O_3^{\bullet-} \quad (64)$$

$$Fe(C_2O_4)_3^{3-} + 2O_3 \longrightarrow 2CO_2 + Fe(C_2O_4)_2^{-} + 2O_3^{\bullet-} \quad (65)$$

The ozonide ion radical gives place to hydroxyl radicals according to:

$$O_3^{\bullet-} + H^{+} \longrightarrow HO_3^{\bullet} \quad (66)$$

$$HO_3^{\bullet} \longrightarrow {}^{\bullet}OH + O_2 \quad (67)$$

Therefore, the simultaneous combinations of ozone-Fenton or ozone-photo-Fenton processes lead to the formation of higher amounts of hydroxyl radicals pointing out a synergistic effect between the processes.

CONCLUSION

From the results shown in this chapter it is concluded the important role of hydrogen peroxide in the so called advanced oxidation processes (AOPs).

First, it should be highlighted the common point of many of the AOPs which present some steps where hydrogen peroxide is formed. Second, hydrogen peroxide then combines with other agent mainly ozone, electrons in the conduction band of illuminated catalysts or with iron (and light) in Fenton or Photo-Fenton processes.

Then, the addition of more hydrogen peroxide to these systems allows high concentrations of hydroxyl radicals to be formed enhancing the rate of oxidation and mineralization of water contaminants in different aquatic environments, wastewater included.

Acknowledgments

Authors thank the Spanish Research General Direction and FEDER and the Universidad de Extremadura for economic support (Projects CTQ2009/13459/C05/05 and Action VII-13, respectively). Also, Dr Ana Rey thanks the University of Extremadura for a postdoctoral research contract.

References

Abate, S.; Centi, G.; Melada, S.; Perathoner, S.; Pinna, F.; Strukul, G. "Preparation, performances and reaction mechanism for the synthesis of H2O2 from H2 and O2 based on palladium membranes". *Catal. Today 104 (2005) 323-328.*

Abate, S.; Lanzafame, P.; Perathoner, S.; Centi, G. "SBA-15 as a support for palladium in the direct synthesis of H2O2 from H2 and O2". *Catal. Today* 169 (2011) 167-174.

Abrahamson, H.B.; Rezvani, A.B.; Brushmiller, J.G. "Photochemical and spectroscopic studies of complexes of iron (III) with citric acid and other carboxilyc acids". *Inorg. Chimica Acta* 226 (1994) 117-127.

Abramović, B.F.; Banić, N.D.; Šojić, D.V. "Degradation of thiacloprid in aqueous solution by UV and UV/H_2O_2 treatments". *Chemosphere* 81 (2010) 114-119.

Addamo, M.; Augugliaro, V.; García-López, E.; Loddo, V.; Marcì, G.; Palmisano, L. "Oxidation of oxalate ion in aqueous suspensions of TiO_2 by photocatalysis and ozonation". *Catal. Today* 107-108 (2005) 612-618.

Agladze, G.R.; Tsurtsumia, G.S.; Jung, B.I.; Kim, J.S.; Gorelishvili, G. "The "in-cell" and "ex-cell" Fenton treatment of phenol, 4-chlorophenol and aniline". *J. Appl. Electrochem.* 37 (2007) 385-393.

Aguinaco, A.; Pocostales, J.P.; García-Araya, J,F.; Beltrán. F.J. "Decomposition of Hydrogen Peroxide in the Presence of Activated Carbons of Different Nature". *J. Chem. Technol. Biotechnol.* 86 (2011) 595-600.

Aguinaco, A.; Beltrán F.J.; García-Araya J.F.; Oropesa, A. "Photocatalytic Ozonation to Remove the Pharmaceutical Diclofenac from Water. Influence of Variables". *Chem. Eng. J.* 189-190 (2012) 275-282.

Agustina, T.E.; Ang, H.M; Vareek; V.K. "A review of synergistic effect of photocatalysis and ozonation on wastewater treatment". *J. Photochem. Photobiol.C: Photochem. Rev.* 6 (2005) 264-273.

Al-Hayek, N.; Eymery, J.P.; Doré, M. "Catalytic oxidation of phenols with hydrogen peroxide. Structural study of iron/alumina and iron-copper/alumina catalysts by Moessbauer spectroscopy. *Water Res.* 19 (1985) 657-666.

Alhamedi, F.H.; Rauf, M.A.; Ashraf, S.S., "Degradation studies of Rhodamine B in the presence of UV/H_2O_2". *Desalination* 239 (2009) 159-166.

Álvarez, P.M.; García-Araya, J.F.; Beltrán, F.; Giraldez, I. "Aqueous Ozone Descomposition in the Presence of Granular Activated Carbons of Different Surface Chemical Properties".17th world congress and exhibition. Strasbourg, France, 2005.

Álvarez, P.M.; García-Araya, J.F.; Beltrán, F.; Giraldez, I.; Jaramillo, J.; Gómez-Serrano, V. "The influence of various factors on aqueous ozone decomposition by granular activated carbons and the development of a mechanistic approach". *Carbon* 44 (2006) 3102-3112.

Andreozzi, R.; D'Apuzzo, A.; Marotta, R. "Oxidation of aromatic substrates in water/goethite slurry by means of hydrogen peroxide". *Water Res.* 36 (2002) 4691-4698.

Anheden, M.; Goswami, D.Y.; Svedberg, G. "Photocatalytic treatment of wastewater from 5-fluorouracil manufacturing". *J. Solar Energy Eng.* 118 (1996) 2-8.

Anipsitakis, G.P.; Dionysiou, D.D. "Radical generation by the interaction of transition metals with common oxidants". *Environ. Sci. Technol.* 38 (2004) 3705-3712.

Bach, A.; Shemer, H.; Semiat, R. "Kinetics of phenol mineralization by Fenton-like oxidation". *Desalination* 264 (2010) 188-192.

Baeza, C.; Knappe, D.R.U. "Transformation kinetics of biochemically active compounds in low-pressure UV Photolysis and UV/H_2O_2 advanced oxidation processes". *Water Res.* 45 (2011) 4531-4543.

Bahnemann, D. "Photocatalytic water treatment: solar energy applications". *Sol. Energy* 77 (2004) 445–459.

Bailey, P. S. "Ozonation in Organic Chemistry". Ed. New York, *Academic Press*, 1978.

Balzani, V.; Carassiti, V. "Photochemistry of coordination compounds". *Academic Press*, London, Chap 10 (1970) 145-192.

Bansal, R.T.; Donnet, J.B. y Stoeckli, F., "Activated Carbon". Ed. Marcel Dekker. New York. 1988.

Barrault, J.; Abdellaoui, M.; Bouchoule, C.; Majesté, A.; Tatibouët, J.M.; Louloudi, A.; Papayannakos, N.; Gangas, N.H. "Catalytic wet peroxide oxidation over mixed (Al-Fe) pillared clays". *Appl. Catal. B Environ.* 27 (2000) 225-230.

Bates, H.G.C.; Uri, N. "Oxidation of aromatic compounds in aqueous solution by free radicals produced by photo-excited electron transfer in iron complexes". *J. Amer. Chem. Soc.* 75 (1953) 2754-2759.

Bautista, P.; Mohedano, A.F.; Gilarranz, M.A.; Casas, J.A.; Rodríguez, J.J. "Application of Fenton oxidation to cosmetic wastewaters treatment". *J. Haz. Mat.* 143 (2007) 128-134.

Bautista, P.; Mohedano, A.F.; Casas, J.A.; Zazo, J.A.; Rodríguez, J.J. "An overview of the application of Fenton oxidation to industrial wastewaters treatment". *J. Chem. Technol. Biotechnol.* 83 (2008) 1323-1338.

Bautista, P.; Mohedano, A.F.; Menéndez, N.; Casas, J.A.; Rodríguez, J.J. "Catalytic wet peroxide oxidation of cosmetic wastewaters with Fe-bearing catalysts". *Catal. Today* 151 (2010) 148-152.

Bautista, P.; Mohedano, A.F.; Casas, J.A.; Zazo, J.A.; Rodríguez, J.J. "Highly stable Fe/gamma-Al2O3 catalyst for catalytic wet peroxide oxidation". *J. Chem. Technol. Biotechnol.* 86 (2011) 497-504.

Baxendale, J.H.; Wilson, J.A. "The photolysis of hydrogen peroxide at high light intensities". *Trans. Faraday Soc.* 53 (1957) 344-356.

Beltrán, F.J. Ozone Reaction Kinetics for Water and Wastewater Systems, Lewis Publisher, Boca Ratón, Florida, 2004.

Beltrán, F.J.; Rivas, F.J.; Montero-de-Espinosa, R. "Iron type catalysts for the ozonation of oxalic acid in water". *Water Res.* 39 (2005) 3553-3564.

Beltrán F.J.; Aguinaco, A.; García-Araya J.F.; Oropesa, A. "Ozone and photocatalytic processes to remove the antibiotic sulfamethoxazole from water". *Water Res.* 42 (2008) 3799-3808.

Beltrán F.J., Aguinaco, A.; García-Araya J.F. "Mechanism and kinetics of sulfamethoxazole photocatalytic ozonation in water". *Water Res.* 43 (2009a), 1359-1369.

Beltrán, F.J.; Pocostales, P.; Alvarez, P.; Oropesa, A. Diclofenac removal from water with ozone and activated carbon". *J. Hazard. Mater.* 163 (2009b) 768-776.

Beltrán, F.J.; Pocostales, P.; Alvarez, P.; Aguinaco, A. "Ozone-activated Carbon Mineralization of 17-α-Ethynylestradiol Aqueous Solutions". *Ozone Sci. Eng.* 31 (2009c) 422-427.

Beltrán, F.J.; Aguinaco, A.; García-Araya, J.F. "Kinetic modelling of TOC removal in the photocatalytic ozonation of diclofenac aqueous solutions". *Appl. Catal. B: Environ.* 100 (2010) 289-298.

Beltrán F.J.; Aguinaco, A.; García-Araya J.F. "Application of ozone involving advanced oxidation processes to remove some pharmaceutical compounds from urban wastewaters". *Ozone Sci. Eng.* 34 (2012) 3-15.

Benítez, F.J.; Acero, J.L.; Real, F.J.; Roldán, G.; Casas, F. "Comparison of different chemical oxidation treatments for the removal of selected pharmaceuticals in water matrices". *Chem. Eng. J.* 168 (2011) 1149-1156.

Benkelberg, H.J.; Warneck, P. "Photodecomposition of Fe(III) hydroxo and sulfato complexes in aqueous solution: Wavelength dependence of OH and SO4- quantum yields". *J. Phys. Chem.* 99 (1995) 5214-5221.

Bianco, B.; De Michelis, I.; Veglio, F. "Fenton treatment of complex industrial wastewater: Optimization of process conditions by surface response method". *J. Haz. Mat.* 186 (2011) 1733-1738.

BŁędzka, D.; Gryglik, D.; Olak, M.; Gębicki, J.L.; Miller, J.S. "Degradation of n-butylparaben and 4-tert-octylphenol in H_2O_2/UV system". *Radiat. Phys. Chem.* 79 (2010) 409-416.

Bossmann, S.H.; Oliveros, E.; Göb, S.; Siegwart, S.; Dahlen, E.P.; Payawan Jr., L.; Straub, M.; Worner, M.; Braun, A.M. "New evidence against hydroxyl radicals as reactive intermediates in the thermal and photochemically enhanced Fenton reactions". *J. Phys. Chem.* A 102 (1998) 5542-5550.

Brillas, E.; Sauleda, R.; Casado, J. "Peroxi-coagulation of aniline in acidic médium using an oxygen diffusion cathode". *J. Electrochem. Soc.* 144 (1997) 2374-2379.

Brillas, E.; Boye, B.; Sirés, I.; Garrido, J.A.; Rodríguez, R.M.; Arias, C.; Cabot, P.; Comninellis, C. "Electrochemical destruction of chlorophenoxy herbicides by anodic oxidation and electro-Fenton using a boron-doped diamond electrode". *Electrochim. Acta* 49 (2004) 4487-4496.

Brillas, E.; Sirés, I.; Oturan, M.A. "Electro-Fenton process and related electrochemical technologies based on Fenton's reaction chemistry". *Chem. Rev.* 109 (2009) 6570-6631.

Bühler, R. E.; Staehelin, J.; Hoigne, J. "The ozone descomposition in water by use radiolysis I. HO_2/O_2^- and HO_3/O_3^- as intermediates". *J. Phys. Chem.* 88 (1984) 2560-2564.

Buxton, G.V.; Greenstock, C.L.; Helman, W.P.; Ross, A.B. "Critical review of data constants for reactions electrons, hydrogen atoms and hydroxyl radicals ($\cdot OH/\cdot O^-$) in aqueous solution". *J. Phys. Chem. Ref.* Data 17 (1988) 513-886.

Cai, R.; Kubota, Y.; Fujishima,A. "Effect of copper ions on the formation of hydrogen peroxide from photocatalytic titanium dioxide particles". *J. Catal.* 219 (2003) 214-218.

Can, Z.S.; Çakır, E. "Treatability of Organic Constituents in the Paşaköy Wastewater Treatment Plant Effluent by O_3 and O_3/H_2O_2". *Ozone Sci. Eng.* 32 (2010) 209-214.

Carriazo, J.G.; Centeno, M.A.; Odriozola, J.A.; Moreno, S.; Molina, R. "Effect of Fe and Ce on Al-pillared bentonite and their performance in catalytic oxidation reactions". *Appl. Catal. A Gen.* 317 (2007) 120-128.

Catrinescu, C.; Teodosiu, C.; Macoveanu, M.; Miehe-Brendlé, J.; Le Dred, R. "Catalytic wet peroxide oxidation of phenol over Fe-exchanged pillared beidellite". *Water Res.* 37 (2003) 1154-1160.

Caudo, S.; Centi, G.; Genovese, C.; Perathoner, S. "Homogeneous versus heterogeneous catalytic reactions to eliminate organics from waste water using H2O2". *Top. Catal.* 40 (2006) 207-219.

Centi, G.; Perathoner, S.; Torre, T.; Verduna, M.G. "Catalytic wet oxidation with H2O2 of carboxylic acids on homogeneous and heterogeneous Fenton-type catalysts". *Catal. Today* 55 (2000) 61-69.

Chang, M.W.; Chen, T.S.; Chern, J.M. "Initial degradation rate of p-nitrophenol in aqueous solution by Fenton reaction". Ind. *Eng. Chem. Res.* 47 (2008) 8533-8541.

Chang, E.E.; Liu, T.Y.; Huang, C.P.; Liang, C.H.; Chiang, P.C. "Degradation of mefenamic acid from aqueous solutions by the ozonation and O_3/UV processes". Sep. Purif. Technol. In Press, *Accepted Manuscript,* 2012.

Chelme-Ayala, P.; El-Din, M.G.; Smith, D.W.; Adams, C.D. "Oxidation kinetics of two pesticides in natural waters by ozonation and ozone combined with hydrogen peroxide". *Water Res.* 45 (2011) 2517-2526.

Chen, J.; Ollis, D.F.; Rulkens, W.H.; Bruning,H. "Photocatalyzed oxidation of alcohols and organochlorides in the presence of native TiO_2 and metallized TiO_2 suspensions. Part (II): Photocatalytic mechanisms". *Water Res.* 33 (1999) 669-676.

Chen, W.S.; Lin, S.Z. "Destruction of nitrotoluenes in wastewater by electro-Fenton oxidation". *J. Hazard. Mat.* 168 (2009) 1562-1568.

Chiou, C.H.; Wu, C.Y.; Juang, R.S. "Influence of operating parameters on photocatalytic degradation of phenol in UV/TiO_2 process". *Chem. Eng. J.* 139 (2008) 322-329.

Choi, H.; Al-Abed, S.R.; Dionysiou, D.D.; Stathatos, E.; Lianos, P. "TiO_2-based advanced oxidation nanotechnologies for water purification and reuse". *Sustain. Sci. Eng.* 2 (2010) 229–254.

Chong, M.N.; Jin, B.; Chow, C.W.K.; Saint, C. "Recent developments in photocatalytic water treatment technology: a review". *Water Res.* 44 (2010) 2997–3027.

Chou, S.S.; Huang, C.P. "Application of a supported iron oxyhydroxide catalyst in oxidation of benzoic acid by hydrogen peroxide". *Chemosphere* 38 (1999) 2719-2731.

Chou, S.S.; Huang, C.P.; Huang, Y.H. "Heterogeneous and homogeneous catalytic oxidation by supported gamma-FeOOH in a fluidized bed reactor: Kinetic approach" *Environ. Sci. Technol. 35* (2001) 1247-1251.

Chu, L.; Anastasio, C. "Formation of hydroxyl radical from the photolysis of hydrogen peroxide in ice". *J. Phys. Chem.* A 109 (2005) 6264-6271.

Clark, N.; Knowles, G. "High purity water using H2O2 and UV radiation". *Effluent Water Treatment J.* 22 (1982) 335-341.

Contescu, A.; Vass, M.; Contescu, C., Putyera, K.;Schwarz, J.A."Acid buffering capacity of basic carbons revealed by their continous pK distribution". *Carbon* 36 (1998) 247-258.

Costa, R.C.C.; Moura, F.C.C.; Ardisson, J.D.; Fabris, J.D.; Lago, R.M. "Highly active heterogeneous Fenton-like systems based on Fe0/Fe3O4 composites prepared by controlled reduction of iron oxides". *Appl. Catal. B Environ.* 83 (2008) 131-139.

Criegee, R. "The course of ozonation of unsaturated compounds." Record of Chemical Progress, 18 (1957) 110-120.

Criegee, R. Justus Liebigs Annalen Chemistry, 564, pag 9-20, 1953.

Crowther, N.; Larachi, F. "Iron-containing silicalites for phenol catalytic wet peroxidation". *Appl. Catal. B Environ.* 46 (2003) 293-305.

Daifullah, A.A.M.; Girgis, B.S. "Removal of some subtituyed phenols activated carbon obtained from agricultural waste". *Water Res.* 32 (1998) 1169- 1177.

De Laat, J.; Le, G.T.; Legube, B. "A comparative study of the effects of chloride, sulfate and nitrate ions on the rates of decomposition of H2O2 and organic compounds by Fe(II)/H2O2 and Fe(III)/H2O2". *Chemosphere* 55 (2004) 715-723.

De Laat.; Le, G.T. "Effects of chloride ions on the iron(III)-catalyzed decomposition of hydrogen peroxide on the efficiency of the Fenton-like oxidation process". *Appl. Catal. B Environ.* 66 (2006) 137-146.

De León, M.A.; Castiglioni, J.; Bussi, J.; Sergio, M. "Catalytic activity of an iron-pillared montmorillonitic clay mineral in heterogeneous photo-Fenton process". *Catal. Today* 133 (2008) 600-605.

Deng, N.; Wu, F.; Luo, F.; Xiao, M. "Ferric citrate-induced photodegradation of dyes in aqueous solutions". *Chemosphere* 36 (1998) 3101-3112.

Dionysiou, D.D.; Suidan, M.T.; Bekou, E.; Baudin, I.; Laı□né, J.M. "Effect of ionic strength and hydrogen peroxide on the photocatalytic degradation of 4-chlorobenzoic acid in water". *Appl. Catal. B: Environ.* 26 (2000) 153-171.

Doré M.,"Chimie des Oxidants et traitament des eaux", *Editorial Lavoisier* Publishing, Paris , 1989.

Du, Y.; Zhou, M.; Lei, L. "Role of the intermediates in the degradation of phenolic compounds by Fenton-like process". *J. Haz. Mat.* 136 (2006) 859-865.

Du, W.; Xu, Y.; Wang, Y. "Photoinduced degradation of orange II on different iron (Hydr)oxides in aqueous suspension: Rate enhancement on addition of hydrogen peroxide, silver nitrate, and sodium fluoride". *Langmuir* 24 (2008) 175-181.

Duarte, F.; Maldonado-Hodar, F.J.; Pérez-Cadenas, A.F.; Madeira, L.M. "Fenton-like degradation of azo-dye orange II catalyzed by transition metals on carbon aerogels". *Appl. Catal. B Environ.* 85 (2009) 139-147.

Duarte, F.; Madeira, L.M. "Fenton and photo-Fenton-like degradation of a textile dye by heterogeneous processes with Fe/ZSM-5 zeolite". *Sep. Sci. Technol.* 45 (2010) 1512-1520.

Duesterberg, C.K,; Waite, T.D. "Process optimization of Fenton oxidation using kinetic modeling". *Environ. Sci. Technol.* 40 (2006) 4189-4195.

Duesterberg, C.K.; Waite, T.D. "Kinetic modeling of the oxidation of p-hydroxybenzoic acid by Fenton's reagent: Implications of the role of quinones in the redox cycling of iron". *Environ. Sci. Technol.* 41 (2007) 4103-4110.

Dusenbury, J.; Cannon F.S. "Advanced Oxident Regeneration Pertaining to Granular Activated Carbon Beds for Air Pollution Control". *Carbon* 34 (1996) 1577-1589.

El-Gohary, F.A.; Badawy, M.I.; El-Khateeb, M.A.; El-Kalliny, A.S. "Integrated treatment of olive mill wastewater (OMW) by the combination of Fenton's reaction and anaerobic treatment". *J. Haz. Mat.* 162 (2009) 1536-1541.

Elmolla, E.; Chaudhuri, M. "Optimization of Fenton process for treatment of amoxicillin, ampicillin and cloxacillin antibiotics in aqueous solution". *J. Haz. Mat.* 170 (2009) 666-672.

Evgenidou, E.; Bizani, E.; Christophoridis, C.; Fytianos, K. "Heterogeneous photocatalytic degradation of prometryn in aqueous solutions under UV–Vis irradiation". *Chemosphere* 68 (2007) 1877-1882.

Fajerwerg, K.; Castan, T.; Foussard, J.N.; Perrard, A.; Debellefontaine, H. "Dependency on some operating parameters during wet oxidation of phenol by hydrogen peroxide with Fe-ZSM-5 zeolite". *Environ. Technol.* 21 (2000) 337-344.

Fallmann, H.; Krutzler, T.; Bauer, R.; Malato, S.; Blanco, J. "Applicability of the photo-Fenton method for treating water containing pesticides". *Catal. Today* 54 (1999) 309-319.

Fan, H.J.; Huang, S.T.; Chung, W.H.; Jan, J.L.; Lin, W.Y.; Chen, C.C. "Degradation pathways of crystal violet by Fenton and Fenton-like systems: Condition optimization and intermediate separation and identification". *J. Haz. Mat.* 171 (2009) 1032-1044.

Faust, B.C.; Hoigné, J. "Photolysis of Fe(III)-hydroxy complexes as sources of OH radicals in clouds, fog and rain". *Atm. Environ.* 24A (1990) 79-89.

Faust, B.C.; Zepp, R.G. "Photochemistry of aqueous iron (III) polycarboxylate complexes-roles in the chemistry of atmospheric and surface waters". *Environ. Sci. Technol.* 27 (1993) 2517-2522.

Feng, J.; Hu, X.; Yue, P.L.; Zhu, H.Y.; Lu, G.Q. (2003a) "A novel laponite clay-based Fe nanocomposite and its photo-catalytic activity in photo-assisted degradation of orange II". *Chem. Eng. Sci.* 58 (2003) 679-685.

Feng, J.; Hu, X.; Yue, P.L.; Zhu, H.Y.; Lu, G.Q. (2003b) "Discoloration and mineralization of Reactive Red HE-3B by heterogeneous photo-Fenton reaction". *Water Res.* 37 (2003) 3776-3784.

Feng, J.; Hu, X.; Yue, P.L. "Novel bentonite clay-based Fe-nanocomposite as a heterogeneous catalyst for photo-Fenton discoloration and mineralization of organge II". *Environ. Sci. Technol.* 38 (2004) 269-275.

Feng, J.; Hu, X.; Yue, P.L. "Discoloration and mineralization of Orange II by using a bentonite clay-based Fe nanocomposite film as a heterogeneous photo-Fenton catalyst". *Water Res.* 39 (2005) 89-96.

Fernández, J.; Bandara, J.; López, A.; Albers, P.; Kiwi, J. "Efficient photo-assisted Fenton catalysis mediated by Fe ion son Nafion membranes active in the abatement of non-biodegradable azo-dye". *Chem. Comm.* (1998) 1493-1494.

Fernández, J.; Dhananjeyan, M.R.; Kiwi, J.; Senuma, Y.; Hilborn, J. "Evidence for Fenton photoassisted processes mediated by encapsulated Fe ions at biocompatible pH values". *J. Phys. Chem.* B 104 (2000) 5298-5301.

Fenton, H.J.H. "Oxidation of tartaric acid in presence of iron". *J. Chem. Soc.* 65 (1894) 899-911.

Ferreira de Oliveira, T.; Chedeville, O.; Fauduet, H.; Cagnon, B. "Use of ozone/activated carbon coupling to remove diethyl phthalate from water: Influence of activated carbon textural and chemical properties". *Desalination* 276 (2011) 359-365.

Fox, M.A.; Dulay, H.T.; "Heterogeneous photocatalysis". *Chem. Rev.* 93 (1993) 341–357.

Fujishima, A.; Zhang, X.; Tryk, D.A. "TiO_2 photocatalysis and related surface phenomena". *Surf. Sci. Rep.* 63 (2008) 515–582.

García-Araya, J.F.; Beltrán F.J.; Aguinaco, A. "Diclofenac removal from water by ozone and photolytic TiO2 catalysed processes". *J. Chem. Technol. Biotechnol.* 85 (2010) 798-804.

Garrido-Ramírez, E.G.; Theng, B.K.G.; Mora, M.L. "Clays and oxide minerals as catalysts and nanocatalysts in Fenton-like reactions-A review". *Appl. Clay Sci*. 47 (2010) 182-192.

Gajovic, A.; Silva, A.M.T.; Segundo, R.A.; Sturm, S.; Jancar, B.; Ceh, M. "Tailoring the phase composition and morphology of Bi-doped goethite-hematite nanostructures and their catalytic activity in the degradation of an actual pesticide using a photo-Fenton-like process". *Appl. Catal. B Environ.* 103 (2011) 351-361.

Georgi, A.; Kopinke, F.D. "Interaction of adsorption and catalytic reactions in water decontamination processes Part I. Oxidation of organic contaminants with hydrogen peroxide catalyzed by activated carbon". *Appl. Catal. B Environ.* 58 (2005) 9-18.

Gimeno, O.; Carbajo, M.; Beltrán, F.J.; Rivas, F.J. "Phenol and substituted phenols AOPs remediation". *J. Hazard. Mater.* 119 (2005) 99-108.

Giráldez, I.; García-Araya, J.F.; Beltrán, F.J. "Activated Carbon Promoted Ozonation of Polyphenol Mixtures in Water:□ Comparison with Single Ozonation" Ind. *Eng. Chem. Res.* 46 (2007) 8241–8247.

Glaze, W.H.; Kang J.W.; Chapin, D.H. "The chemistry of water treatment processes involving ozone, hydrogen peroxide and ultraviolet radiation". *Ozone Sci. Eng.* 9 (1987) 335–352.

Gogate, P.R.; Pandit, A.B. "A review of imperative technologies for wastewater treatment. I: oxidation technologies at ambient conditions". *Adv. Environ. Res.* 8 (2004) 501-551.

Goi, A.; Trapido, M. "Hydrogen peroxide photolysis, Fenton reagent and photo-Fenton for the degradation of nitrophenols: a comparative study. *Chemosphere* 46 (2002) 913-922.

Gokulakrishnan, N.; Pandurangan, A.; Sinha, P.K. "Catalytic wet peroxide oxidation technique for the removal of decontaminating agents ethylenediaminetetraacetic acid and oxalic acid from aqueous solution using efficient Fenton type Fe-MCM-41 mesoporous materials". *Ind. Eng. Chem. Res.* 48 (2009) 1556-1561.

Goldstein, S.; Aschengrau, D.; Diamant, Y.; Rabani, J. "Photolysis of aqueous H_2O_2: Quantum yield and applications for polychromatic UV actinometry in photoreactors". *Environ. Sci. Technol.* 41 (2007) 7486-7490.

Goldstein, S.; Rabani, J. "The ferrioxalate and iodide-iodate actinometers in the UV region". *J. Photochem. Photobiol. A Chem.* 193 (2008) 50-55.

Gong, J.; Liu, Y.; Sun, X. "O_3 and UV/O_3 oxidation of organic constituents of biotreated municipal wastewater". *Water Res.* 42 (2008) 1238-1244.

González-Bahamon, L.F.; Hoyos, D.F.; Benítez, N.; Pulgarin, C. "New Fe-immobilized natural bentonite plate used as photo-Fenton catalyst for organic pollutant degradation". *Chemosphere* 82 (2011) 1185-1189.

González-Olmos, R.; Roland, U.; Toufar, H.; Kopinke, F.D.; Georgi, A. "Fe-zeolites as catalysts for chemical oxidation of MTBE in water with H2O2". *Appl. Catal. B Environ.* 89 (2009) 356-364.

Goslich, R.; Dillert, R.; Bahnemann, D. "Solar water treatment principles and reactors". *Water Sci. Technol.* 35 (1997) 137–148.

Grcic, I.; Vujevic, D.; Sepcic, J.; Koprivanac, N. "Minimization of organic content in simulated industrial wastewater by Fenton type processes: A case study". *J. Haz. Mat.* 170 (2009) 954-961.

Guélou, E.; Barrault, J.; Fournier, J.; Tatibouët, J.M. "Active iron species in the catalytic wet peroxide oxidation of phenol over pillared clays containing iron". *Appl. Catal. B Environ.* 44 (2003) 1-8.

Guittonneau, S.; De Laat, J.; Dore, M.; Duguet, J.P.; Bonnel, C. "Comparative study of the photodegradation of aromatic compounds in water by UV and H2O2/UV". *Environ. Technol. Lett.* 9 (1988) 1115-1128.

Guo, L.; Chen, F.; Fan, X.; Cai, W.; Zhang, J. "S-doped alpha-Fe2O3 as a highly active heterogeneous Fenton-like catalyst towards the degradation of acid orange 7 and phenol". *Appl. Catal. B Environ.* 96 (2010) 162-168.

Guzman-Perez, C.A.; Soltan, J.; Robertson, J. "Kinetics of catalytic ozonation of atrazine in the presence of activated carbon". *Sep. Purif. Technol.* 79 (2011) 8-14.

Haber, F.; Weiss, J. "The catalytic decomposition of hydrogen peroxide by iron salts". *Proc. Roy. Soc. Lond.* A 147 (1934) 332-351.

Hassler J.W., "Activated carbon". Ed. Leonard Hill, London. 1967.

He, J.; Tao, X.; Ma, W.H.; Zhao, J.C. (2002a) "Heterogeneous photo-fenton degradation of an azo dye in aqueous H2O2/iron oxide dispersions at neutral pHs". *Chem. Lett.* 1 (2002) 86-87.

He, J.; Ma, W.; Zhao, J.; Yu, J.C. (2002b). "Photooxidation of azo dye in aqueous dispersions of H2O2/α-FeOOH". *Appl. Catal. B Environ.* 29 (2002) 211-220.

He, J.; Ma, W.; Song, W.; Zhao, J.; Qian, X.; Zhang, S.; Yu, J.C. "Photoreaction of aromatic compounds at α-FeOOH/H2O interface in the presence of H2O2: Evidence for organic-goethite surface complex formation". *Water Res.* 39 (2005) 119-128.

Heckert, E.G.; Seal, S.; Self, W.T. "Fenton-like reaction catalyzed by the rare earth inner transition metal cerium". *Environ. Sci. Technol.* 42 (2008) 5014-5019.

Herrmann, J.M. "Fundamentals and misconceptions in photocatalysis". *J. Photochem. Photobiol.* A. 216 (2010) 85–93.

Herrmann, J.M. "Heterogeneous photocatalysis: fundamentals and applications to the removal of various types of aqueous pollutants". *Catal. Today* 53 (1999) 115–129.

Homem, V.; Santos, L. "Degradation and removal methods of antibiotics from aqueous matrices. A Review". *J. Environ. Manag.* 92 (2011) 2304-2347.

Hrvoje, K.; Loncaric, B.A.; Koprivanac, N. "Fenton type processes for minimization of organic content in coloured wastewaters: Part I: Processes optimization". *Dyes and Pigments* 74 (2007) 380-387.

Hsing, H.J.; Chiang, P.C.; Chang, E.E.; Chen, M.Y. "The decolorization and mineralization of Acid Orange 6 azo dye in aquous solution by advanced oxidation processes: A comparative study". *J. Hazard. Mat.* 141 (2007) 8-16.

Huang, C.P.; Dong, C.H.; Tang, Z."Advanced Chemical Oxidation: Its Present Role and Future Potencial in Hazardous Waste Treatment". *Waste Management* 13 (1993) 361-377.

Huang, H.H.; Lu, M.C.; Chen, J.N. "Catalytic decomposition of hydrogen peroxide and 2-chlorophenol with iron oxides". *Water Res.* 35 (2001) 2291-2299.

Huang, C.P.; Huang, Y.H. "Comparison of catalytic decomposition of hydrogen peroxide and catalytic degradation of phenol by immobilized iron oxides". *Appl. Catal. A Gen.* 346 (2008) 140-148.

Im, J.K.; Cho, I.H.; Kim, S.K.; Zoh, K.D. "Optimization of carbamazepine removal in $O_3/UV/H_2O_2$ system using a response surface methodology with central composite design". *Desalination* 285 (2012) 306-314.

Ince, N.H. "Critical effect of hydrogen peroxide in photochemical dye degradation". *Water Res.* 33 (1999) 1080-1084.

Iurascu, B.; Siminiceanu, I.; Vione, D.; Vicente, M.A.; Gil, A. "Phenol degradation in water through a heterogeneous photo-Fenton process catalyzed by Fe-treated laponite". *Water Res.* 43 (2009) 1313-1322.

Jain, R.; Shrivastava, M. "Photocatalytic removal of hazardous dye cyanosine from industrial waste using titanium dioxide". *J. Hazard. Mater.* 152 (2008) 216-220.

Jans, U.; Hoigné, J. "Activated carbon and carbon black catalysed transformation of aqueous ozone into OH-Radicals". *Ozone Sci. Eng.*20 (1998) 67-90.

Jeong, J.; Yoon, J. "Dual roles of $CO_2^{\cdot-}$ for degrading synthetic organic chemicals in the photo/ferrioxalate system". *Water Res.* 38 (2004) 3531-3540.

Ju, L.; Chen, Z.; Fang, L.; Dong, W.; Zheng, F.; Shen, M. "Sol-gel synthesis and photo-Fenton-like catalytic activity of EuFeO3 nanoparticles". *J. Am. Cer. Soc.* 94 (2011) 3418-3424.

Jung, Y.J.; Kim, W.G.; Yoon, Y.; Kang, J.W.; Hong, Y.M.; Kim, H.W. "Removal of amoxicillin by UV and UV/H_2O_2 processes". *Sci. Total Environ.* 420 (2012) 160-167.

Kasiri, M.B.; Aleboyeh, H.; Aleboyeh, A. "Degradation of Acid Blue 74 using Fe-ZSM5 zeolite as a heterogeneous photo-Fenton catalyst". *Appl. Catal. B Environ.* 84 (2008) 9-15.

Khalil, L.B.; Girgis, B.S.; Tawfik, T.A. "Descomposition of H_2O_2 on activated carbons obtained from olives stones". *J. Chem. Technol. Biotechnol.* 76 (2001), p.1132-1140.

Kiwi, J.; Denisov, N.; Gak, Y.; Ovanesyan, N.; Buffat, P.A.; Suvorova, E.; Gostev, F.; Titov, A.; Sarkisov, O.; Albers, P.; Nadtochenko, V. "Catalytic Fe3+ clusters and complexes in Nafion active in photo-Fenton processes. High resolution electron microscopy and femtosecond studies". *Langmuir 18 (2002) 9054-9066.*

Ku, Y.; Lin, H.S. "Decomposition of phorate in aqueous solution by photolytic ozonation". *Water Res.* 36 (2002) 4155-4159.

Kuma, K.; Nakabayashi, S.; Matsunaga, K. "Photoreduction of Fe(III) by hidroxycarboxilyc acids in seawater". *Water Res.* 29 (1995) 1559-1569.

Kusic, H.; Koprivanac, N.; Bozic, A.L. "Minimization of organic pollutant content in aqueous solution by means of AOPs: UV- and ozone-based technologies. *Chem. Eng. J.* 123 (2006) 127–137.

Kusic, H.; Bozic, A.L.; Koprivanac, N. "Fenton type processes for minimization of organic content in coloured wastewaters: Part I: Processes optimization". *Dyes and Pigments* 74 (2007) 380-387.

Lam, F.L.Y.; Hu, X. "A high performance bimetallic catalyst for photo-Fenton oxidation of Orange II over a wide pH range". *Catal. Comm.* 8 (2007) 2125-2129.

Langlais, B.; Reckhow, D.A.; Brink, D.R. Ozone in Water Treatment. Application and Engineering, Lewis Publisher, New York, 1991.

Lau, T.K.; Chu, W.; Graham, N. "Reaction pathways and kinetics of butylated hydroxyanisole with UV, ozonation, and UV/O_3 processes". *Water Res.* 41 (2007) 765-774.

Lee, C.; Yoon, J.; von Gunten, U." Oxidative degradation of N-nitrosodimethylamine by conventional ozonation and the advanced oxidation process ozone/hydrogen peroxide". *Water Res.* 411 (2007) 581-590.

Lee, H.; Shoda, M. "Removal of COD and color from livestock wastewater by the Fenton method". *J. Haz. Mat.* 153 (2008) 1314-1319.

Lee, C.; Sedlak, D.L. "A novel homogeneous Fenton-like system with Fe(III)-phosphotungstate for oxidation of organic compounds at neutral pH values". *J. Mol. Catal. A Chem.* 311 (2009) 1-6.

Legrini, O.; Oliveros, E.; Braun, A.M. "Photochemical processes for water treatment". *Chem. Rev.* 93 (1993) 671-698.

Lei, L.; Gu, L.; Zhang, X; Su, Y. "Catalytic oxidation of highly concentrated real industrial wastewater by integrated ozone and activated carbon". *Appl. Catal. A: Gen.*, 327 (2007) 287–294.

Leitzke, A.; von Sonntag, C. "Ozonolysis of Unsaturated Acids in Aqueous Solution: Acrylic, Methacrylic, Maleic, Fumaric and Muconic Acids". *Ozone Sci. Eng.* 31 (2009) 301-308.

Li, F.B.; Chen, J.J.; Liu, C.S.; Dong, J.; Liu, T.X. "Effect of iron oxides and carboxylic acids on photochemical degradation of bisphenol A". *Biol. Fert. Soils* 42 (2006) 409-417.

Liang, X.L.; Zhong, Y.H.; Zhu, S.Y.; Ma, L.Y.; Yuan, P.; Zhu, J.; He. H.; Jiang, Z. "The contribution of vanadium and titanium on improving methylene blue decolorization through heterogeneous UV-Fenton reaction catalyzed by their co-doped magnetite". *J. Haz. Mat.* 199 (2012) 247-254.

Lin, S.S.; Gurol, M.D. "Catalytic decomposition of hydrogen peroxide on iron oxide: kinetics, mechanism, and implications". *Environ. Sci. Technol.* 32 (1998) 1417-1423.

Liou, R.M.; Chen, S.H.; Hung, M.Y.; Chin, H.; Lay, J. "Fe(III) supported on resin as effective catalyst for the heterogeneous oxidation of phenol in aqueous solution". *Chemosphere* 59 (2005) 117-125.

Liu, X.; Chen, Z.; Wang, L.; Wu, Y.; Garoma, T. "Degradation of polycyclic musk HHCB in water by O_3, UV, and UV/O_3 ".*J. Photochem. Photobiol.*, A: Chem. 230 (2012) 1-9.

Lu, M.C. "Oxidation of chlorophenols with hydrogen peroxide in the presence of goethite". *Chemosphere* 40 (2000) 125-130.

Lucas, M.S.; Peres, J.A. "Removal of COD from olive mill wastewater by Fenton's reagent: Kinetic study". J. Haz. Mat. 168 (2009) 1253-1259.

Lucas, M.S.; Peres, J.A.; Puma, G.L. "Treatment of winery wastewater by ozone-based advanced oxidation processes (O_3, O_3/UV and O_3/UV/H_2O_2) in a pilot-scale bubble column reactor and process economics". *Sep. Purif. Technol.* 72 (2010) 235-241.

Lüking, F.; Köser, H.; Jank, M.; Ritter, A. "Iron powder, graphite and activated carbon as catalysts for the oxidation of 4-chlorophenol with hydrogen peroxide in aqueous solution" *Water Res.* 32 (1998) 2607-2614.

Magario, I.; García-Einschlag, F.S.; Rueda, E.H.; Zygadlo, J.; Ferreira, M.L. "Mechanisms of radical generation in the removal of phenol derivatives and pigments using different Fe-based catalytic systems". *J. Mol. Catal. A Chem.* 352 (2012) 1-20.

Malato, S.; Fernández-Ibáñez, P.; Maldonado, M.I.; Blanco, J.; Gernjak, W. "Decontamination and disinfection of water by solar photocatalysis: Recent overview and trends". *Catal. Today* 147 (2009) 1-59.

Mantzavinos, D. "Removal of benzoic acid derivatives from aqueous effluents by the catalytic decomposition of hydrogen peroxide". *Proc. Saf. Environ. Protect.* 81 (2003) 99-106.

Martínez, F.; Calleja, G.; Melero, J.A.; Molina, R. "Heterogeneous photo-Fenton degradation of phenolic aqueous solutions over iron-containing SBA-15 catalyst". *Appl. Catal. B Environ.* 60 (2005) 181-190.

Martínez, F.; Pariente, M.I.; Melero, J.A.; Botas, J.A.; Gómez, E. (2007a) "Catalytic wet peroxidation of phenol in a fixed bed reactor". *Water Sci. Technol.* 55 (2007) 75-81.

Martínez, F.; Melero, J.A.; Botas, J.A.; Pariente, M.I.; Molina, R. (2007b) "Treatment of phenolic effluents by catalytic wet hydrogen peroxide oxidation over Fe2O3/SBA-15 extruded catalyst in a fixed-bed reactor". *Ind. Eng. Chem. Res.* 46 (2007) 4396-4405.

Martínez, F.; Calleja, G.; Melero, J.A.; Molina, R. (2007c) "Iron species incorporated over different silica supports for the heterogeneneous photo-Fenton oxidation of phenol". *Appl. Catal. B Environ* 70 (2007) 452-460.

Martínez, F.; Pariente, M.I.; Melero, J.A.; Botas, J.A. "Catalytic wet peroxide oxidation process for the continuous treatment of polluted effluents on a pilot plant scale". *J. Adv. Oxid. Technol.* 11 (2008) 65-74.

Martinez-Nieto, L.; Hodaifa, G.; Rodríguez, S.; Giménez, J.A.; Ochando, J. "Degradation of organic matter in olive-oil mill wastewater through homogeneous Fenton-like reaction". *Chem. Eng. J.* 173 (2011) 503-510.

Massa, P.; Dafinov, A.; Cabello, F.M.; Fenoglio, R. "Catalytic wet peroxide oxidation of phenolic solutions over Fe2O3/CeO2 and WO3/CeO2 catalyst systems". *Catal. Comm.* 9 (2008) 1533-1538.

Menéndez, J.A.; Suarez, D.; Fuente, E.; Montes- Morán, M.A., "Contribution of pyrone-type structures to carbon basicity: Theoretical evaluation of the pK_a of cluster models". *Carbon* 37 (1999) 1002-1006.

Miller, R.M.; Singer, G.M.; Rosen, J.D.; Bartha, R. "Sequential degradation of chlorophenols by photolytic and microbial treatment". *Environ. Sci. Technol.* 22 (1988) 1215-1219.

Mills, A.; Lee, K. "A web-based overview of semiconductor photochemistry based on current commercial applications". *J. Photochem. Photobiol. A.* 152 (2002) 233–247.

Molina, C.B.; Casas, J.A.; Zazo, J.A.; Rodríguez, J.J. "A comparison of Al-Fe and Zr-Fe pillared clays for catalytic wet peroxide oxidation". *Chem. Eng. J.* 118 (2006) 29-35.

Molina C.B.; Zazo, J.A.; Casas, J.A.; Rodríguez, J.J. "CWPO of 4-CP and industrial wastewater with Al-Fe pillared clays". *Water Sci. Technol.* 61 (2010) 2161-2168.

Monteagudo, J.M.; Durán, A.; San Martín, I.; Aguirre, M. "Effect of continuous addition of H2O2 and air injection on ferrioxalate-assisted solar photo-Fenton degradation of Orange II". *Appl. Catal. B Environ.* 89 (2009) 510-518.

Mrowetz, M.; Selli, E. "Photocatalytic degradation of formic and benzoic acids and hydrogen peroxide evolution in TiO_2 and ZnO water suspensions. *J. Photochem. Photobiol.: Chem.* 180 (2006) 15-22.

Muñoz, M.; de Pedro, Z.; Casas, J.A.; Rodríguez, J.J. "Assessment of the generation of chlorinated byproducts upon Fenton-like oxidation of chlorophenols at different conditions". *J. Haz. Mat.* 190 (2011) 993-1000.

Mvula, M.; von Sonntag, C. "Ozonolysis of Phenols in Aqueous Solution". *Org. Biomol. Chem.* 1 (2003) 1749-1756.

Navalon, S.; Álvaro, M.; García, H. "Heterogeneous Fenton catalysts based on clays, silicas and zeolites". *Appl. Catal. B Environ.* 99 (2010) 1-26.

Nawrocki, J.; Kasprzyk-Hordern, B. "The efficiency and mechanisms of catalytic ozonation". *Appl. Catal. B Environ.* 99 (2010) 27-42.

Neyens, E.; Baeyens, J. "A review of classic Fenton's peroxidation as an advanced oxidation technique". *J. Hazard. Mater.* 98 (2003) 33-50.

Nicole, I.; De Laat, J.; Dore, M.; Duguet, J.P.; Bonnel, C. "Utilisation du rayonnement ultraviolet dans le traitement des eaux: mesure du flux photonique par actinometrie chimique au peroxyde d'hydrogene". *Water Res.* 24 (1990) 157-168.

Nieto, L.M.; Hodaifa, G.; Rodríguez, S.; Giménez, J.A.; Ochando, J. "Degradation of organic matter in olive-oil mill wastewater through homogeneous Fenton-like reaction". *Chem. Eng. J.* 173 (2011) 503-510.

Noorjahan, A.; Kumari, V.D.; Subrahmanyam, A.; Panda, L. "Immobilized Fe(III)-HY: an efficient and stable photo-Fenton catalyst". *Appl. Catal. B Environ.* 57 (2005) 291-298.

Oliveira, L.C.A.; Silva; C.N.; Yoshida, M.I.; Lago, R.M. "The effect of H_2 treatment on the activity of activated carbon for the oxidation of organic contaminants in water and the H_2O_2 decomposition". *Carbon* 42 (2004) 2279-2284.

Oliveira, L.C.A.; Goncalves, M.; Guerreiro, M.C.; Ramalho, T.C.; Fabris, J.D.; Pereira, M.C. Sapag, K. "A new catalyst material based on niobia/iron oxide composite on the oxidation of organic contaminants in water via heterogeneous Fenton mechanisms". *Appl. Catal. A Gen.* 316 (2007) 117-124.

Oliveira, L.C.A.; Ramalho, T.C.; Souza, E.F.; Goncalves, M.; Oliveira, D.Q.L.; Pereira, M.C.; Fabris, J.D. "Catalytic properties of goethite prepared in the presence of Nb on oxidation reactions in water: Computational and experimental studies". *Appl. Catal. B Environ.* 83 (2008) 169-176.

Oller, I.; Malato, S.; Sánchez-Pérez, J.A.; Maldonado, M.I.; Gernjak, W.; Pérez-Estrada, L.A.; Muñoz, J.A.; Ramos, C.; Pulgarín, C. "Pre-industrial-scale combined solar photo-fenton and immobilized biomass activated-sludge biotreatment". *Ind. Eng. Chem. Res.* 46 (2007) 7467-7475.

Ollis, D.F. "Contaminant degradation in water". *Environ. Sci. Technol.* 19 (1985) 480-484.

Ollis, D.F.; Pelizzetti, E.; Serpone, N. "Photocatalyzed destruction of water contamination". *Environ. Sci. Technol.* 25 (1991) 1523-1529.

Ovejero, G.; Sotelo, J.L., Martínez, F.; Gordo, L. "Novel heterogeneous catalysts in the wet peroxide oxidation of phenol". *Water Sci. Technol.* 44 (2001) 153-160.

Pan, X.M.; Schuchmann, M.N.; von Sonntag, C. "Oxidation of benzene by the OH radical. A product and pulse radiolysis study in oxygenated aqueous solution". *J. Chem. Soc. Perkin Trans.* 2 (1993) 289–297.

Panizza, M.; Cerisola, G. "Removal of organic pollutants from industrial wastewater by electrogenerated Fenton's reagent". *Water Res.* 35 (2001) 3987-3992.

Panizza, M.; Cerisola, G. "Electro-Fenton degradation of synthetic dyes". *Water Res.* 43 (2009) 339-344.

Papadopoulos, A.E.; Fatta, D.; Loizidou, M. "Development and optimization of dark Fenton oxidation for the treatment of textile wastewaters with high organic load". *J. Haz. Mat.* 146 (2007) 558-563.

Pariente, M.I.; Melero, J.A.; Botas, J.A.; Velegraki, T.; Xekoukoulotakis, N.; Mantzavinos, D. "Heterogeneous photo-Fenton oxidation of benzoic acid in water: Effect of operating conditions, reaction by-products and coupling with biological treatment". *Appl. Catal. B Environ.* 85 (2008) 24-32.

Parra, S.; Guasaquillo, I.; Enea, O.; Mielczarski, E.; Mielczarki, J.; Albers, P.; Kiwi-Minsker, L.; Kiwi, J. "Abatement of an azo dye on structured C-Nafion/Fe-ion surfaces by photo-Fenton reactions leading to carboxylate intermediates with a remarkable biodegradability increase of the treated solution". *J. Phys. Chem.* B 107 (2003) 7026-7035.

Parra, S.; Nadtotechenko, V.; Albers, P.; Kiwi, J. "Discoloration of Azo-dyes at biocompatible pH-values through an Fe-histidine complex immobilized on nafion via Fenton-like processes". *J. Phys. Chem.* B 108 (2004) 4439-4448.

Perathoner, S.; Centi, G. "Wet hydrogen peroxide catalytic oxidation (WHPCO) of organic waste in agro-food and industrial streams". *Top. Catal.* 33 (2005) 207-224.

Pérez-Estrada, L; Maldonado, M.I.; Gernjak, W.; Agüera, A.; Fernández-Alba, A.R.; Ballesteros, M.M.; Malato, S. "Decomposition of diclofenac by solar driven photocatalysis at pilot plant scale". *Catal. Today* 101 (2005) 219-226.

Pérez-Moya, M.; Graells, M.; Buenestado, P.; Mansilla, H.D. "A comparative study on the empirical modeling of photo-Fenton treatment process performance". *Appl. Catal. B Environ.* 84 (2008) 313-323.

Peternel, I.; Koprivanac, N.; Kusic, H. "UV-based processes for reactive azo dye mineralization". *Water Res.* 40 (2006) 525-532.

Peyton, G.R. Significance and Treatment of Volatile Organic Compounds in Water Supplies, Lewis Publisher, Chelsea, p. 313, 1990.

Peyton, G.R.; Glaze, W.H. "Destruction of pollutants in water with ozone in combination with ultraviolet radiation. 3. Photolysis of aqueous ozone". *Environ. Sci. Technol.* 22 (1988) 761–767.

Piera, E.; Calpe, J.C.; Brillas, E.; Doménech, X.; Peral, J. "2,4-Dichlorophenoxyacetic acid degradation by catalyzed ozonation: TiO2/UVA/O3 and Fe(II)/UVA/O3 systems". *Appl. Catal. B Environ.* 27 (2000) 169-177.

Pignatello, J.J. "Dark and photoassisted iron(3+)-catalyzed degradation of chlorophenolxy herbicides by hydrogen peroxide". *Environ. Sci. Technol.* 26 (1992) 944-951.

Pignatello, J.J.; Oliveros, E.; MacKay, A. "Advanced oxidation processes for organic contaminant destruction based on the Fenton reaction and related chemistry". *Crit. Rev. Environ. Sci. Technol.* 36 (2006) 1-84.

Pimentel, M.; Oturan, N.; Dezotti, M.; Oturan, M.A. "Phenol degradation by advanced electrochemical oxidation process electro-Fenton using a carbón felt cathode". *Appl. Catal. B Environ.* 83 (2008) 140-149.

Popiel, S.; Nalepa, T.; Dzierżak, D.; Stankiewicz, R.; Witkiewicz, Z. "Rate of dibutylsulfide decomposition by ozonation and the O_3/H_2O_2 advanced oxidation process". *J. Hazard. Mater.* 164 (2009) 1364-1371.

Popiel, S.; Witkiewicz, Z.; Chrzanowski, M. "Sulfur mustard destruction using ozone, UV, hydrogen peroxide and their combination". *J. Hazard. Mater.* 153 (2008) 37-43.

Prieto-Rodríguez, L.; Oller, I.; Zapata, A.; Agüera, A.; Malato, S. "Hydrogen peroxide automatic dosing based on disolved oxygen concentration during solar photo-Fenton". *Catal. Today* 161 (2011) 247-254.

Pupo-Nogueira, R.F.; Trovó, A.G.; da Silva, M.R.A.; Villa, R.D.; Oliveira, M.C. "Fundamentos e aplicações ambientais dos procesos Fenton e foto-Fenton". *Quim. Nova* 30 (2007) 400-408.

Rakowski, S.; Cherneva, D. "Kinetics and mechanism of the reaction of ozone with aliphatic alcohols". *Int. J. Chem. Kinetics* 22 (1990) 321-329.

Ramírez, J.H.; Maldonado-Hodar, F.J.; Pérez-Cadenas, A.F.; Moreno-Castilla, C.; Costa, C.A.; Madeira, L.M. 2007a "Azo-dye orange II degradation by heterogeneous Fenton-like reaction using carbon-Fe catalysts". *Appl. Catal. B Environ.* 75 (2007) 312-323.

Ramírez, J.H.; Costa, C.A.; Madeira, L.M.; Mata, G.; Vicente, M.A.; Rojas-Cervantes, M.L.; López-Peinado, A.J.; Martín-Aranda, R.M. 2007b. "Fenton-like oxidation of Orange II solutions using heterogeneous catalysts based on saponite clay". *Appl. Catal. B Environ.* 71 (2007) 44-56.

Ramírez, H.J.; Vicente, M.A.; Madeira, L.M. "Heterogeneous photo-Fenton oxidation with pillared clay-based catalysts for wastewater treatment: A review". *Appl. Catal. B Environ.* 98 (2010) 10-26.

Rao, Y.F.; Chu, W. "A new approach to quantify the degradation kinetics of linuron with UV, ozonation and UV/O_3 processes ". *Chemosphere* 74 (2009) 1444-1449.

Rao, Y.F.; Chu, W. "Degradation of linuron by UV, ozonation, and UV/O_3 processes—Effect of anions and reaction mechanism". *J. Hazard. Mater.* 180 (2010) 514-523.

Rey, A.; Faraldos, M.; Bahamonde, A.; Casas, J.A.; Zazo, J.A.; Rodríguez, J.J. "Role of the activated carbón surface on catalytic wet peroxide oxidation". *Ind. Eng. Chem. Res.* 47 (2008) 8166-8174.

Rey, A.; Faraldos, M.; Casas, J.A.; Zazo, J.A.; Bahamonde, J.A.; Rodríguez, J.J. "Catalytic wet peroxide oxidation of phenol over Fe/AC catalysts: Influence of iron precursor and activated carbon surface". *Appl. Catal. B Environ* 86 (2009) 69-77.

Rey, A.; Bahamonde, A.; Casas, J.A.; Rodríguez, J.J. "Selectivity of hydrogen peroxide decomposition towards hydroxyl radicals in the CWPO over Fe/AC catalysts". *Water Sci. Technol.* 61 (2010) 2769-2778.

Rey, A.; Zazo, J.A.; Casas, J.A.; Bahamonde, A.; Rodríguez, J.J. "Influence of the structural and surface characteristics of activated carbón on the catalytic decomposition of hydrogen peroxide". *Appl. Catal. A Gen.* 402 (2011) 146-155.

Rigg, T.; Taylor, W.; Weiss, J. "The rate constant of the reaction between hydrogen peroxide and ferrous ions". *J. Chem. Phys.* 22 (1954) 575-577.

Rivas, F.J.; Beltrán, F.J.; Gimeno, O.; Carvalho, F. "Fenton-like oxidation of landfill leachate". *J. Environ. Sci. Health A.* 38 (2003) 371-379.

Rivas, F.J.; Navarrete, V.; Beltrán, F.J.; García-Araya, J.F. "Simazine Fenton's oxidation in a continuous reactor". *Appl. Catal. B Environ.* 48 (2004) 249-258.

Rivera-Utrilla, J.; Sánchez-Polo, M.; Prados-Joya, G.; Ferro-García, M.A.; Bautista-Toledo, I. "Removal of tinidazole from waters by using ozone and activated carbon in dynamic regime". *J. Hazard. Mater.* 174 (2010) 880-886.

Rodríguez, E.; Núñez, B.; Fernández, G.; Beltrán, F.J. "Effects of some carboxylic acids on the Fe(III)/UVA photocatalytic oxidation of muconic acid in water". *Appl. Catal. B Environ.* 89 (2009) 214-222.

Rodríguez, E.M.; Fernández, G.; Álvarez, P.M.; Beltrán, F.J. "TiO2 and Fe(III) photocatalytic ozonation processes of a mixture of emergent contaminants of water". *Water Res.* 46 (2012) 152-166.

Rosales, E.; Pazos, M.; Longo, M.A.; Sanromán, M.A. "Electro-Fenton decoloration of dyes in a continuous reactor: A promising technology in colored wastewater treatment". *Chem. Eng. J.* 155 (2009) 62-67.

Rosal, R.; Rodríguez, A.; Perdigón-Melón, J.A.; Petre,A.; García-Calvo, E. "Oxidation of dissolved organic matter in the effluent of a sewage treatment plant using ozone combined with hydrogen peroxide (O_3/H_2O_2)". *Chem. Eng. J.* 149 (2009) 311-318.

Rosario-Ortiz, F.L.; Wert, E.C. Snyder, S.A. "Evaluation of UV/H_2O_2 treatment for the oxidation of pharmaceuticals in wastewater". *Water Res.* 44 (2010) 1440-1448.

Sahoo, M.K.; Sinha, B.; Marbaniang, M.; Naik, D.B. "Degradation and mineralization of Calcon using UV_{365}/H_2O_2 technique: Influence of pH". *Desalination* 280 (2011) 266-272.

Sánchez, L.; Peral, J.; Domènech, X. "Aniline degradation by combined photocatalysis and ozonation". *Appl. Catal. B: Environ.* 19 (1998) 59-65.

Sánchez-Polo, M.; Rivera-Utrilla, J.; Prados-Joya, G.; Ferro-García, M.A.; Bautista-Toledo, I. "Removal of pharmaceutical compounds, nitroimidazoles, from waters by using the ozone/carbon system". *Water Res.* 42 (2008) 4163-4171.

Santiago, J.; Agüera, A.; Gómez-Ramos, M.M.; Fernández Alba, A.R.; García-Calvo, E.; Rosal, R. "Oxidation by-products and ecotoxicity assessment during the photodegradation of fenofibric acid in aqueous solution with UV and UV/H_2O_2". *J. Hazard. Mater.* 194 (2011) 30-41.

Santos, V.P.; Pereira, M.F.R.; Faria, P.C.C.; Orfao, J.J.M. "Decolourisation of dye solutions by oxidation with H2O2 in the presence of modified activated carbons". *J. Haz. Mat.* 162 (2009) 736-742.

Santos, A.; Yustos, P.; Rodríguez, S.; Romero, A. "Mineralization lumping kinetic model for abatement of organic pollutants using Fenton's reagent". *Catal. Today* 151 (2010) 89-93.

Sauleda, R.; Brillas, E. "Mineralization of aniline and 4-chlorophenol in acidic solution by ozonation catalyzed with Fe2+ and UVA light". *Appl. Catal. B Environ.* 29 (2001) 135-145.

Serpone, N.; Pelizzetti, E. Photocatalysis, Fundamentals and Applications. Wiley-Interscience, New York, 1989.

Siedlecka, E.M.; Mrozik, W.; Kaczynski, Z.; Stepnowski, P. "Degradation of 1-butyl-3-methylimidazolium chloride ionic liquid in a Fenton-like system". *J. Haz. Mat.* 154 (2008) 893-900.

Sirés, I.; Oturan, N.; Oturan, M.A.; Rodríguez, R.M.; Garrido, J.A.; Brillas, E. "Electro-Fenton degradation of antimicrobials triclosan and triclocarban". *Electrochim. Acta* 52 (2007) 5493-5503.

Sirés, I.; Guivarch, E.; Oturan, N.; Oturan, M.A. "Efficient removal of triphenylmethane dyes from aqueous medium by in situ electrogenerated Fenton's reagent at carbon-felt cathode". *Chemosphere* 72 (2008) 592-600.

Song, S.; Liu, Z.; He, Z.; Li, Y.; Chen, J.; Li, C. "Degradation of the biocide 4-chloro-3,5-dimethylphenol in aqueous medium with ozone in combination with ultraviolet irradiation: Operating conditions influence and mechanism". *Chemosphere* 77 (2009) 1043-1051.

Song, S.; Xu, L.J.; He, Z.Q.; Chen, J.M.; Xiao, X.Z.; Yan, B. "Mechanism of the photocatalytic degradation of C.I. Reactive Black 5 at pH 12.0 using $SrTiO_3/CeO_2$ as the catalyst". *Environ. Sci. Technol.* 41 (2007) 5846–5853.

Sotelo, J.L.; Ovejero, G.; Martínez, F; Melero, J.A.; Milieni, A. "Catalytic wet peroxide oxidation of phenolic solutions over a LaTi1-xCuxO3 perovskite catalyst". *Appl. Catal. B Environ.* 47 (2004) 281-294.

Staehelin, J.; Hoigné, J. "Decomposition of ozone in water: rate of initiation by hydroxide ion and hydrogen peroxide." *Environ. Sci. Technol.* 16 (1982) 676-680.

Staehelin, J.; Bühler, R. E.; Hoigné, J. "The ozone descomposition in water by use radiolysis II. HO_4 as chain intermediates." *J. Phys. Chem.* 88 (1984) 5999-6004.

Sun, H.H.; Sun, S.P.; Fan, M.H.; Guo, H.Q.; Quiao, L.P.; Sun, R.X. "A kinetic study on the degradation of p-nitroaniline by Fenton oxidation process". *J. Haz. Mat.* 148 (2007) 172-177.

Suri, R.P.S.; Liu, J.; Hand, D.W.; Crittenden, J.C.; Perram, D.L.; and Mullins, M.E. "Heterogeneous Photocatalytic Oxidation of Hazardous Organic Contaminants in Water". *Water Environ. Fed.* 65 (1993) 665-673.

Tang, W.; Huang, C. "2,4-Dichlorophenol oxidation kinetics by Fenton's reagent". *Environ. Technol.* 17 (1996) 1371-1378.

Tekbas, M.; Yatmaz, H.C.; Bektas, N. "Heterogeneous photo-Fenton oxidation of reactive azo dye solutions using iron exchanged zeolite as a catalyst". *Microp. Mesop. Mat.* 115 (2008) 594-602.

Teresa, Z.; Gunther, G.; Fernando, H. "Chemical oxygen demand reduction in coffee wastewater through chemical flocculation and advanced oxidation processes". *J. Environ. Sci.* 19 (2007) 300–305.

Tomiyasu, H., Fukutomi, H. et al. "Kinetics and mechanism of ozone decomposition in basic aqueous solution." *Inorg. Chem.* 24 (1985) 2962-2966.

Trovo, A.G.; Melo, S.A.S.; Nogueira, R.F.P. "Photodegradation of the pharmaceuticals amoxicillin, bezafibrate and paracetamol by the photo-Fenton process-Application to sewage treatment plant effluent". *J. Photochem. Photobiol. A Chem.* 198 (2008) 215-220.

Turchi, S.C.; Ollis, D.F. "Photocatalytic degradation of organic water contaminants: mechanisms involving hydroxyl radical attack". *J. Catal.* 122 (1990) 178-192.

Ulmann, M.; de Tacconi, N.R.; Augustynski, J. "Behavior of surface peroxo species in the photoreactions at titanium dioxide". *J. Phys. Chem.* 90 (1986) 6523-6530.

Uyguner, C.S.; Bekbolet, M. A review on the photocatalytic degradation of humic substances. In: Nikolau, A., Selcuk, H., Rizzo, L. (Eds.), Control of Disinfection By-Products in Drinking Water Systems. NOVA Science Publishers Inc., New York, pp. 419–446 (Chapter 7.4), 2007.

Vidic, R.D., Tessmer, C.H.; Uranowski, L.J. "Impact of surface properties of activated carbons on oxidative coupling of phenolic compounds". *Carbon.* 35 (1997) 1349-1359.

Voelker, B.M.; Sulzberger, B. "Effects of fulvic acid on Fe(II) oxidation by hydrogen peroxide". Environ. Sci. Technol. 30 (1996) 1106-1114.

Wang, G.S.; Hsieh, S.T.; Hong, C.S. "Destruction of humic acid in water by UV light – catalyzed oxidation with hydrogen peroxide". *Water Res.* 34 (2000) 3882–3887.

Wang, S.; Shiraishi, F.; Nakano, K. "A synergistic effect of photocatalysis and ozonation on decomposition of formic acid in an aqueous solution". *Chem. Eng. J.* 87 (2002) 261-271.

Wang, S. "A comparative study of Fenton and Fenton-like reaction kinetics in decolourisation of wastewater". *Dyes and Pigm.* 76 (2008) 714-720.

Watts, R.J.; Sarasa, J.; Loge, F.J.; Teel, A.L. "Oxidative and reductive pathways in manganese-catalyzed Fenton's reactions". *J. Environ. Eng. ASCE* 131 (2005) 158-164.

Wu, J.J.; Yang, J.S.; Muruganandham, M.; Wu, C.C. "The oxidation study of 2-propanol using ozone-based advanced oxidation processes". *Sep. Purif. Technol.* 62 (2008) 39-46.

Xia, M.; Long, M.; Yang, Y.; Chen, C.; Cai, W.; Zhou, B. "A highly active bimetallic oxides catalyst supported on Al-containing MCM-41 for Fenton oxidation of phenol solution". *Appl. Catal. B Environ.* 110 (2011) 118-125.

Xu, M.; Wang, Q.; Hao, Y. "Removal of organic carbon from wastepaper pulp effluent by lab-scale solar photo-Fenton process". *J. Haz. Mat.* 148 (2007) 103-109.

Xu, B.; Chen, Z.; Qi, F.; Ma, J.; Wu, F. "Comparison of N-nitrosodiethylamine degradation in water by UV irradiation and UV/O_3: Efficiency, product and mechanism". *J. Hazard. Mater.* 179 (2010) 976-982.

Yalfani, M.S.; Contreras, S.; Medina, F.; Sueiras, J. "Phenol degradation by Fenton's process using catalytic in situ generated hydrogen peroxide". *Appl. Catal. B Environ.* 89 (2009) 519-526.

Yalfani, M.S.; Contreras, S.; Llorca, J.; Domínguez, M.; Sueiras, J.E.; Medina, F. "Simultaneous in situ generation of hydrogen peroxide and Fenton reaction over Pd-Fe catalysts". *Phys. Chem. Chem. Phys.* 12 (2010) 14673-14676.

Yalfani, M.S.; Contreras, S.; Medina, F.; Sueiras, J.E. (2011a). "Hydrogen substitutes for the in situ generation of H2O2: An application in the Fenton reaction". *J. Haz. Mat.* 192 (2011) 340-346.

Yalfani, M.S.; Georgi, A.; Contreras, S.; Medina, F.; Kopinke, F.D. (2011b). "Chlorophenol degradation using a one-pot reduction-oxidation process". *Appl. Catal. B Environ.* 104 (2011) 161-168.

Yasunaga, N.; Hirotsuji, J. "Efficient Decomposition of Trichloroethylene (TCE) in Groundwater by Ozone-Hydrogen Peroxide Treatment". *Ozone Sci. Eng.* 30 (2008) 127-135.

Yoon, J.; Lee, Y.; Kim, S. "Investigation of the reaction pathway of OH radicals produced by Fenton oxidation in the conditions of wastewater treatment". *Water Sci. Technol.* 44 (2001) 15-21.

Yu, Y.H.; Ma, J.; Hou, Y.J. "Degradation of 2,4-dichlorophenoxyacetic acid in water by ozone-hydrogen peroxide process". *J. Environ. Sci.* 18 (2006) 1043-1049.

Yuan, F.; Hu, C.; Hu, X.; Qu, J.; Yang, M. "Degradation of selected pharmaceuticals in aqueous solution with UV and UV/H_2O_2". *Water Res.* 43 (2009) 1766-1774.

Zapata, A.; Oller, I.; Rizzo, L.; Hilgert, S.; Maldonado, M.I.; Sánchez-Pérez, J.A.; Malato, S. "Evaluation of operating parameters involved in solar photo-Fenton treatment of wastewater: Interdependence of initial pollutant concentration, temperature and iron concentration". *Appl. Catal. B Environ.* 97 (2010) 292-298.

Zarei, M.; Niaei, A.; Salari, D.; Khataee, A. "Application of response surface methodology for optimization of peroxi-coagulation of textile dye solution using carbon nanotube PTFE cathode". *J. Hazard. Mat.* 173 (2010) 544-551.

Zazo, J.A.; Casas, J.A.; Mohedano, A.F.; Rodríguez, J.J. "Catalytic wet peroxide oxidation of phenol with a Fe/active carbon catalyst". *Appl. Catal. B Environ.* 65 (2006) 261-268.

Zazo, J.A.; Casas, J.A.; Molina, C.B.; Quintanilla, A.; Rodríguez, J.J. "Evolution of ecotoxicity upon Fenton's oxidation of phenol in water". *Environ. Sci. Technol.* 41 (2007) 7164-7170.

Zazo, J.A.; Casas, J.A.; Mohedano, A.F.; Rodríguez, J.J. (2009a) "Semicontinuous Fenton oxidation of phenol in aqueous solution. A kinetic study". *Water Res.* 43 (2009) 4063-4069.

Zazo, J.A.; Fraile, A.F.; Rey, A.; Bahamonde, A.; Casas, J.A.; Rodríguez, J.J. (2009b) "Optimizing calcination temperatura of Fe/activated carbón catalysts for CWPO". *Catal. Today* 143 (2009) 341-346.

Zhang, Y.; Dou, X.M.; Liu, J.; Yang, M.; Zhang, L.P.; Kamagata, Y. "Decolorization of reactive brilliant red X-3B by heterogeneous photo-Fenton reaction using an Fe-Ce bimetal catalyst". *Catal. Today* 126 (2007) 387-393.

Zhang, Z.; Feng, Y.; Liu, Y.; Sun, Q.; Gao, P.; Ren, N. 2010a "Kinetic degradation model and estrogenicity changes of EE_2 (17α-ethinylestradiol) in aqueous solution by UV and UV/H_2O_2 technology". *J. Hazard. Mater.* 181 (2010) 1127-1133.

Zhang, G.; Gao, Y.; Zhang, Y.; Guo, Y. 2010b "Fe2O3-Pillared rectorite as an efficient and stable Fenton-like heterogeneous catalyst for photodegradation of organic contaminants". *Environ. Sci. Technol.* 44 (2010) 6384-6389.

Zou, L.; Zhu, B. "The synergistic effect of ozonation and photocatalysis on color removal from reused water". *J. Photochem. Photobiol. A: Chem.* 196 (2008) 24-32.

Zuo, Y.; Hoigné, J. "Photochemical decomposition of oxalic, glyoxalic and pyruvic-acid catalyzed by iron in atmospheric waters". *Atmosph. Environ.* 28 (1994) 1231-1239.

In: Hydrogen Peroxide
Editors: Gilberto Aguilar and Raphael A. Guzman
ISBN: 978-1-62257-414-8

Chapter 2

ELECTROCHEMICAL APPROACH TO QUANTIFY CELLULAR HYDROGEN PEROXIDE AND MONITOR ITS RELEASE PROCESS FROM LIVING CELLS

Ping Wu, Hui Zhang and Chenxin Cai
Jiangsu Key Laboratory of New Power Batteries,
College of Chemistry and Materials Science, Nanjing Normal University, China

ABSTRACT

Hydrogen peroxide (H_2O_2) is one of significant representative of reactive oxygen species (ROS), it can express diverse physiological and pathological consequences within living systems in a concentration-dependent manner. At its low concentration, H_2O_2 functions as the second messenger for several growth factors, cytokines, and signal transduction. Whereas high concentration of H_2O_2 is a source of oxidative damage to cellular equilibrium leading to aging, Alzheimer's and related neurodegenerative diseases, cardiovascular disorders, even cancer. This chapter introduces the methods to detect cellular H_2O_2, including colorimetry, electrochemical methods, chemiluminescence, fluorescence, and electron spin resonance, etc., and mainly describes the electrochemical approach to quantify the level of H_2O_2 in various cell lines and monitor the release process of H_2O_2 from the living cells. Electrochemical detection approaches depicted in this chapter are based on electrochemical reduction of H_2O_2 catalyzed by enzymes (including proteins) and nanomaterials. Horseradish peroxidase (HRP), and hemoglobin (Hb) were selected as model enzyme and protein, nitrogen-doped graphene (N-graphene) was chosen as a representative of nanomaterials to catalyze H_2O_2. The preparation, immoblization, characterization, and application of the above mentioned catalysts to detect cellular H_2O_2, to study kinetics on the exocytotic process, and to evaluate the whole efficiency of the process in real-time are introduced. These results presented in this chapter open a new way for quantification cellular H_2O_2 level and also develop a new platform for a reliable collection of kinetic information on cellular H_2O_2 release, which could be potentially useful in study of downstream biological effects in physiology and pathology.

1. INTRODUCTION

All primary animal cells are aerobic, i.e., oxygen is involved in their essential cellular functions. The stepwise reduction of oxygen in a biological system will generate a number of ROS (reactive oxygen species) including superoxide anion ($O_2^{\bullet -}$), hydroxyl radical (·OH), hydrogen peroxide (H_2O_2), hypochlorous acid (HOCl), and peroxynitrite ($ONOO^-$), etc [1]. They are regarded as a series of potentially harmful species in some system [2] as well as considered as biochemical mediators in cellular pathology which may be involved in a number of pathological events [3].

H_2O_2 is the most important representative of ROS studied in cellular environments. Emerging evidences support it can exert diverse pathological and/or physiological effects in living systems and participate in physiological processes in a concentration dependent manner. H_2O_2 plays a physiological role as a second messenger for several growth factors, cytokines, signal transduction, and signal amplification. For example, H_2O_2 regulates for a number of oxidative stress-related states [4], H_2O_2 bursts trigger mitogen-activated protein (MAP) kinase and nuclear factor κB (NF-κB) pathways that affect cell proliferation and cell death [5,6]. On the other hand, at its high concentrations, it can cause damage to biological macromolecules and bring about major various pathological events ranging from aging [7-9], Alzheimer's and related neurodegenerative diseases [10,11], cardiovascular disorders [12,13], even to cancer [14-16]. In normal cells, H_2O_2 is scavenged primarily by catalase and secondly by peroxidases (glutathione peroxidase), which catalyze, respectively, its fast disproportionation into oxygen and water or into oxidized organic substrates and water, so that its concentration is usually maintained at nanomolar steady-state levels [17-20]. However, catalase and peroxidases are inhibited by their substrate, at high concentrations of hydrogen peroxide (typically a few tens to a few hundred millimolars) [21,22]. Thus, the balance of cellular environment will be broken, then aerobic cells might remain unprotected against H_2O_2 whenever their concentrations rise rapidly at levels significantly beyond normal physiological ones. Moreover, H_2O_2 is extremely cytotoxic to living cells among all ROS, since its high stability, good penetrability and long lifetime allows it to penetrate into almost any cellular compartment for potentially inducing various harmful biological modifications such as peroxidation of cell membrane lipids, DNA bases, and backbone hydroxylation [23]. Therefore, the determination of cellular H_2O_2 could lead to a better understanding of the clinical consequences of the enhancement in H_2O_2 concentration as well as assisting in studies designed to elucidate the biological effect of H_2O_2 in cells.

A variety of techniques including colorimetry [6.24], electrochemical methods [25-27], chemiluminescence [28-31], fluorescence [32-35], and electron spin resonance [36,37], etc., have been developed for assay of cellular H_2O_2. Among these methods, amperometric technique has attracted much attention in both *in vivo* and *ex vivo* biological application because it possesses good temporal resolution for rapid and real-time measurements, offering great potential to image the dynamic release process of H_2O_2 from cells. Moreover, this kind of technique can also provide much more precise kinetic information on the exocytotic process itself and allow evaluation in real-time of the whole efficiency of the process [38].

This chapter introduces the methods to detect cellular H_2O_2, including colorimetry, electrochemical methods, chemiluminescence, fluorescence, and electron spin resonance, etc., and mainly describes the electrochemical approach to quantify the level of H_2O_2 in various cell lines and monitor the release process of H_2O_2 from the living cells. Electrochemical detection approaches depicted in this chapter are based on electrochemical reduction of H_2O_2 catalyzed by enzymes (including proteins) and nanomaterials. Horseradish peroxidase (HRP), and hemoglobin (Hb) were selected as model enzyme and protein, nitrogen-doped graphene (N-graphene) was chosen as a representative of nanomaterials to catalyze H_2O_2. The preparation, immoblization, characterization, and application of the above mentioned catalysts to detect cellular H_2O_2, to study kinetics on the exocytotic process, and to evaluate the whole efficiency of the process in real-time are introduced. These results presented in this chapter open a new way for quantification cellular H_2O_2 level and also develop a new platform for a reliable collection of kinetic information on cellular H_2O_2 release, which could be potentially useful in study of downstream biological effects in physiology and pathology.

2. Detection Techniques for Cellular H_2O_2

There has been various detection techniques developed for assay for cellular H_2O_2, such as colorimetry, electron spin resonance, fluorescence, chemiluminescence, and electrochemical methods, etc.

2.1. Colorimetry

Colorimetric detection method of H_2O_2 is based on spectrophotometer record on the absorption of the reaction product of H_2O_2 and color reagent that contains a dye. For examples, colorimetric H_2O_2 kit (Sigma, Ltd) employs a xylenol orange in an acidic solution with sorbitol and ammonium iron sulfate that reacts to produce a purple color proportional to the concentration of H_2O_2 in the sample. Titanium-based assays (Ti-PAPS reagents) were developed for spectrophotometric detection of H_2O_2 [39].

The Fox assay performed on ferrous ion oxidation in the presence of the ferric ion indicator xylenol orange under acidic conditions [40]. Tanner and co-workers studied the detection of H_2O_2 through a reaction with pyridine-2,6-dicarboxylic acid and vanadate(V) in acidic solution to form a orange-red complex chelate complex, oxo-peroxo-pyridine-2,6-dicarboxylato vanadate (V) detectable at 432 nm [41].

Recently, Luo and co-workers developed a detection method based on oxidation of methyl orange using an iron-catalyzed Fenton reaction system under acidic conditions [42]. Moreover, horseradish peroxidase (HRP)-catalyzed reaction is one of the most popular enzymatic assays used for determination of H_2O_2 [43], Zhou et al, used this peroxidase system for determination of H_2O_2. H_2O_2 reacted with 4-aminoantipyrine and phenol to produce a stable red product in the presence of peroxidase, measured the absorbance of the red product at 505 nm.

Although these methods are sensitive, they are limited by interference from metal ions, restriction with pH environments of reaction system, and inhibition from chelators. In addition, if intracellular dyes are used, which capture the analyte inside the cells, will largely increase the complication of detection.

2.2. Electron Spin Resonance

Electron spin resonance (ESR) technique depends on the fact that certain atomic systems have a permanent magnetic moment. Transitions among the energy levels of the magnetic system influenced by the surrounding atoms and by external magnetic fields levels can be detected by monitoring the power absorbed from an alternating magnetic field, or by absorption of light. ESR spectroscopy is used for studying chemical species that have one or more unpaired electrons, such as organic and inorganic free radicals or inorganic complexes possessing a transition metal ion.

During the generation process of cellular H_2O_2, H_2O_2 formed by $O_2^{\bullet-}$ disproportionation, and it may act as a potential source of hydroxyl radicals $OH^{\bullet}$ through Fenton reaction in the presence of Fe(II) free ions or metalloenzymes. Therefore, ESR can be used as an investigative tool for the study of H_2O_2-related ROS radicals, since the radicals typically produce an unpaired spin on the molecule from which an electron is removed. For Park et al, used ESR spectroscopy to study $OH^{\bullet}$ radicals and H_2O_2 scavenging activity from Sargassum thunbergii [44]; Khachatryan reported the yields of ·OH, $O_2^{\bullet-}$, and H_2O_2 generated by EPFR of 2-monochlorophenol associated with copper oxide-containing silica particles in phosphate buffered saline (PBS) solutions using a 5,5-dimethyl-1-pyrroline-N-oxide (DMPO) spin-trapping agent in conjunction with ESR spectroscopy [45]. In these results, ESR is unique technique that specifically and directly ''see'' unpaired spins of free radicals. However, ESR detection is an indirect determination of ROS via quantification of products of their reactions. A significant limitation of this technique is its relatively low sensitivity especially under aqueous conditions at ambient temperature and ESR technique requires a specialized and expensive ESR spectrometer.

2.3. Fluorescence

Fluorescent probes combined with fluorescence spectroscopy are well suited to detection H_2O_2 in cells. Usually, fluorescent probes are non-fluorescent (or weakly fluorescent) compounds, but which can yield fluorescent products upon reaction with ROS H_2O_2. Several types of reagents have been used as probes for examining for H_2O_2 detection, including dihydro analogues of fluorescent dyes (e.g. 2',7'-dichlorod ihydrofluorescein (DCFH), Amplex Red, dihydrorhodamine123), phosphine-based fluorophores, lanthanide coordination complexes, and chromophores with ROS-cleavable protecting groups [46-50]. For instances, Chang and his co-workers developed a selective, cell-permeable probe Peroxyfluor-1 for H_2O_2 in living cells. They also report a new series of monoboronate probes that respond to H_2O_2 over a range of rationally tunable visible excitation and emission wavelengths. Three derivatives, peroxyfluor-3 (PF3), peroxy yellow 1(PY1), and peroxy orange 1 (PO1), are sensitive enough to image H_2O_2 signals produced in RAW 264.7 macrophages and A431 cells during immune response and growth factor stimulation, respectively. However, limitations of currently available H_2O_2-responsive probes include interfering background fluorescence from other ROS, the need for an external activating enzyme, lack of water solubility or compatibility, and/or excitation profiles in the ultraviolet region, which can damage living samples and cause interfering auto-fluorescence from native cellular species.

2.4. Chemiluminescence

Chemiluminescence (CL) is observed when the electronically excited product of an exoergic reaction relaxes to its ground state with emission of photons, and it can be defined in simplistic terms: chemical reactions that emit light (ultraviolet, visible or infra-red radiation). H_2O_2 can generate electronically excited products, which emit the weak CL during their decay to the ground state. Although it is not strong enough to detect the light emission directly by CL techniques, it can be enhanced by CL reagents, such as luminol and lucigenin. In general, in the presence of horseradish peroxide (HRP) and H_2O_2, luminol will produce a strong, durable blue light with an emission maximum at 425 nm. Kawamato and his group has performed systematic studies of the luminal- H_2O_2-HRP assay for H_2O_2 [51,52]. Li et al. have reported hemoglobin (Hb) as another model enzyme to detect H_2O_2 [53]. The metal-ion-catalyzed CL oxidation of luminal by H_2O_2 in alkaline solution has also been studied extensively to determine H_2O_2, such as Co(II) and Cu(II) [54,55]. Moreover, peroxyoxalate (PO) and many PO compounds including bis[2-(3,6,9-trioxadecarbonyloxycarbonyl)-4-nitrophenyl] oxalate (TDPO) and bis(2,4,6-trichlorophenyl) oxalate (TCPO), et al are one type of sensitized CL for assay H_2O_2 [56]. Although, CL can provide an effective detection and quantification of ROS with high sensitivity, wide linear range, and considerable reduction of background noise, the presence of CL reagents will make the detection system more complex.

2.5. Electrochemical Methods

Electrochemical detection of H_2O_2 has received considerable attention and become an attractive method for in vivo and ex vivo biological application. Compared with other methods, the electrochemical method can offer the direct, rapid, and real-time measurements with no need for using the poor stability and easy photobleaching fluorescent probe molecule, adding chemiluminescence reagents, or detecting under rigorous environment conditions. Electrochemical measurements of a cellular H_2O_2 generally rely on the reduction/oxidation of the molecules released at the electrode surface located close to the cell membrane. The recorded currents thus evidence the nature of the molecules released and their quantities as well as the dynamics of the release itself. Two main electrochemical techniques have been used, amperometry and cyclic voltammetry.

In amperometry, the working electrode is held at a constant potential at which the molecules under investigation are reducible/oxidizable. It ensues a continuous recording of the reduction/oxidation current as a function of the time without severe contamination by capacitive current. Therefore, amperometry detection method can possess the best temporal resolution for measurements, offer great potential to image the dynamic release process of H_2O_2.

Moreover, it can also provide much more precise kinetic information on the exocytotic process itself and allow evaluation of the whole efficiency of the process in real-time. In the following, the amperometric detection method to assay cellular H_2O_2 in our previous work is illustrated.

3. Electrochemical Detection Based on Electrochemical Reduction of H_2O_2 Catalyzed by Protein/Enzymes

Electrochemical approaches for the determination of H_2O_2 are often based on the direct oxidation of H_2O_2 on the electrode surface. However, the oxidation of H_2O_2 usually requires a relatively high positive potential (usually over +0.6 V versus SCE).

Many other electroactive species coexisting in the biological fluids can also be oxidized at such a high potential, and their electrochemical signals thus severely affect the selectivity of the assay. Alternatively, H_2O_2 can be electrochemically detected based on its reduction because the electrochemical reduction of H_2O_2 that occurred at a relatively negative potential does not give rise to the interference from those common interfering species [57,58]. Therefore, we depict amperometry detection method based on heme proteins or enzymes can catalyze the reduction of H_2O_2 with high sensitivity and selectivity in this chapter [59-63]. We chose HRP and Hb as models of redox protein and enzyme, hydroxyapatite (HAP) and attapulgite (attapulgite) nanostructures with large surface area and fine biocompatibility as the protein/enzyme immobilization matrix. In the following paragraph, we describe the fabrication and characterization of three novel nanohybrid (HRP-HAP, HRP-attapulgite and Hb-attapulgite) and their application in assay of cellular H_2O_2.

3.1. Fabrication of Protein/Enzyme-Based Biosensors

Generally, enzyme-based biosensors are developed by immobilizing protein/ enzymes on the surface of the suitable materials or on the electrode surface because the direct use of enzymes in solution suffers the poor stability [64,65]. To keep the intrinsic secondary structure of the enzyme and enhance the efficiency of the biosensor, it is critical to choose a biocompatible material for immobilizing the enzymes. HAP adopts the ideal composition $Ca_{10}(PO_4)_6(OH)_2$ that is thermally stable and contains both acid and base sites, including hydroxylenriched channels [66].These structural features allow HAP to have good biological properties such as non-toxicity, and good biocompatibility [67], etc. The TEM image on shape of HAP is displayed in Figure 1A. The TEM image indicates that the HAP nanostructures have a needle-like morphology with an average size of 20–25 nm in width and 160–180 nm in length. The HAP nanostructures have a rough surface characteristic. This feature is very useful for the nanostructures being used for substrate to immobilize the enzymes because the rough surface can significantly enhance the enzymes loading and therefore, increasing the electrochemical response. And Attapulgite, a natural nanostructure clay, has a special laminated chain structure with specific fibrous morphology (as shown in Figure 1B), high surface area, good thermal resistance, moderate absorptive capacity, and fine biological affinity, providing much potential for diverse applications such as catalyst supports, nanocomposites, pharmaceutical industry and environmental absorbents [68]. HRP and Hb (three dimensional structures of enzyme molecules are shown in Figure 2) were selected as amodel enzyme/protein to fabricate the H_2O_2 enzyme-based biosensor due to their high electrocatalytic activity toward the reduction of H_2O_2, known structure, and commercial availability [69].

(A) 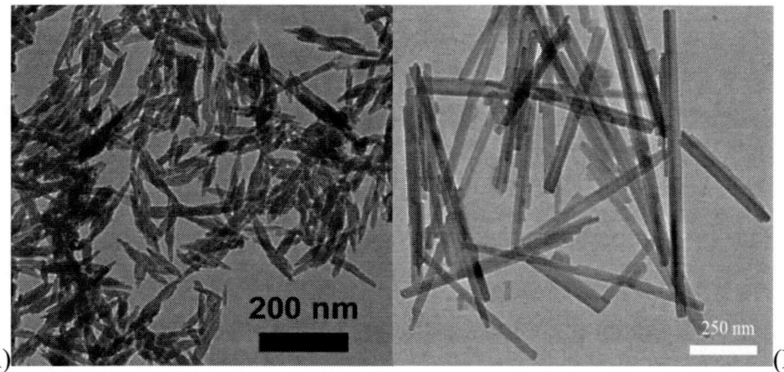 (B)

Figure 1. TEM images of HAP(A) and attapulgite (B), respectively. Panel A is cited from Ref. 26, and Panel B is cited from Ref. 27.

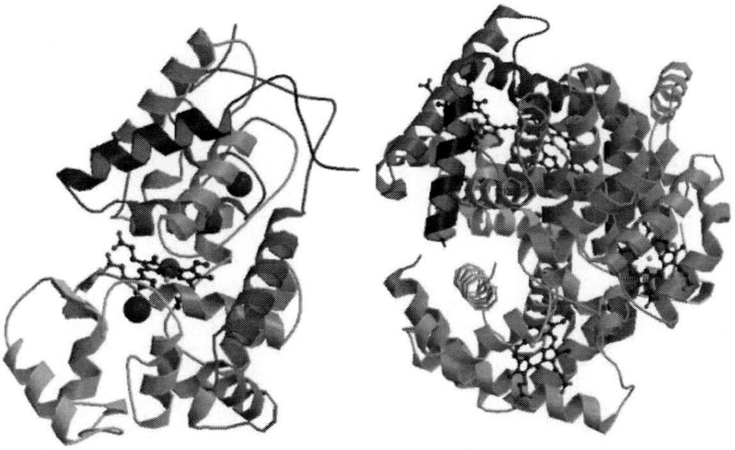

(A) (B)

Figure 2. The structure of HRP and Hb (adapted from protein data bank).

For the fabrication HRP-HAP nanohybid, HAP (10 mg) was dispersed into HRP solution (2 mL, 5 mg mL^{-1} in pH 7.4 PBS) by continuously stirring at 4 ºC for 2 h. HRP was assembled on the surface of the negatively charged HAP via electrostatic interaction. After that, the mixture was centrifuged and the HRP-HAP nanohybrids were collected by removing the supernatant.

The nanohybrids were thoroughly washed with PBS to remove the loosely assembled enzyme molecules.

To fabricate the HRP-HAP/GC electrode, the HRP-HAP nanohybrid was dispersed into PBS (pH 7.4) to form a homogeneous suspension (5 mg mL^{-1}). Then, the suspension (4 mL) was cast onto the surface of the glassy carbon (GC, 3 mm in diameter, CH Instruments) electrode with a microsyringe, and the solvent was allowed to evaporate at ambient temperature before use. The HRP-HAP/GC electrode was stored at 4 ºC when not in use.

By similar procedures, HRP-attapulgite/GC, Hb-attapulgite/GC, HRP/GC, Hb/GC, HAP/GC, and attapulgite/GC electrode were fabricated.

3.2. Characterization of Protein/Enzyme-Based Biosensors

Several techniques including scanning electron microscopic (SEM), atomic force microscopy (AFM), FTIR, UV-Vis, Circular dichroic (CD) spectroscopy have been used for characterization of the preparation of protein/enzyme-based biosensors.SEM images can provide information on the assembly of HRP on the surface of the HAP nanostructures. The image indicates that the needle-like nanostructures distribute almost uniformly on the entire surface of the GC electrode, exhibiting a special three-dimensional porous structure (Figure. 3a), which can significantly increase the effective surface of the electrode for loading biomolecules. The image of the HRP-HAP nanohybrids present in Figure. 3b shows that the size of the HRP-HAP nanohybrids is slightly larger than that of HAP nanostructures, indicating the assembly of HRP on the surface of HAP. Moreover, the morphology of HRP-HAP nanohybrids on the electrode surface has a similar three-dimensional structure to that of HAP nanostructures. Such a structure is expected to be very attractive for the detection of substrates because each of the HRP-HAP nanohybrids is fully and easily accessible to substrates, and thus can be used as an electrochemical sensing unit, yielding a high signal-to-noise ratio for electrochemical determination.AFM images can reflect the change of size and shape of immobilizing material prior or upon protein/enzyme loading more intuitively. Take Hb-attapulgite nanohybrids as an instance, attapulgite present in AFM image shows a needle-like morphology with an average size of 20–50nm in width and 0.5–1μm in length (Figure 4A). Upon immobilizing Hb, the morphology of attapulgite changes significantly. Some small and uniformly distributed islandlike nanostructures are clearly observed at the surface of attapulgite.

Moreover, the thickness of attapulgite has a significant increase from 50 nm to 65 nm with the assembly of Hb (Figure 4B), indicating the formation of Hb-attapulgite hybrid. Spectroscopy techniques are powerful tools to characterize the immobilization of protein/enzyme onto nanomaterials as well as test whether the secondary conformation of protein/enzyme has changed during the immobilization process.

In this chapter, we selected the HRP-attapulgite nanohybrids as an example to describe loading protein/enzyme on the surface of immobilizing matrix characterized by UV-vis, CD, and FTIR spectroscopy.

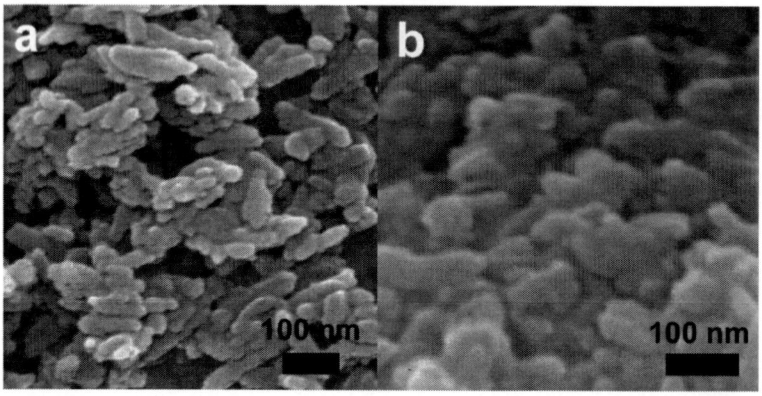

Figure 3. SEM images of HAP nanostructures (a) and HRP-HAP nanohybrids(b). Cited from Ref. 26.

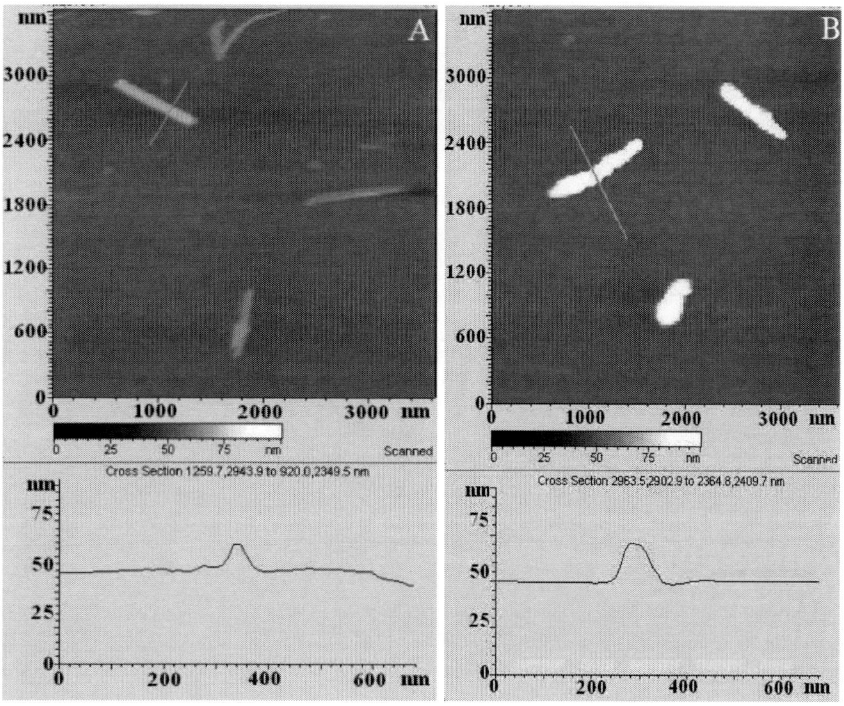

Figure 4. Typical AFM images and the cross–sectional analysis of attapulgite with (B) and without (A) assembly of Hb.

The UV-vis spectrum of attapulgite shows a feature-less curve (curve a, Figure 5A), however, HRP-attapulgite shows the Soret band at 403 nm (curve c, Figure 5A), indicative of the effective assembly of HRP on the attapulgite nanostructures.

The features of the peak are identical with those of native HRP dissolving in solution (403 nm, curve b, Figure 5A), demonstrating that no significant denaturation occurred to the enzyme after being assembled onto the surface of the nanostructures. This conclusion can be further supported by CD spectra. The spectrum in the far-UV region (200–250 nm) corresponds to the peptide $n \rightarrow \pi^*$ electronic transition and it is sensitive to the conformational changes of proteins [70].

The spectra of HRP aqueous solution in this region exhibit two negative peaks at ca. 207 and 222 nm (curve a, Figure 5B), respectively, which are the characteristic features of heme proteins [71]. The CD spectrum of HRP-attapulgite (curve b, Figure 5B) in this region is also characterized by two negative peaks, whose positions are almost the same as those depicted in curve a (Figure 5B), indicating that the structure and the conformation of HRP after being confined on the surface of attapulgite nanostructures are essentially the same as those in its natural state.

IR spectrum of attapulgite shows the bands of 989, 1198 cm^{-1} and 483, 1035cm^{-1} (curve b in Figure 6), which are ascribed to the stretch and in–plane vibrations of Si–O bond, respectively [72-74]. The bands at 3619 and 3550 cm^{-1} are attributed to stretching mode of the surface –OH on attapulgite [72-74].

The deformation vibration mode of adsorbed water appears at 1650 cm^{-1}[75]. After the assembly of HRP, the major IR bands of protein appear at 1653 cm^{-1} (amide I, the C=O stretching vibrations) and 1540 cm^{-1} (amide II, the N–H bending with contribution of the C–N stretching vibrations) [76-79], respectively (curve c in Figure 6). The shape and position of these bands are very similar to those of the native HRP (1652 and 1539 cm^{-1}, curve a in Figure 6), indicating that HRP has been effectively assembled on the surface of attapulgite nanostructures.

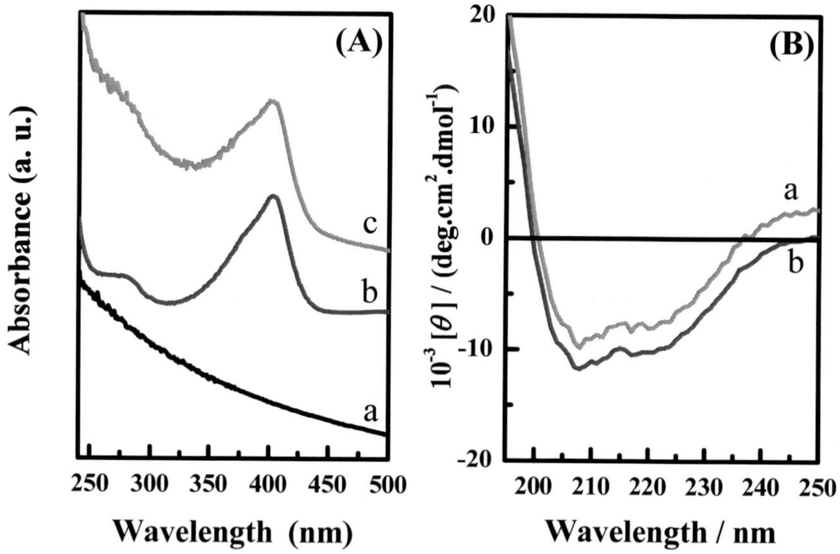

Figure 5. Panel A: UV–vis spectra of attapulgite nanostructures (a), HRP in solution (0.5mg mL^{-1} (b), and the HRP-attapulgite nanohybrids (c). Panel B: CD spectra of HRP in solution (a) and HRP-attapulgite nanohybrids (b) in the far-UV region. Curve (a) is displayed by two units' shifts for clarity. Cited from Ref. 27.

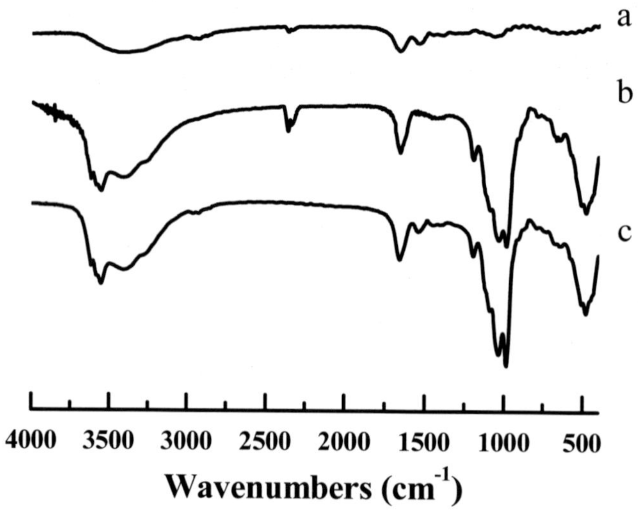

Figure 6. IR spectra of HRP (a), attapulgite (b), and the HRP-attapulgite hybrid (c). Cited from Ref. 27.

Moreover, these results also demonstrate that HRP molecules can retain their native secondary conformation after assembly. This conclusion is in good agreement with that obtained from UV and CD spectroscopy.

3.3. Electrocatalytic Reduction of H_2O_2 Catalyzed by Protein/Enzyme

Enzyme/redox protein is one sort of catalysts which takes some advantages over other inorganic or organic catalysts, such as high efficiency of catalysis, high affinity to substrate, mild reaction conditions, and easy of regulation. Usually enzymes catalyze only a single type of reaction, and often they work only on one or a few substrate compounds under mild conditions. However, chemical catalysts often require high temperature and/or pressure to function.

For example, nitrogen can be reduced to ammonia industrially by the Haber process, catalyzed by iron at 500° C. and at 300 atmospheres pressure of N_2. In contrast, the same reaction is carried out enzymatically at 25°C and less than 1 atmosphere pressure of N_2. These gentle conditions of temperature, pressure, and pH characterize enzymatic catalysis. Moreover, enzymes are often regulated. An enzyme's effectiveness can be altered without changing the concentration of the enzyme; on the other hand, the effectiveness of a chemical catalyst is generally determined by its overall concentration. Therefore, we discussed the use of enzyme (take HRP as a representative of enzyme model) to reduce H_2O_2 as an electrocatalyst in this chapter.

The electrocatalytical reduction toward H_2O_2 by HRP was firstly studied through the voltammetric responses on the HRP-attapulgite/GC electrode. At the absence of H_2O_2, the voltammetric response of the electrode is characterized by a pair of well–defined redox peak with the cathodic (E_{pc}) and anodic (E_{pa}) peak potential of ca. –360 and –295 mV (at 100 mV s^{-1}, in N_2–saturated PBS), respectively (curve a, panel A, Figure 7), which can be ascribed to the direct electron transfer (DET) reaction of HRP assembled on the nanostructures. This result indicates that attapulgite can facilitate the effective DET between HRP and the electrode surface in spite of the large molecular structure of HRP.

Upon addition of H_2O_2 (0.2 mM), the shape of the cyclic voltammogram of the electrode changes significantly, being characterized by a large cathodic peak at ca.–410 mV (curve b, panel A) and complete disappearance of the anodic peak. The increase of H_2O_2 concentration leads to the enhancement of the cathodic current (not shown here). These characteristics are the typical features of electrocatalytic reactions.

The reduction of H_2O_2 starts at a potential of ca.–100 mV, and the observed reduction current increases quite rapidly during negative potential scanning and reaches the highest catalytic current at ca. –410 mV.

The results presented in panels B, C, and D (Figure 7) show that the reduction current can only be observed at the potential more negative than –400 mV, demonstrating that electrocatalytic activity of bare GC, attapulgite, or HRP immobilized on the GC electrode toward the reduction of H_2O_2 is low.

Therefore, it can be concluded that HRP–attapulgite nanohybrids presents a good electrocatalyst for the reduction of H_2O_2, indicating that HRP assembled on the attapulgite nanostructures retains its intrinsic and good electrocatalytic activities towards its substrates.

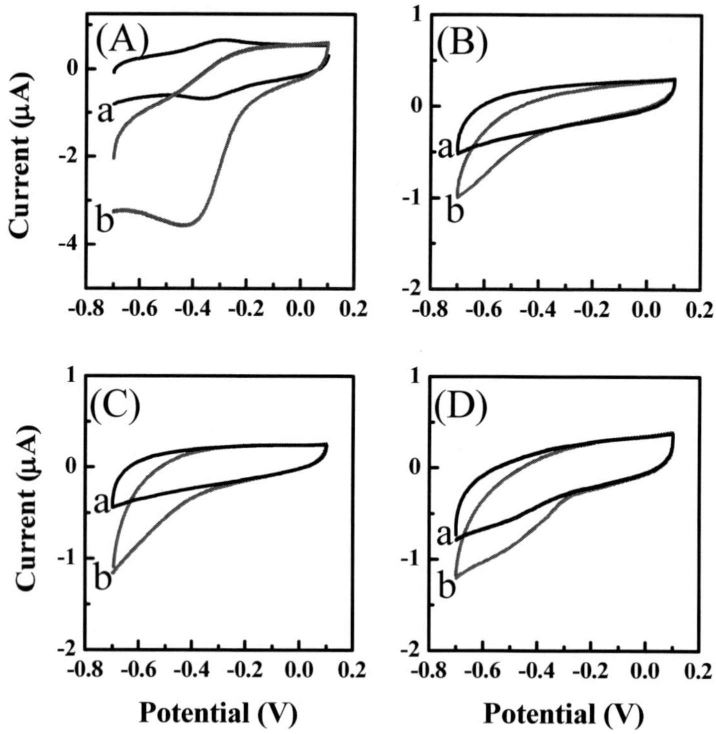

Figure 7. Voltammetric responses of the HRP-attapulgite/GC (panel A), bare GC (panel B), attapulgite/GC (panel C), and HRP/GC electrodes (panel D) in N_2–saturated PBS (0.1 M, pH 7.4) with (curve b) and without (curve a) 0.2 mM H_2O_2. Scan rate was 100 mV s^{-1}. Cited from Ref. 27.

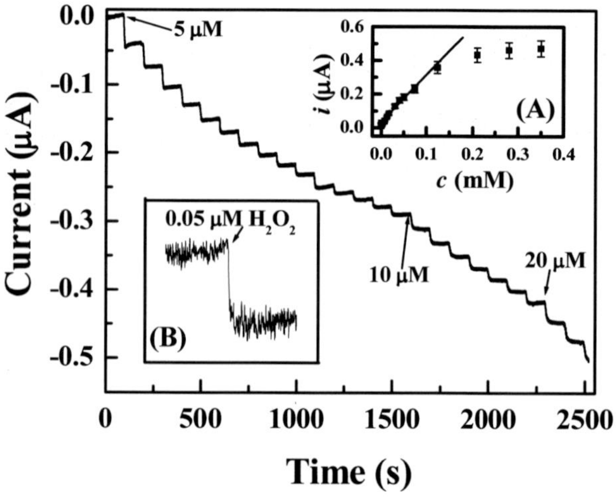

Figure 8. Amperometric responses of the HRP–attapulgite/GC electrode to the successive addition of H_2O_2 in PBS (0.1 M, pH 7.4) at an applied potential of –400 mV. The insets show the dependence of the response of the electrode on H_2O_2 concentration under the optimal conditions (inset A), and the relationship of the signal-to-noise ratio at S/N of 3 (inset B). Cited from Ref. 27.

The electrocatalytic ability of HRP to H_2O_2 was then evaluated by amperometric techniques. Figure 8 depicts the typical steady state amperometric response of the HRP–attapulgite/GC electrode to the successive addition of H_2O_2 in PBS (pH 7.4) solution at a detection potential of –400 mV, which was selected by considering the voltammetric response of H_2O_2 at the electrode and the noise at different detection potentials (Figure 9A). As shown in Figure 9A, the responses of 0.2 mM H_2O_2 at the HRP-attapulgite/GC electrode at different detection potentials were recorded and compared. The potential range was from 0 to –500 mV with a settled interval of 100 mV. Basically, the response keeps on rising when the potential is moved from 0 to –400 mV and then reaches a plateau at more negative potentials. However, a more negative potential leads to increase in noise.

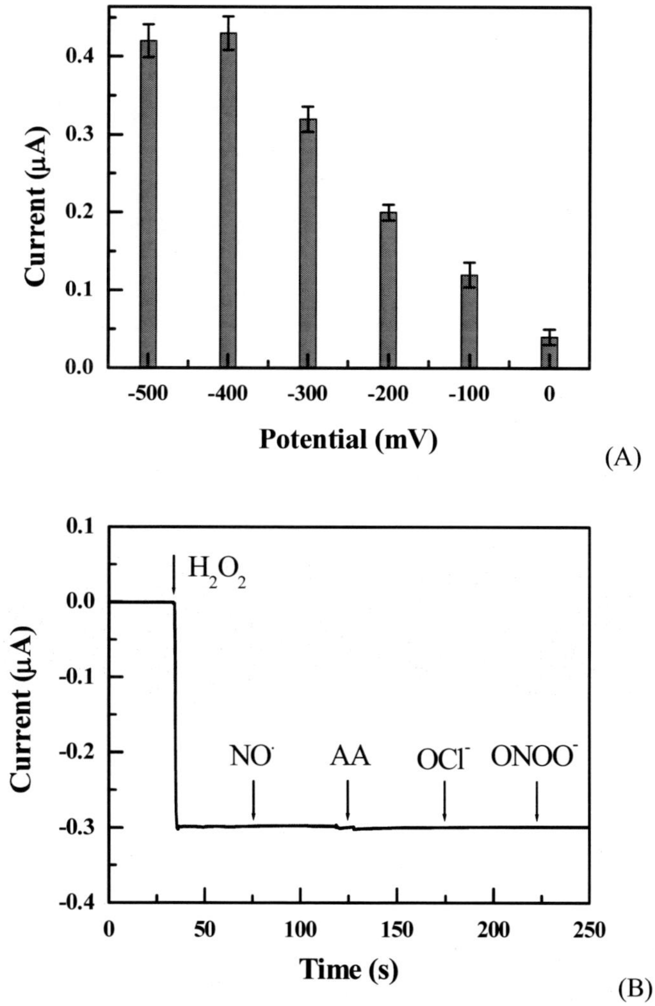

Figure 9. Panel (A), Dependence of the electrocatalytic response of the HRP–attapulgite/GC biosensor toward the reduction of 0.2 mM H_2O_2 in N_2-satrauted PBS (0.1 M, pH 7.4). Panel (B), Amperometric response of H_2O_2 (0.1 mM), NO· (0.1 mM), AA (0.1 mM), OCl^- (0.1 mM), and $ONOO^-$ (0.1 mM) at the HRP–attapulgite/GC electrode under an applied potential of –400 mV (versus SCE). Cited from Ref. 27.

Therefore, a detection potential of –400 mV is selected. Immediately after the addition of H_2O_2, the response increases and reaches 95% of the steady state value within ca. 2, suggesting that the electrode responds rapidly to the change of the substrate concentration.

The electrode linearly responds to H_2O_2 at lower concentration and attains a saturation level at a higher concentration, representing a typical Michaelis–Menten enzymatic reaction characteristic (inset A, Figure 8). The response displays a good linear range from 0.2 to 150 μM with a correlation coefficient of 0.997 and a slope of (3.8 ± 0.2) μA mM^{-1}. Therefore, the sensitivity is calculated to be (55.8 ± 2.9) μA mM^{-1} cm^{-2}. The apparent Michaelis-Menten constant (K_M^{app}) is estimated to be (0.09 ± 0.02) mM, the detection limit is estimated to be ca. (0.05 ± 0.01) μM (at a signal/noise of 3, inset B, Figure 8) implying that attapulgite can provide a good biocompatible microenvironment for maintaining enzymatic activity. This is an advantage of the developed biosensor compared with those reported previously.The selectivity and anti-interference advantages of the enzyme catalyst are demonstrated by recording the response of some common coexisting ROS and compounds in biological system such as OCl^-, NO·, $ONOO^-$, and AA, at the biosensor and compared with that of H_2O_2 (Figure 9B). While a well–defined H_2O_2 response is observed at the biosensor, biological relevant ROS and compounds result a negligible signal, largely due to the use of low operating potential in the detection. Hence, a highly selective response to H_2O_2 is obtained without the use of a perm-selective membrane or enzymatic preoxidation. This is another advantage of the developed catalyst. These results indicate that the suitability of the catalyst to monitor the released process and measure the flux of H_2O_2 from cells.

3.4. Quantification H_2O_2 in Living Cells

Macrophages (RAW 264.7) are key cells in the immune system playing an important role in host protection against a wide range of tumors and microorganisms by phagocytosis, antigen presentation, and the production of cytokines, nitrogen species (RNS), and ROS involved in the destruction of pathogens [80]. Therefore, RAW 264.7 cells were chose as the model cells for study in this chapter.

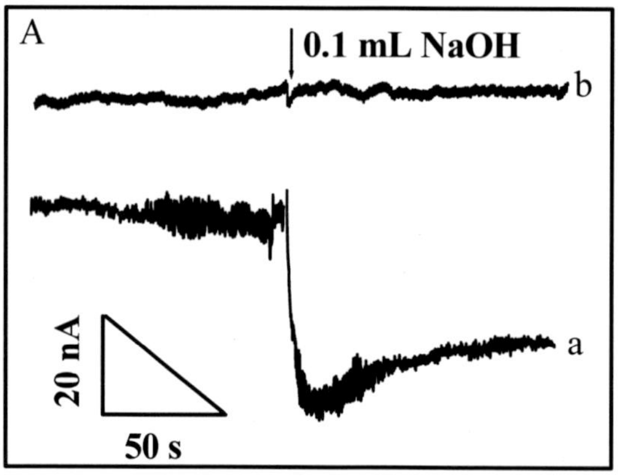

Figure 10. Amperometric response of intracellular H_2O_2 at Hb-attapulgite/GC electrode without NaOH (curve a) and with NaOH (curve b).

The concentrations of H_2O_2 in RAW 264.7 murine macrophage cells were measured by recording the current response on reduction of H_2O_2 electrocatalyzed by the developed Hb-attapulgite nanostructures. Cells were lysed with 0.3 M KOH, and then the lysate was sonicated. The samples were centrifuged at 18000 rpm for 30 min and the supernatant was filtered through a 0.22 μm Nylon filter. The concentration of H_2O_2 in the filtrates was measured by recording the response currents of H_2O_2 reduction at the Hb-attapulgite/GC electrode under a constant potential (–400 mV, versus SCE). The results indicated that the H_2O_2 concentration in the cell is ca. (8.0 ± 0.4) nM (Figure 10), which is in agreement with that reported previously based on fluorescence method [81], suggests that the developed electrode can be used for determination of the concentration of H_2O_2 in cells.

3.5. Monitoring the Release Process of H_2O_2 from Cells

When the cells were stimulated, respiratory burst occurred with H_2O_2 as the end product. The endogenous generation of H_2O_2 in the RAW 264.7 cells induced by PMA was evaluated and the flux of H_2O_2 releasing from the cells was measured with the use of the developed electrode by amperometry since this technique can provide much more precise kinetic information on the exocytotic process itself and allows evaluation in real-time of the whole efficiency of the exocytotic process [82,83].

Curve (a) in Figure 11A depicts amperometric response obtained at the HRP–attapulgite/GC electrode located near the cell in PBS (pH 7.4) at a detection potential of –400 mV. After a very short period (less than 2 s) upon injection of PMA, the response increases suddenly (curve a, Figure 11B), indicative of a large amount of H_2O_2 releasing from the cells through the expansion of the fusion pore [84,85].

Then a stable current reaches, featuring the end of the event. The current decays gradually to zero within ca.200 s (not shown here), indicating H_2O_2 releasing from the cells has been completely consumed or diffused away from the electrode surface. To verify the response depicted in curve a (Figure 11A) is only due to the response of H_2O_2 releasing from the cells, 50 μL of catalase solution (300 U mL^{-1}), which was known as a selective scavenger of H_2O_2 [86] were injected into the solution before the addition of PMA. The response of H_2O_2 at the electrode disappears (curve b, Figure 11A) since the releasing H_2O_2 was scavenged by catalase, demonstrating that the addition of PMA can only induce the release of H_2O_2 from the RAW 264.7 cells and cannot lead to the release of other ROS or RNS. Otherwise, the response of the electrocatalytic reduction of other ROS such as O_2 at the electrode should be observed because catalase cannot scavenge O_2. A maximum current of (47.3 ± 1.6) nA is obtained at ca. 17 s after injection of PMA.

This current corresponds to the H_2O_2 amount of (8.2 ± 0.5) nmol, which was calculated based on the calibration curve depicted in inset (A) of Figure 8. Therefore, the mean flux of H_2O_2 releasing from the cells is evaluated to be (48.2 ± 2.9) amol $cell^{-1}$ s^{-1}.

These observations substantially demonstrate that the developed biosensor represents a new biosensing platform for the reliable determination of cellular H_2O_2 and could be potentially useful for further physiological and pathological studies.

The effects of PMA loading on the amount of H_2O_2 generated in RAW264.7 cells and the flux of the H_2O_2 releasing from the cells can also be studied using the developed method.

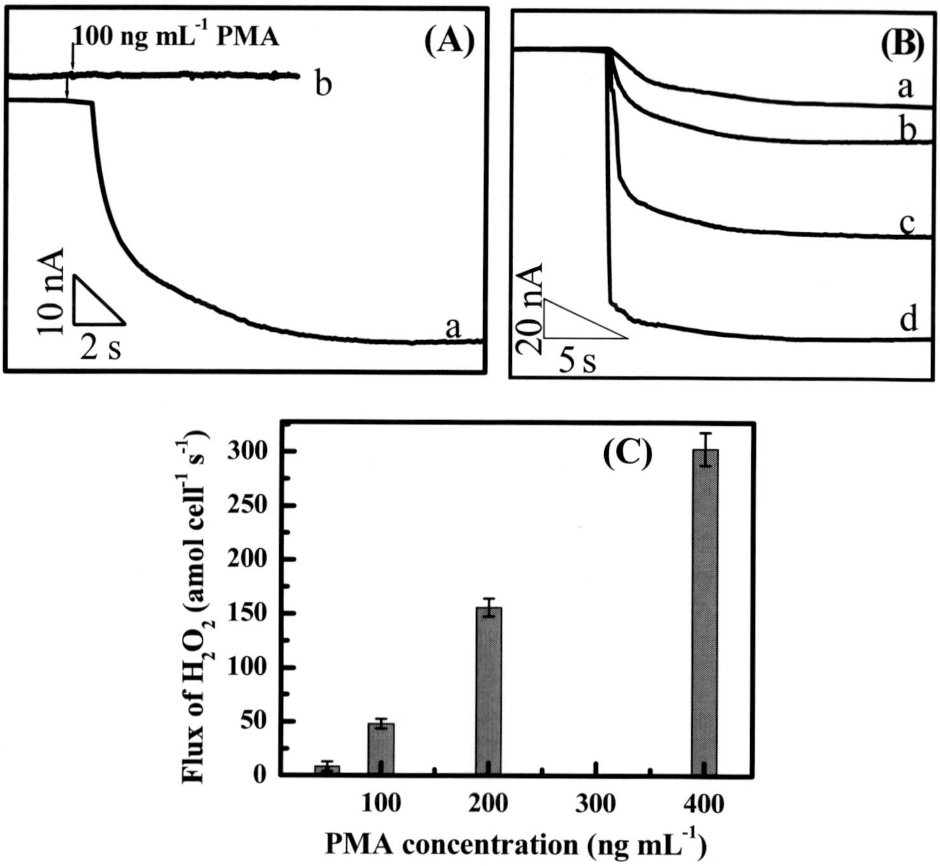

Figure 11. (Panel A), typical amperometric responses of the HRP–attapulgite/GC electrode for the reduction of H_2O_2 released from RAW 264.7 macrophage cells induced by PMA (100 ng mL^{-1}) with (curve b) and without (curve a) addition of 50 μL of catalase (300 U mL^{-1}). (Panel B), the time course of H_2O_2 released from the cells induced by 50 (a), 100 (b), 200 (c), and 400 ng mL^{-1} of PMA (d). (Panel C), dependence of the flux of H_2O_2 releasing from the cells on the concentration of PMA. Cited from Ref. 27.

Figure 11B depicts the time course of H_2O_2 releasing from the cells induced by different concentration of PMA. The release rate of H_2O_2 is enhanced by the increase of PMA loading as reflected by the slope of the curve (a) to (d) in the rising periods, demonstrating the exocytosis develops rapidly with stimulation of large amount of PMA. Moreover, the amount of H_2O_2 releasing from the cells significantly increases with enhancement of the PMA loading.

The mean flux of H_2O_2 from the cells also increases with the increasing of PMA concentration (Figure 11C), showing a PMA concentration–dependent trend.

The developed electrode can be also used for measuring the flux of H_2O_2 releasing from the cells induced by a variety of stimuli such as fMLP, ADP, and AA, which were reported to induce the generation of H_2O_2 in cells [87,88].

The amount of H_2O_2 generated in the cells and the flux of H_2O_2 from the cells are significantly dependent on the stimuli (Figure 12).

The amount of H_2O_2 generated in the cells increases in the order of ADP, AA, fMLP, and PMA (each at the concentration of 400 ng mL^{-1}, Figure 12A). The results presented in Figure 12 also demonstrate that the cells response rapidly to the stimulation, and the generation and the release of H_2O_2 from the cells are fast.

Accordingly, the flux of the H_2O_2 from the cells is also dependent on the stimuli. The maximum flux is obtained when PMA is used as a stimulator, and the flux is the lowest in the case of ADP (Figure 12B). Therefore, the developed electrode could be used for studying the H_2O_2 generation in cells as well as quantifying the flux of H_2O_2 releasing from cells.

These observations substantially demonstrate that the electrode represents a new biosensing platform for a reliable collection of kinetic information of cellular H_2O_2 release and could be potentially useful for oxidative stress biosensors in study of downstream biological effects of various stimuli in physiology and pathology.

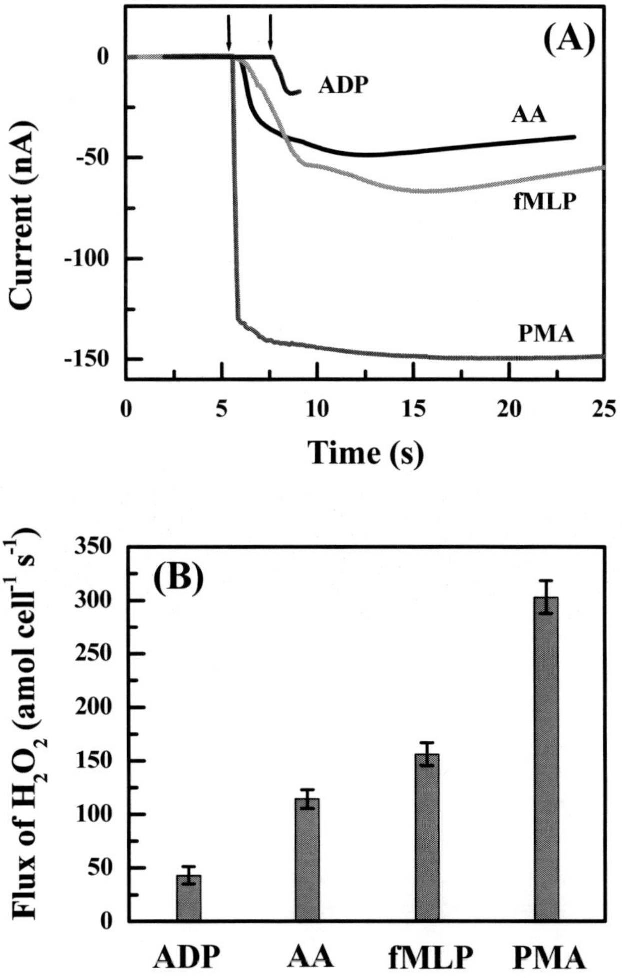

Figure 12. The time course (panel A) and the flux of H_2O_2 (panel B) released from the cells induced by 400 ng mL^{-1} ADP, AA, fMLP, and PMA, respectively. The arrow indicates the addition of stimuli. Cited from Ref. 27.

4. Electrochemical Detection Based on Electrochemical Reduction of H_2O_2 Catalyzed by Nanomaterials

Interest in nanomaterial with high electrocatalytic activity to H_2O_2 is growing rapidly in recent years. Graphene, a two-dimensional form of carbon with atoms arranged in a honeycomb lattice, exhibits various interesting physical properties, large surface area, and high chemical stability, all of which could be utilized for potential applications in nanoelectronics, nanophotonics, sensor devices, and green energy technologies [89-106]. Thus, graphene was selected as representative of nanomaterials for acting as electrocatalyst to catalyze H_2O_2 in this chapter. To better application of graphene to catalyze H_2O_2, modulation of its electrical characteristics is of great technological importance. Chemical doping the graphene with other elements such as nitrogen, boron, and chlorine atoms, etc., is a promising way to achieve this goal [92,92,107-112]. It has been demonstrated that nitrogen-doping can significantly increase the electron conductivity and the electron-donor property of graphene [99,110,113], and enhance binding ability of graphene with guest molecules [101,102]. In the following paragraph, nitrogen-doping graphene (denoted as N-graphene) was discussed as electrocatalyst towards reduction of H_2O_2.

4.1. Synthesis of the N-Graphene

N-graphene was synthesized by a facile method using graphite as a starting material. Graphite powder (3 g, 99.99995%, 325 mesh, Alfa Aesar) was put into a mixture of concentrated H_2SO_4 (12 mL), $K_2S_2O_8$ (2.5 g), and P_2O_5 (2.5 g). The solution was heated to 80 °C and kept stirring for 4.5 h by using oil-bath. Next, the mixture was cooled to room temperature and diluted with double distilled water (0.5 L) and left overnight. Then, the product was obtained by filtering with use of 0.2 micron nylon film and washed with double distilled water to remove the residual acid. The product was dried ambient temperature.

The pre-oxidized graphite was then re-oxidized by Hummers' method [114]. The pretreated graphite powder was put into cold (0 °C) concentrated H_2SO_4 (120 mL). Then, $KMnO_4$ (15 g) was added gradually under stirring and the temperature of the mixture was kept to be below 20 °C by ice-bath. Successively, the mixture was stirred at 35 °C for 2 h, and then diluted with double distilled water (0.5 L) by keeping the temperature at 50 °C. After adding all of 250 mL of double distilled water, the mixture was stirred for 2 h, and then additional water (1 L) was added. Shortly after the dilution, 30% H_2O_2 (20 mL) was added drop by drop, and the color of the mixture changed into brilliant yellow along with bubbling. The mixture was filtered and washed with 1:10 HCl aqueous solution (1 L) to remove metal ions followed by double distilled water (1 L) to remove the acid. The resulting solid was dried in air and diluted to make graphite oxide dispersion (0.5% w/w). Finally, it was purified by dialysis for one week to remove the remaining metal species.

Exfoliation was carried out by sonicating graphite oxide dispersion (0.1 mg/mL) under ambient condition for 20 min to obtain GO. The resulting homogeneous yellow-brown dispersion was used for reduction. The reduction reaction was carried out by adding hydrazine (1.2 mL) into the dispersion of GO sheets (60 mg of GO in 50 mL of water). After sonicated for 1 h and kept stirring for 24 h at 50 °C, nitrogen-doped graphene (N-graphene) sheets were obtained by filtration of the product and drying in vacuum.

4.2. Characterization of the N-Graphene

Transmission electron microscopic (TEM) image shows that the synthesized N-graphene is rippled and resembles crumpled silk veil waves (Figure. 13A), and atomic force microscopic (AFM) image indicates the apparent thickness of ~1 nm (Figure. 13B and C), suggesting the single sheet nature of N-graphene is obtained under our experimental conditions. To further obtain insight into the distribution of C, N, and O in the as-synthesized N-graphene, elemental analysis was carried out with scanning TEM (STEM) (Figure. 13D-G). Figure.13D shows a HAADF (high-angle annular dark-field) micrograph of a selected area of N-graphene. From the elemental mapping image of C, O, and N shown in Figure.13E-G, it can be concluded that C, O, and N distribute on the surface of the prepared N-graphene. The energy-dispersive X-ray (EDX) spectrum (not shown here) also indicates the presence of C, O, and N in the prepared sample. These results suggest that the N-graphene has been synthesized in our experimental conditions.

The formation of N-graphene characterized by FTIR spectroscopy was also depicted. A broad peak appeared at about 3440 cm^{-1} in both the IR spectra of GO and N-graphene is ascribed to the O–H stretching mode of interacted water [115] (Figure. 14A).

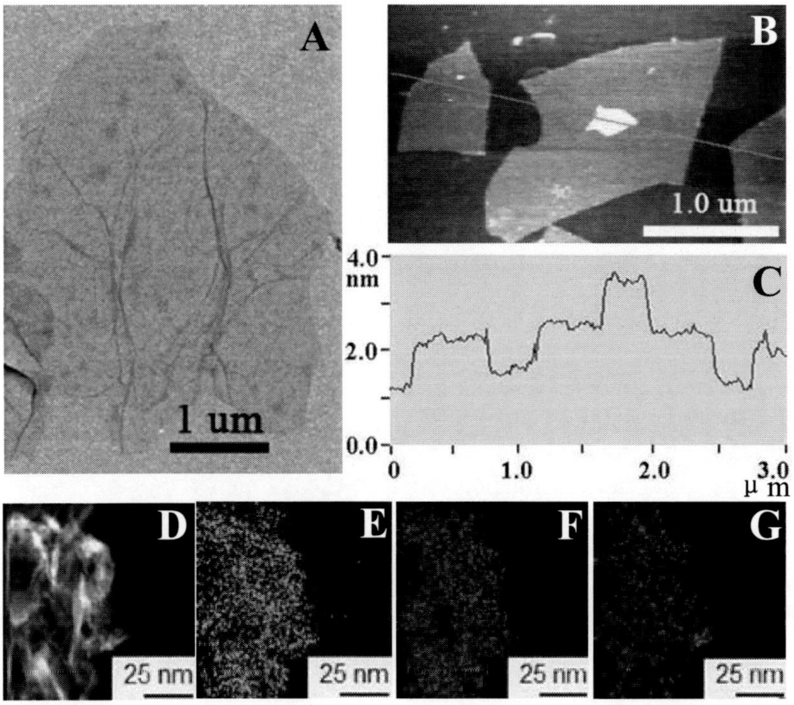

Figure 13. Typical TEM (A) and AFM image (B) and related cross-sectional analysis (C) of the prepared N-graphene. (D) HAADF STEM image of the N-graphene, and the corresponding elemental mapping for C (E), O (F), and N (G), respectively. Cited from Ref. 103.

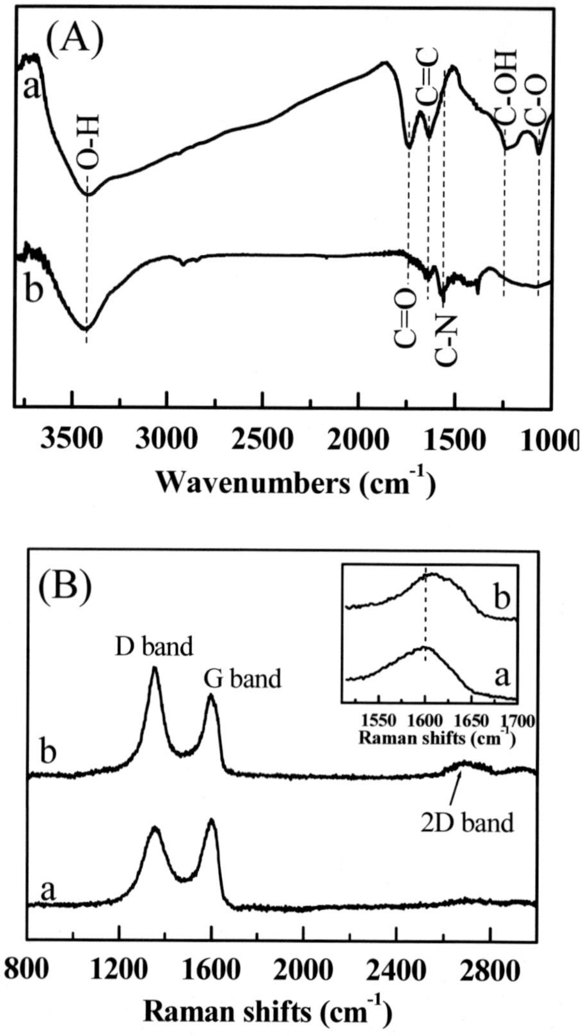

Figure 14. FTIR (A) and Raman spectra (B) of GO (curve a) and N-graphene (curve b). The inset in (B) shows the Raman spectra of G band for GO and N-graphene in a different scale. Cited from Ref. 103.

In the IR spectrum of GO (curve a in Figure. 14A), the characteristic band for the stretching vibration of C=O (1750 cm^{-1}), the sp^2 hybridized C=C (1640 cm^{-1}), and C–OH groups (1233 cm^{-1}), and the deformation mode of C–O groups (1060 cm^{-1}) can be clearly identified [116-118]. After reduced by hydrazine, the intensity of these characteristic bands has a significant decrease or complete disappearance, indicating the removal of these oxygen-containing groups. However, a new peak is certainly identified at about 1572 cm^{-1} (curve b in Figure.14A), which can be assigned to sp^2 bonded C–N [116], demonstrating the formation of the C–N bond in the reduction of GO with hydrazine.

The N-graphene studied by Raman spectroscopy which is an effective tool to detect the doping effect of graphene was also discussed [96]. The spectrum of GO show three features in the 800–3000 cm^{-1} region: a D band at ~1350 cm^{-1}, a G band at ~1590 cm^{-1}, and a weak 2D band at ~2680 cm^{-1} (curve a in Figure. 14B). A strong D band, a signature for disorder, appearing at the spectrum of GO indicates a high density of defects and vacancies in the GO plane due to the harsh oxidation by strong acid in the preparation [100].

After reduced by hydrazine, the characteristics of the Raman spectrum were similar to those of GO (curve b in Figure. 14B), indicative of preserving the basic structural properties of the graphene sheet upon the nitrogen doping. However, the intensities of D and 2D bands have certain increase, suggesting the increase of defects with the nitrogen doping (curve b) [100]. Moreover, it can be found that the G band of N-graphene has a slight blue shift in comparison with that of GO (the inset in Figure 14B), and a broad shoulder (centered at ~1627 cm^{-1}) considered as another Raman feature (D' band) induced by defect also appears.

In addition, the intensity ratio of D band to G band (I_D/I_G), which has been used to evaluate the defect in graphene [119], increases from 0.92 for GO to 1.34 for N-graphene, demonstrating nitrogen doping leads to the increase of the defects in graphene.

The ratio of D to G bands can also be used to estimate the crystalline size (L_a) of graphene. We calculated L_a of GO and N-graphene based on Tuinstra–Koenig relation:

$$L_a(\text{in nm})=(2.4\times10^{-10})\lambda^4(I_D/I_G)^{-1} \quad (1)$$

where λ is the Raman excitation wavelength (λ = 514.5 nm). La of GO and N-graphene is estimated to be ~18.2 nm and 12.5 nm, respectively. The crystallite size of N-graphene obtained here becomes decreased as anticipated since the doped graphene usually exhibits a decrease in the average size of sp^2 domains [99].

Detailed information of elemental composition of GO and N-graphene obtained from X-ray photoelectron spectroscopy (XPS) was described in this chapter. The survey scan spectrum for GO shows C1s (at ~284.7 eV) and O1s peaks (at ~532.7 eV) [120] with C/O atomic ratio of about 1.9 (curve a, Figure. 15A). After reduced with hydrazine, the peak of O1s decreases significantly (curve b, Figure. 15A) and the C/O ratio increases to be ~8.3, which is much higher than that of GO, suggesting the reduction of GO to graphene. In addition, an N1s peak at ~400 eV is clearly observed in spectrum of N-graphene (curve b) [121,122], and the N/C atomic ratio estimated from XPS peaks is ~4.5%. However, the N1s signal is too low to be certainly detected in the spectrum of GO (curve a). These results certainly confirm the simultaneous incorporation of N atom into graphene during the reduction of GO. Deconvoluting the N1s peak can obtain more detailed information on nitrogen functional groups. It is widely established that incorporation of nitrogen atoms into the carbon materials can lead to formation of different kinds of nitrogen functionalities such as pyrrolic and pyridinic groups, amine, and nitrile etc. The nitrogen atom can also substitute the carbon atom forming the substitutional or graphitic nitrogen [123,124]. The existence of these functionalities depends on the experimental conditions. For N-doped graphene and carbon nanotubes (CNTs), it was generally accepted that the main types of N-containing groups were pyridinic (refers to N atoms at the edge of graphene planes, each of which is bonded to two carbon atoms and donates one p-electron to the aromatic π system,), pyrrolic (refers to N atoms that are bonded to two carbon atoms and contribute the π system with two p-electron), graphitic (also called quaternary N or substituted N, in which N atoms are incorporated into the graphene layer and replace carbon atom within a graphite plane), and oxidized nitrogen (oxidized pyridinic nitrogen, which is bonded two carbon atoms and one oxygen atom) [125-127]. Moreover, the IR spectrum presented in Figure. 14A (curve b) indicates that the prepared N-graphene does not contain amine (should be a sharp peak at

~3300–3500 cm^{-1}) and nitrile (the peak should be at ~2220–2260 cm^{-1}) groups. Therefore, we deconvoluted the N1s peak into pyridinic, pyrrolic, graphitic, and oxidized nitrogen.

The four components center at ~398.6, 400.2, 401.3, and 402.8 eV [127] (Figure. 15B), respectively. The surface atomic compositions of pyridinic, pyrrolic, graphitic, and oxidized nitrogen are ~28%, 49%, 19%, and 4%, respectively.

To eliminate the possibility of the N signals in the prepared samples is raised by physisorbed hydrazine or N adatoms, we dispersed the as-synthezied N-graphene in various solvents like DMF (*N,N*-dimethylformamide), acetone, and H_2O by sonication and then collected for further XPS measurements. XPS showed no change in the N-doping level before and after sonication, suggesting formation of covalent C–N bonds in the prepared samples instead of physisorbed hydrazine or N adatoms.The oxygen functionalities and their distribution on the surface of GO and N-graphene were also identified by deconvoluting the C1s peak into its relative components. Two main peaks for GO at ~284.6 and 286.8 eV are ascribed to C–C and C–O (includes epoxide and hydroxyl) bonds (Figure. 15C) [128,129], respectively. The peaks at ~287.6 and 288.9 eV are assigned to C=O and O–C=O, respectively. The C1s peak of N-graphene exhibits the same oxygen functionalities as those have been assigned for GO (Figure. 15D), while some peak intensities (for example C–O, O–C=O) are much lower than those in GO, indicating considerable deoxygenation by the reduction process.

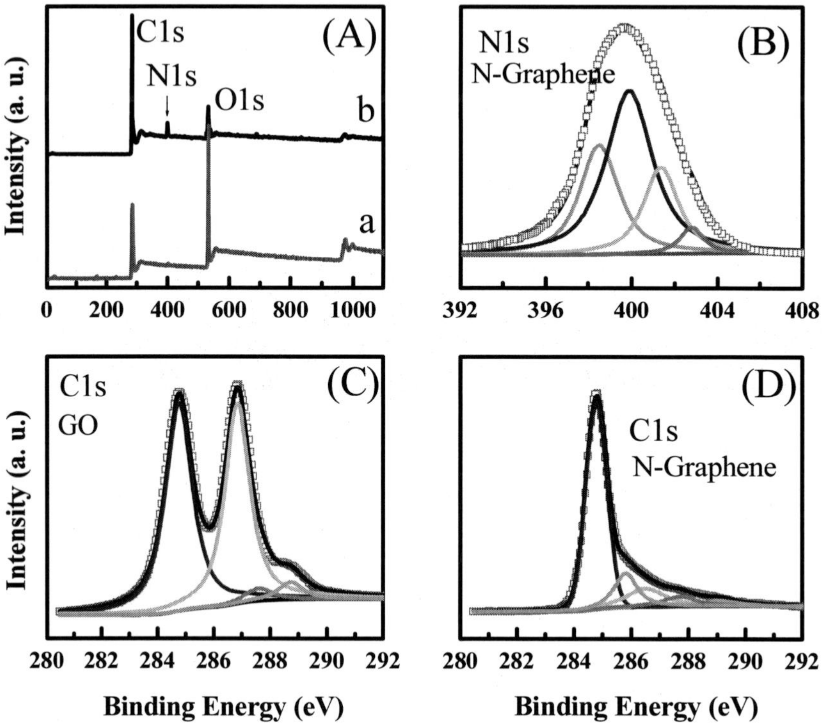

Figure 15. (A) The XPS spectra of GO (a) and N-graphene (b). (B), (C), and (D) are XPS spectra of the N1s in N-graphene (B), C1s in GO (C), and C1s in N-graphene (D), respectively, and their related curve-fitted components. Cited from Ref. 103.

Besides those oxygen functionalities, the C–N peak (~285.8 eV) [130,131] can be certainly identified at C1s peak of N-graphene (Figure. 15D), further demonstrating the doping of the graphene with nitrogen atoms under the reduction of hydrazine. These results demonstrate that N-graphene can certainly be synthesized in the reduction of GO with use of hydrazine as a reducing agent at low temperature.

4.3. Enhanced Electrocatalytic Activity to H_2O_2 Reduction

To evaluate the effects of the N-doping on the electrocatalytic characteristics of graphene, we deposited the prepared N-graphene on the glassy carbon (GC) electrode to form N-graphene/GC electrode and then studied the electrocatalytic reduction of H_2O_2 in phosphate buffer (PBS, 0.1 M, pH 7.4) with use of this electrode. A featureless voltammetric current within the potential range of interest is observed for N-graphene in nitrogen-saturated PBS (curve a, Figure. 16A). In contrast, a large cathodic current emerges in voltammetric curve upon adding 0.2 mM H_2O_2 in PBS (curve b). This cathodic current is ascribed to the catalytic reduction of H_2O_2 by N-graphene.

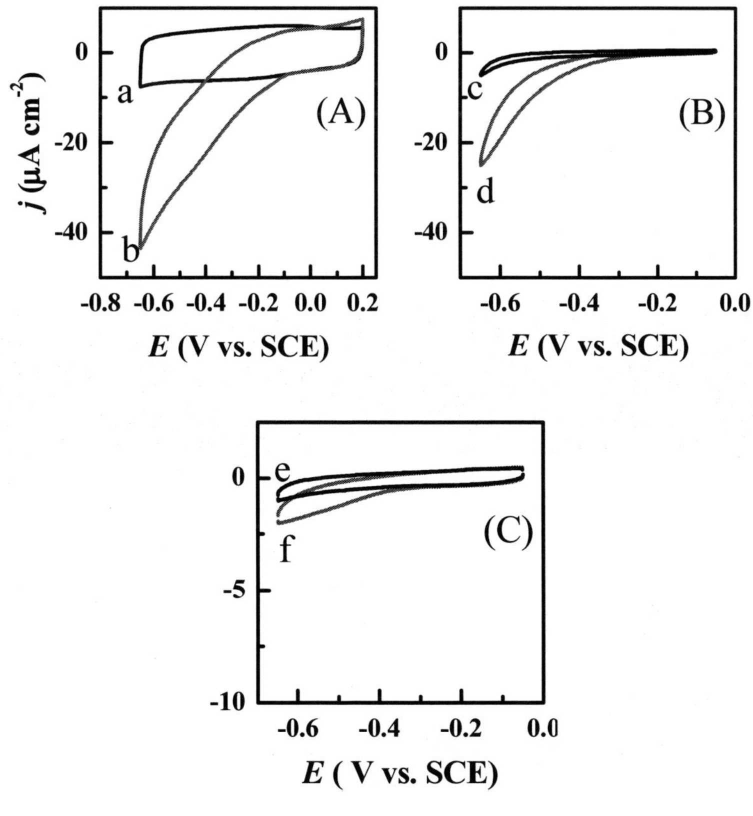

Figure 16. Voltammetric response of electrocatalytic reduction of 0.2 mM H_2O_2 by N-graphene (A), N-free graphene (B), and bare GC (C). The responses were recorded in N_2-saturated PBS (0.1 M, pH 7.4) with (curves b, d, and f) and without (curves a, c, and e) 0.2 mM H_2O_2. Scan rate was 100 mV/s. Cited from Ref. 103.

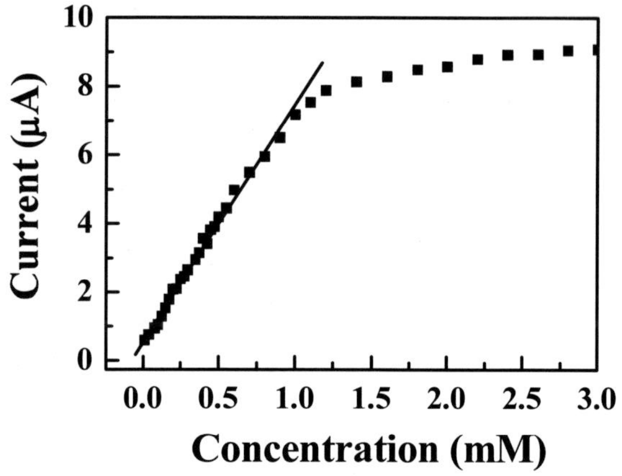

Figure 17. Dependence of the steady state current response of the electrocatalytic reduction of H_2O_2 by N-graphene on the concentration of H_2O_2.

The starting potential of H_2O_2 reduction is about –100 mV (vs SCE), which is ~230 and 280 mV, respectively, more positive than those obtained with N-free graphene (about –330 mV, curve d in Figure. 16B) and bare GC electrode (about –380 mV, curve f in Figure. 16C).

Moreover, the cathodic current for the reduction of 0.2 mM H_2O_2 catalyzed by N-graphene is twice and 25 times more than those catalyzed by N-free graphene and bare GC, respectively.

These results indicate the superior electrocatalytic activity of N-graphene toward H_2O_2 reduction.

The H_2O_2 concentration-dependent electrocatalytic current was also estimated to establish the basis for the assay of cellular H_2O_2. The electrocatalytic response of H_2O_2 at N-graphene recorded at –400 mV, which was selected by considering the response and the noise at different detection potentials, displays a good linear range of 0.5 μM–1.2 mM (Figure. 17). The detection limit is estimated to be ~(0.05 ± 0.01) μM (at a signal/noise ratio of 3). Moreover, N-graphene has good selectivity and anti-interference ability for the reduction of H_2O_2 because the common coexisting ROS and compounds in biological systems such as OCl^-, NO•, $ONOO^-$, and AA, do not affect the electrocatalytic reduction of H_2O_2 by N-graphene.

The catalytic sites in N-graphene are fairly stable since continuous catalysis (1000 s) at –400 mV only causes a slight decrease in the responses.

This is an advantage of the N-graphene used as an electrocatalyst for the H_2O_2 reduction compared with those employing enzymes as catalyst since enzymes usually undergo a denaturation.

Furthermore, graphene has another advantage over other forms of carbon materials, such as carbon nanotubes (CNTs), when it is used for biological measurements. The needle-like CNTs may more easily penetrate the cell membrane, resulting in greater cell membrane damage [132].

Due to its flat shape structure, graphene hardly damages the cell membrane and has a better biocompatibility than CNTs [133]. These results indicate the suitability of N-graphene for measurements of cellular H_2O_2.

4.4. Explanation on Enhanced Electrocatalytic Activity

It has been reported that N-doped carbon nanostructures such as N-doped CNTs [131], ordered mesoporous graphitic carbon [134], and silk-derived carbon [135] can improve their electrocatalytic activity. However, the contribution of the doped N atoms to the enhanced electrocatalytic activity of these carbon materials is not well defined and still remains controversial.

Dai has recently reported that the carbon atoms adjacent to nitrogen dopants in CNTs possess a substantially higher positive charge density, which results in an enhanced adsorption of O_2 and reactive intermediates that proceed to accelerate the reduction reaction of O_2. To explain the enhanced electrocatalytic activity of the prepared N-graphene toward H_2O_2 reduction, the distribustion of the charge density of the carbon atoms and the energy gap of LUMO (the lowest unoccupied molecular orbital) to HOMO (the highest occupied molecular orbital) were calculated by B3LYP hybrid DFT method for pure graphene and N-graphene with pyridinic, pyrrolic, and graphitic nitrogen-doped structure, respectively (the oxidized nitrogen-doping graphene was not calculated because the amount of this type of N in our prepared N-graphene is less than 4%).

For the pure graphene, the carbon atoms near its H-terminated end show negative charge due to hydrogen bonding, but the negative charge is very weak (Figure. 18A).

In this case, carbon atoms a few hexagons away from the H-terminated end become almost neutral.

For the N-graphene, the N atom is quite negative, with a charge of –0.690, –0.736, and –0.874 for pyridinic (Figure. 18B), graphitic (Figure. 18C), and pyrrolic nitrogen (Figure. 18D), respectively.

This is a consequence of N atom having a more electronegative and strong electron-accepting ability than carbon. Most of the compensating positive charge is distributed on the adjacent carbon atoms in graphene plane, therefore, the charges of the carbon atoms (marked with green) adjacent N atom in N-graphene have significant enhancement in comparison with those in the case of pure graphene.

In pyridinic N-doping graphene, the charges of the three carbon atoms adjacent N atom are 0.07, 0.150, and 0.169 (the second neighbor of the N atom), respectively (Figure. 18B). In the graphitic N-doping graphene, the charges of the three carbon atoms adjacent N atom are 0.284, 0.282, and 0.254, respectively (Figure. 18C).

In pyrrolic N-doping graphene, it is interesting that a much higher positive charges of 0.314 and 0.319 are found on the carbon atoms in a pentagon ring adjacent to the N atom (Figure. 18D).

The optimized geometry of the pyridinic and graphitic N-doping graphene is as nearly perfect as the ideal pure graphene but it has significant convexes near the N atom in pyrrolic N-doping graphene.

The nitrogen-induced charge enhancement of the carbon atoms adjacent of the N atom and charge delocalization could be favorable for H_2O_2 chemisorptions, effectively weaken the O–O bond to facilitate the reduction reaction as in the case of the O_2 reduction on the N-doping CNTs. Moreover, the presence of N atom can also enhance the ability of N-graphene to donate electrons, which is advantageous for the reduction reactions. Therefore, the N-graphene exhibits higher electrocatalytic activity than N-free graphene.

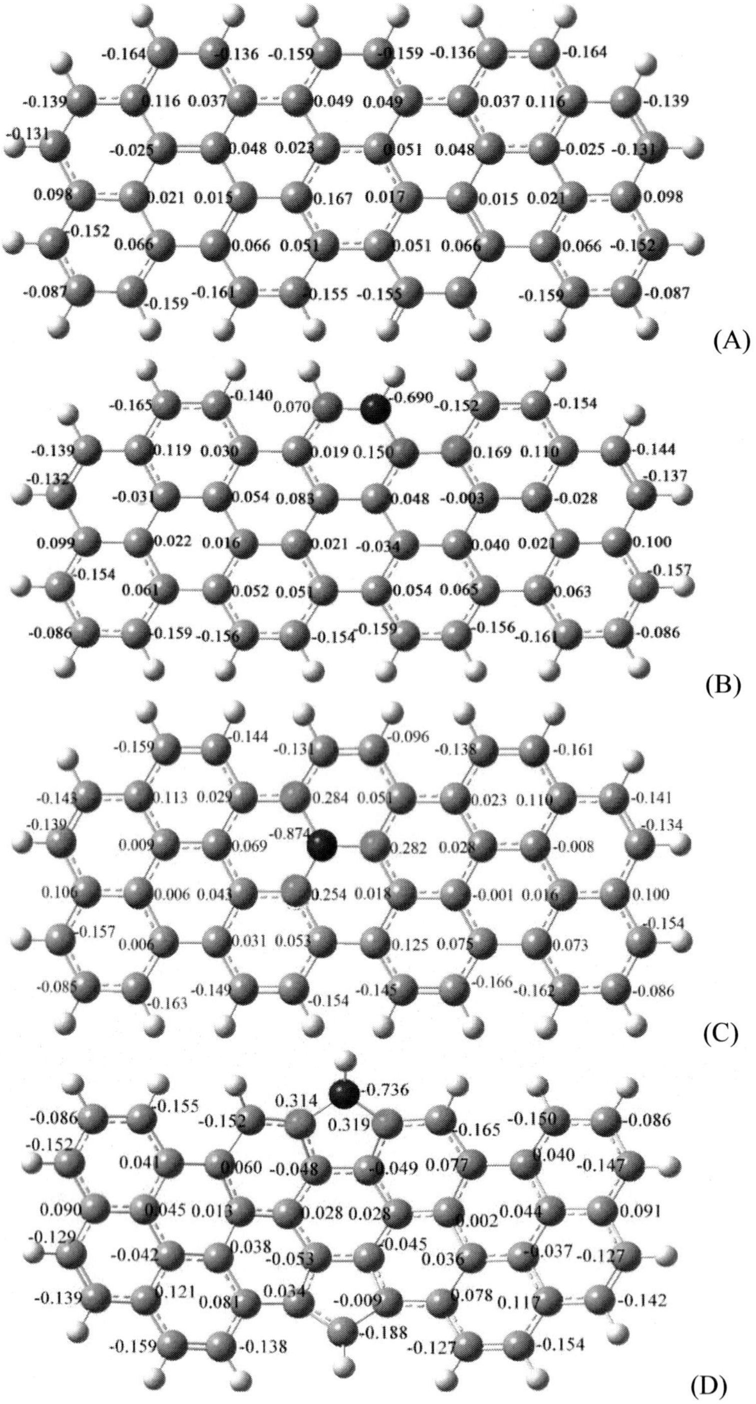

Figure 18. Charge distributions on the pure (A) and N-graphene with pyridinic (B), graphitic (C), and pyrrolic nitrogen-doped structure (D), respectively, in their optimized forms. The gray balls are carbon atoms, the blue balls are nitrogen atoms, and the smaller white balls are hydrogen atom. The green balls are those carbon atoms having significant enhancement in charge density after N doping. Cited from Ref. 103.

The high electrocatalytic activity of N-graphene can also be explained from the level of its chemical reactivity. The energy gap of HOMO-LUMO can be used as a simple indicator of kinetic stability [136]. A small HOMO-LUMO gap implies low kinetic stability and high chemical reactivity, because it is energetically favorable to add electrons to a high-lying LUMO, to extract electrons from a low-lying HOMO, and so to form the activated complex of any potential reaction [137]. We have calculated the HOMO-LUMO gap for pure graphene and N-graphene with pyridinic, pyrrolic, and graphitic N structure, respectively. For pure graphene, the gap is ~2.77 eV, which agrees with the results for similar size graphene calculated by Shemella et al. [138] and Zhang et al [137], respectively. After N doping, the HOMO-LUMO gap reduces to be ~1.40, 1.20, and 1.39 eV for pyridinic, pyrrolic, and graphitic N-doping graphene, respectively.

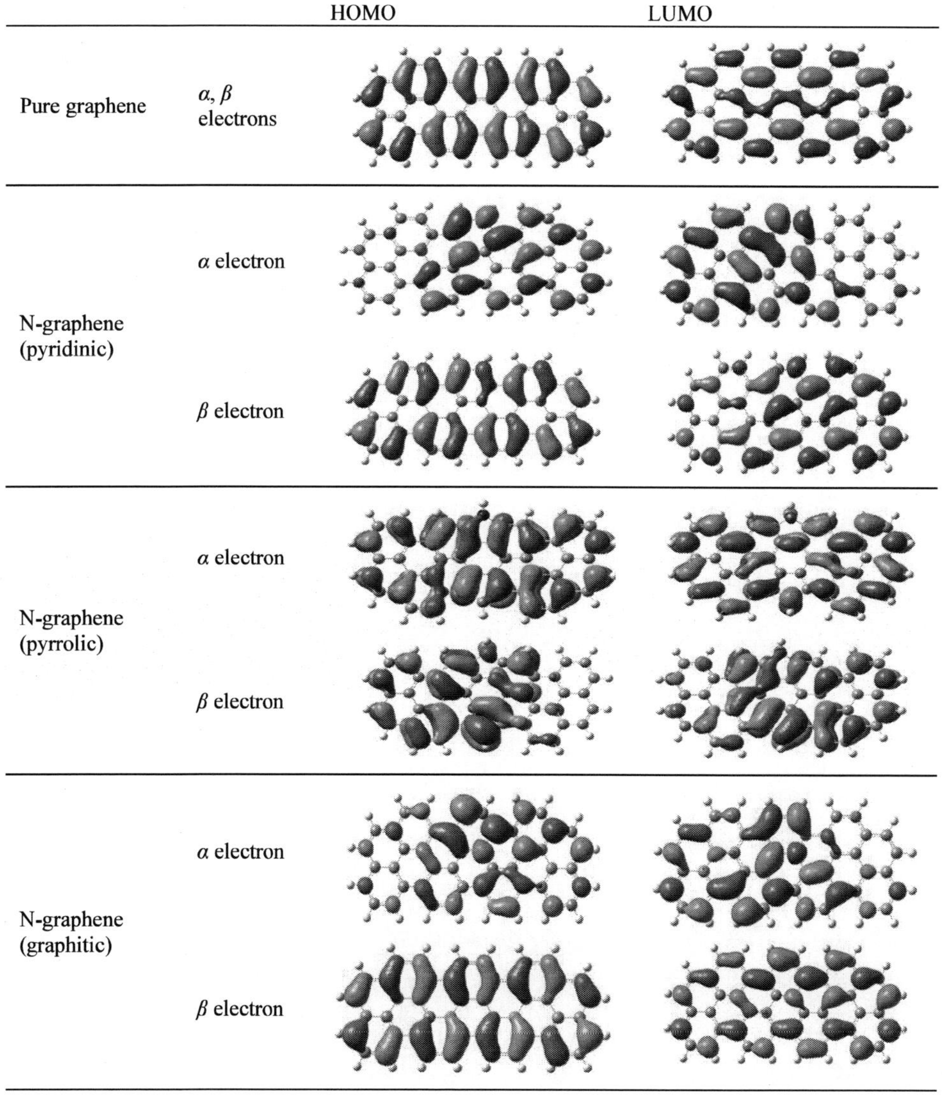

Figure 19. HOMO and LUMO spatial distributions of α electron and β electron for pure graphene and N-graphene with pyridinic, pyrrolic, and graphitic nitrogen-doped structure, respectively.

Hence, the chemical reactivity of the N-graphene is significantly improved because the electrons are more easily excited from valence bands to conduction bands.

We also obtained the HOMO and LUMO spatial distributions of α and β electrons for pure graphene and N-graphene with pyridinic, pyrrolic, and graphitic N structures, respectively (Figure. 19). For the pure graphene, both HOMO and LUMO spatial distribution are delocalized, with the same spatial distribution for α electron and β electrons. However, the HOMO and LUMO of α electron are localized distribution in the N-graphene. Obviously, N-doping introduces an unpaired electron, which causes the localized distribution of the molecule orbital. As a result, the chemical reactivity of the N-graphene is significantly enhanced.

4.5. Measurements of the Dynamic Process of H_2O_2 Releasing from Living Cells

Several studies suggested that nanomaterials elicit toxicity in cell cultures and in rodent models [132,139]. Therefore, the cytotoxicity of the prepared N-graphene was evaluated before it was used as catalyst for measuring the H_2O_2 releasing from living cell. MTT (3-(4,5-dimethlthiazol-2-yl)-2,5-diphenyltetrazolium bromide) assay, a common method of evaluating the adverse effect of nanomaterials in cell culture, was used to measure the mitochondra function of the selected cells (neutrophil, RAW 264.7 macrophage, and MCF-7 cells). After cultured with N-graphene (with the concentration of 2 μg/mL) for 24 h, the metabolic activities of the neutrophil, RAW 264.7 macrophage, and MCF-7 cells almost remained unaltered (Figure. 20A), indicative of the good biocompatibility of the N-graphene.

After verified the biocompatibility of the prepared N-graphene, we used it as platform to monitor the induced releasing process of H_2O_2 from neutrophil by amperometry. The neutrophil is a key cell in the immune system playing an important role in host protection against a wide range of tumors and microorganisms [140]. This cell is extremely sensitive to a variety of chemotactic stimulants [141]. As the result of stimulation by PMA, respiratory burst occurred with H_2O_2 as the end product inside cells. The endogenous generation of H_2O_2 in the cells was evaluated and the flux of H_2O_2 releasing from the cells was measured. Curve (a) in Figure. 20B depicts amperometric response obtained at the N-graphene/GC electrode located near the cells in PBS (pH 7.4) at a potential of −400 mV. After a very short period (less than 2 s) upon injection of PMA (with final dose of 100 ng/mL), the response increases suddenly, indicative of a large amount of H_2O_2 releasing from the cells. Then a stable current can be reached (~12 s), featuring the end of the event.

To verify the response depicted in curve (a) is only due to the response of H_2O_2 releasing from the cells, 50 μL of catalase solution (300 U/mL), an well-known selective scavenger of H_2O_2, was injected into the solution before the addition of PMA.

Then the response of H_2O_2 at the electrode disappears (curve b) since the releasing H_2O_2 was scavenged by catalase, demonstrating that the appearing response current was attributed to the release of H_2O_2 from the cells. Otherwise, the response of the electrocatalytic reduction of other ROS such as O_2 at the electrode should be observed because catalase cannot scavenge O_2.

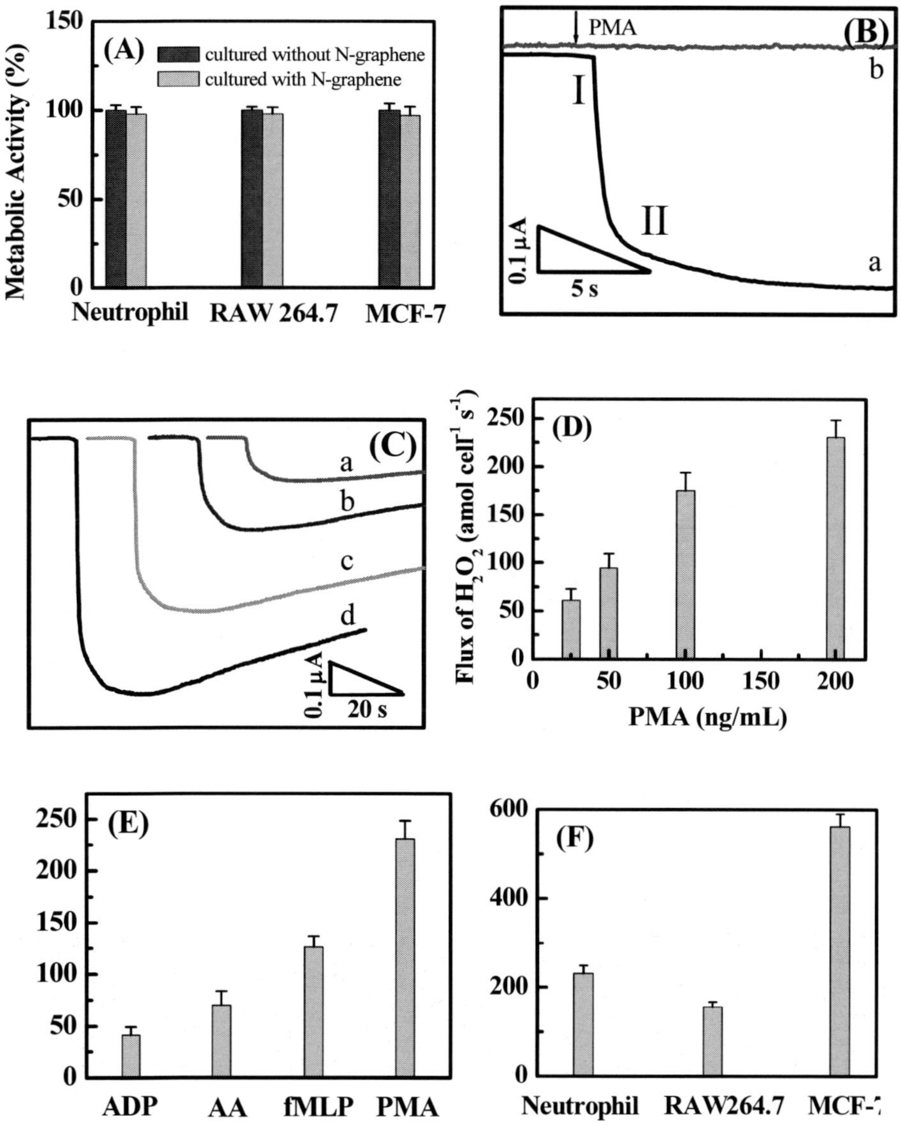

Figure 20. (A) Effects of the prepared N-graphene on metabolic activities of neutrophil, RAW 264.7 macrophage, and MCF-7 cells. (B) Typical amperometric responses of the N-graphene/GC electrode for the reduction of H_2O_2 releasing from neutrophils induced by PMA (100 ng/mL) with (b) and without (a) addition of 50 μL of catalase (300 U/mL). (C) The time course of H_2O_2 releasing from neutrophils induced by 25 (a), 50 (b), 100 (c), and 200 ng/mL of PMA (d), respectively. (D) Dependence of the flux of H_2O_2 releasing from neutrophils on the PMA dose. (E) Comparison of the flux of H_2O_2 releasing from neutrophils induced by 200 ng/mL ADP, AA, fMLP, and PMA, respectively. (F) Comparing the flux of H_2O_2 releasing from neutrophils, RAW 264.7 macrophage, and MCF-7 cells, respectively, induced by 200 ng/mL PMA. Cited from Ref. 103.

We also quantitatively assayed cellular H_2O_2 based on the H_2O_2 concentration-dependent electrocatalytic current. A maximum current of ~0.56 μA is obtained after injection of PMA, corresponding to $(2.1 \pm 0.2) \times 10^{-15}$ mol of H_2O_2 releasing from each cell, which was estimated based on the calibration curve depicted in the Figure.17 and the number of cells used in the measurements (~1 × 10^{-6} cells). This value agrees well with that reported previously (2 × 10^{-15} mol) [141]. Therefore, the mean flux of H_2O_2 releasing from each cell is evaluated to be about (175 ± 17) amol cell^{-1} s^{-1}. These results suggest that the N-graphene could be used as an effctive electrocatalyst in determining the releasing flux of H_2O_2 from cells and studying its dynamic process.

The effects of PMA dose on the amount of H_2O_2 generated and release from the cells can also be evaluated using the developed method. Figure. 20C depicts the time course of H_2O_2 releasing from the cells induced by different dose of PMA.

The release rate of H_2O_2 is enhanced by the increase of PMA dose as reflected by the slope of the curve (a) to (d) (Figure. 20C) in the rising periods, demonstrating the exocytosis develops rapidly with stimulation of high dose of PMA.

Moreover, the flux of H_2O_2 releasing from the cells significantly increases with enhancement of the PMA dose (Figure. 20D), showing a PMA dose-dependent trend. Please note that the slight decrease of the response with time in Figure. 20C reflects the consumption of the H_2O_2 due to the electrocatalytic reduction and/or its diffusion away from the electrode surface (the time scales used in Figure. 20B and C are different).

The developed approach can be also used for evaluating the dependence of H_2O_2 flux releasing from the cells on application of different stimuli such as fMLP (*N*-formylmethionyl-leucylphenylalanine), ADP (adenosine 5'-diphosphate), and AA (ascorbic acid) The amount of H_2O_2 generated in the cells and the flux of H_2O_2 from the cells are significantly dependent on the stimuli (Figure. 20E). The flux of H_2O_2 from the cells increases in the order of ADP, AA, fMLP, and PMA (each at a concentration of 200 ng/mL), indicative of a stimuli type-dependent manner.

The developed approach can be further used for evaluating the flux of H_2O_2 releasing from other cells such as RAW 264.7 macrophage and MCF-7 cells. After induced by 200 ng/mL PMA, the released H_2O_2 from RAW 264.7 macrophage and MCF-7 cells was measured using N-graphene and comparied with that from neutrophils (Figure. 20F).

The highest flux is obtained on MCF-7 cells (~560 ± 30 amol cell^{-1} s^{-1}), while the lowest flux is obtained on RAW 264.7 macrophage (~150 ± 10 amol cell^{-1} s^{-1}) among the three studied cells, showing the cells type-dependent manner.

4.6. Information on the Mechanism of H_2O_2 Release from Cells

The information on the mechanism of H_2O_2 releasing from cells can also be inferred from the time-dependent response (curve a, Figure. 20B). After stimulated by PMA, the H_2O_2 is generated rapidly in the vesicles by the reducation of the intracellular O_2 via a cascade of signaling events including the increase of protein kinase C expression and activation of the nicotinamide adenine dinucleatide phosphate (NADPH) oxidase, then the vesicles dock to the cell membrane via SNARs (soluble N-ethylamleimide sensitive fusion protein attachment receptors) complexes [82] evolving to formation of a fusion pore, through which little H_2O_2 begin to diffuse out (step I, Figure. 20B). Because of the formation of the fusion pore, the ionic exchanges between the vesicular matrix and extracellular medium occur spontanetously [82].

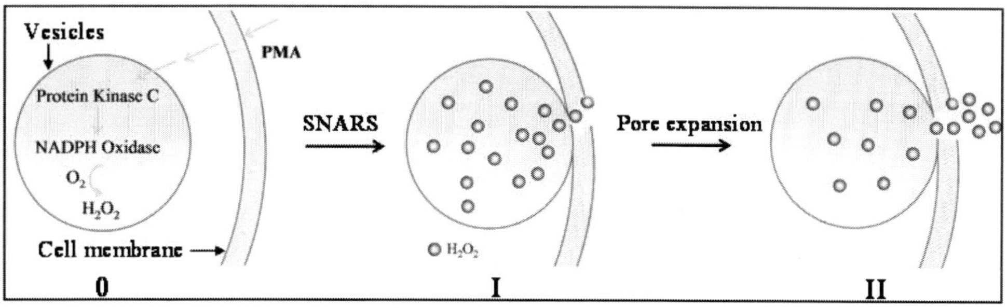

Figure 21. Schematic illustration of the generation of H_2O_2 in the cells induced by PMA and the releasing processes. Step 0, generation of H_2O_2 in cells; step I, formation of fusion pore; and step II, expansion of fusion pore. Cited from Ref. 103.

This process provokes the expansion of the fusion pore, leading to release a larger flux of the H_2O_2 into extracellular medium (step II). After that, the current decays gradually to zero (not shown here), indicating H_2O_2 releasing from the cells has been completely consumed or diffused away from the electrode surface.

The generation of H_2O_2 in the cells induced by PMA and the releasing processes are schematically illustrated in Figure. 21.

CONCLUSIONS

In summary, electrochemical method to quantify the level of cellular H_2O_2 and monitor its release process from living cells has been proposed in this chapter.

The developed method performed the detection based on the electrochemical reduction of H_2O_2. The fabrication, characterization, and electrocatalysis for the reduction of H_2O_2 of HRP-HAP, HRP-attapulgite, Hb-attapulgite, and N-graphene were discussed.

The electrochemical approach based on these electrocatalyst developed to detect the concentration of cellular H_2O_2 and study its dynamic release process from RAW 264.7 cells and neutrophil cells was depicted. The study in this chapter provides a reliable collection of kinetic information of cellular H_2O_2 release and could be potentially useful for oxidative stress biosensors in study of downstream biological effects of various stimuli in physiology and pathology.

ACKNOWLEDGEMENTS

The work presented in this chapter is supported by the National Natural Science Foundation of China (20905036 and 21175067), the Research Fund for the Doctoral Program of Higher Education of China (20103207110004), the Natural Science Foundation of Jiangsu Province (BK2011779), the Foundation of the Jiangsu Education Committee (09KJA150001, 09KJB150006, and 10KJB150009), the Foundation of Jiangsu Provincial Key Laboratory of Palygorskite Science and Applied Technology (HPK201005 and HPK201102), the Program for Outstanding Innovation Research Team of Universities in Jiangsu Province, and the Priority Academic Program Development of Jiangsu Higher Education Institutions.

REFERENCES

[1] Acik, M; Lee, G; Mattevi, C; Pirkle, A; Wallace, RM; Chhowalla, M; Cho, K; Chabal, Y. The role of oxygen during thermal reduction of graphene oxide studied by infrared absorption spectroscopy. *J. Phys. Chem. C,* 2011, 115, 19761–19781.

[2] Aihara, JI. Reduced HOMO–LUMO gap as an index of kinetic stability for polycyclic aromatic hydrocarbons. *J. Phys. Chem. A,* 1999, 103, 7487–7495.

[3] Amatore, C; Arbault, S; Bonifas, I; Bouret, Y; Erard, M; Ewing, AG; Sombers, LA. Correlation between vesicle quantal size and fusion pore release in chromaffin cell exocytosis. *Biophys. J.,* 2005, 88, 4411–4420.

[4] Amatore, C; Arbault, S; Guille, M; Lemaître, F. Electrochemical monitoring of single cell secretion: vesicular exocytosis and oxidative stress. *Chem. Rev.,* 2008, 108, 2585–2621.

[5] Amatore, C; Arbault, S; Koh, AC. Simultaneous detection of reactive oxygen and nitrogen species released by a single macrophage by triple potential–step chronoamperometry. *Anal. Chem.,* 2010, 82, 1411–1419.

[6] Arnes, BN; Shigenagra, MK. In Molecular Biology of Free Radical ScaVenging Systems. Edition: Scandalios, JS, New York: Cold Spring Harbor Laboratory Press, 1992.

[7] Arnous, A; Petrakis, C; Makris, DP; Kefalas, P. Detection of hydrogen peroxide with chemiluminescent micelles. *J. Pharmacol. Toxicol. Methods,* 2002, 48, 171–177.

[8] Arrigo, R; Havecker, M; Schlogl, R; Su, DS. Dynamic surface rearrangement and thermal stability of nitrogen functional groups on carbon nanotubes. *Chem. Commun.,* 2008, 40, 4891–4893.

[9] Bartosz, G. Use of spectroscopic probes for detection of reactive oxygen species. *Clin. Chim. Acta,* 2006, 368, 53–76.

[10] Behera, SK. Enhanced rate performance and cyclic stability of Fe_3O_4-graphene nanocomposites for Li ion battery anodes. *Chem. Commun.,* 2011, 47, 10371–10373.

[11] Belyanskaya, L; Weigel, S; Hirsch, C; Tobler, U; Krug, HF; Wick, P. Effects of carbon nanotubes on primary neurons and glial cells. *Neurotoxicology,* 2009, 30, 702–711.

[12] Borgmann, S. Electrochemical quantification of reactive oxygen and nitrogen: challenges and opportunities. *Anal. Bioanal. Chem.,* 2009, 394, 95–105.

[13] Cai, CX; Chen, J. Direct electrochemistry of horseradish peroxidase at a carbon nanotube electrode. *Acta Chim. Sin.,* 2004, 62, 335–340.

[14] Carlsson, GH; Nicholls, P; Svistunenko, D; Berglund, GI; Hajdu, J. Complexes of horseradish peroxidase with formate, acetate, and carbon monoxide. *Biochemistry,* 2005, 44, 635–642.

[15] Chance, B. Primary compounds of catalase and peroxidase. *Arch. Biochem. Biophys.,* 1952, 41, 404–415.

[16] Chance, B; Maehly, AC. Assay ofcatalase and peroxidases. *Methods Enzymol.,* 1955, 2, 764–775.

[17] Chang, MC; Pralle, A; Isacoff, EY; Chang, CJ. A selective, cell–permeable optical probe for hydrogen peroxide in living cells. *J. Am. Chem. Soc.,* 2004, 126, 15392–15393.

[18] Chen, W; Li, S; Chen, C; Yan, L. Self-assembly and embedding of nanoparticles by in situ reduced graphene for preparation of a 3D graphene/nanoparticle aerogel. *Adv. Mater.,* 2011, 23, 5679–5683.

[19] Chow, RH; Vonruden, L; Neher, E. Delay in vesicle fusion revealed by electrochemical monitoring of single secretory events in adrenal chromaffin cells. *Nature,* 1992, 356, 60–63.

[20] Dequaire, M; Limoges, B; Moiroux, J; Saveant, JM. Mediated electrochemistry of horseradish peroxidase. *J. Am. Chem. Soc.,* 2002, 124, 240–253.

[21] Dickinson, BC; Huynh, C; Chang, CJ. A palette of fluorescent probes with varying emission colors for imaging hydrogen peroxide signaling in living cells. *J. Am. Chem. Soc.,* 2010, 132, 5906–5915.

[22] DiMauro, S; Schon, EA. Mitochondrial disorders in the nervous system. *Ann. Rev. Neurosci.,* 2008, 31, 91–123.

[23] Droge, W; Schipper, HM. Oxidative stress and aberrant signaling in aging and cognitive decline. *Aging Cell,* 2007, 6, 361–370.

[24] Ferrari, AC. Graphene: New bridge between condensed matter physics and quantum electrodynamics. *Solid State Commun.,* 2007, 143, 47–57.

[25] Fike, CD; Slaughter, JC; Kaplowitz, MR; Zhang, Y; Aschner, JL. Reactive oxygen species from NADPH oxidase contribute to altered pulmonary vascular responses in piglets with chronic hypoxia–induced pulmonary hypertension. *Am. J. Physiol. Lung Cell Mol. Physiol.,* 2008, 295, L881–L888.

[26] Finkel, T. Oxidant signals and oxidative stress. *Curr. Opin. Cell Biol.,* 2003, 15, 247–254.

[27] Finkel, T. Redox-dependent signal transduction. *FEBS Lett.,* 2000, 476, 52–54.

[28] Finkel, T; Serrano, M; Blasco, MA. The common biology of cancer and ageing. *Nature,* 2007, 448, 767–774.

[29] Gammon, WJ; Kraft, O; Reilly, AC; Holloway, BC. Experimental comparison of N(1s) X–ray photoelectron spectroscopy binding energies of hard and elastic amorphous carbon nitride films with reference organic compounds. *Carbon,* 2003, 41, 1917–1923.

[30] Geim, K; Novoselov, KS. The rise of graphene. *Nat. Mater.,* 2007, 6, 183–191.

[31] Gong, K; Du, F; Xia, Z; Dustock, M; Dai, L. Nitrogen-doped carbon nanotube arrays with high electrocatalytic activity for oxygen reduction. *Science,* 2009, 323, 760–764.

[32] Gong, XC; Li, QL; Xu, KH; Liu, X; Li, HM; Chen, ZZ; Tong, LL; Tang, B; Zhong, HB. Potent method for the simultaneous determination of glutathione and hydrogen peroxide in mitochondrial compartments of apoptotic cells with microchip electrophoresis–laser induced fluorescence. *Electrophoresis,* 2009, 30, 1983–1990.

[33] Greenway, GM; Leelasattarathkul, T; Liawruangrath, S; Wheatley, RA; Youngvises, N. Ultrasound–enhanced flow injection chemiluminescence for determination of hydrogen peroxide. *Analyst,* 2006, 131, 501–508.

[34] Griffith, AW; Cooper, JM. Single-cell measurements of human neutrophil activation using electrorotation. *Anal. Chem.,* 1998, 70, 2607–2612.

[35] Guissani, A; Henry, YA. EPR characterization of molecular targets for NO· in mammalian cells. *J. Mol. Struct.,* 1997, 404, 257–265.

[36] Guo, B; Liu, Q; Chen, E; Zhu, H; Fang, L; Gong, JR. Controllable N-doping of graphene. *Nano Lett.*, 2010, 10, 4975–4980.

[37] Guo, CX; Zheng, XT; Lu, ZS; Lou, XW; Li, CM. Biointerface by cell growth on layered graphene-artificial peroxidase-protein nanostructure for in situ quantitative molecular detection. *Adv. Mater.*, 2010, 22, 5164–5167.

[38] Guo, HL; Wang, XF; Qian, YD; Wang, FB; Xia, XH. A green approach to the synthesis of graphene nanosheets. *ACS Nano,* 2009, 3, 2653–2659.

[39] He, WF.; Liu, G; Chen, X; Lu, JC; Abe, H; Huang, KX; Manabe, M; Kodama, H. Inhibitory effects of ginsenosides from the root of Panax ginseng on stimulus-induced superoxide generation, tyrosyl or serine/threonine phosphorylation, and translocation of cytosolic compounds to plasma membrane in human neutrophils. *J. Agric. Food Chem.,* 2008, 56, 1921–1927.

[40] Hempel, SL; Buettner, GR; O'Malley, YQ; Wessels, DA; Flaherty, DM. Dihydrofluorescein diacetate is superior for detecting intracellular oxidants: comparison with 2',7'–dichlorodihydrofluorescein diacetate, 5(and 6)–carboxy–2',7'–dichlorodihydrofluorescein diacetate, and dihydrorhodamine 123. *Free Rad. Biol. Med.,* 1999, 27, 146–159.

[41] Henmi, T; Yoshinaga, N. Alteration of imogolite by dry grinding. *Clay Miner.,* 1981, 16, 139–149.

[42] Houstis, N; Rosen, ED; Lander, ES. Reactive oxygen species have a causal role in multiple forms of insulin resistance. *Nature,* 2006, 440, 944–948.

[43] Hu, YJ; Wu, P; Yin, YJ; Zhang, H; Cai, CX. Effects of structure, composition, and carbon support properties on the electrocatalytic activity of Pt-Ni-graphene nanocatalysts for the methanol oxidation. *Appl. Catal. B: Environmental,* 2012, 111-112, 208–217.

[44] Hu, YJ; Zhang, H; Wu, P; Zhang, H; Zhou, B. Cai, CX. Bimetallic Pt-Au nanocatalysts electrochemically deposited on graphene and their electrocatalytic characteristics towards oxygen reduction and methanol oxidation. *Phys. Chem. Chem. Phys.*, 2011, 13, 4083–4094.

[45] Hummers, WS; Offeman, RE. Preparation of graphite oxide. *J. Am. Chem. Soc.,* 1958, 80, 1339–1340.

[46] Irace, G; Bismuto, E; Savy, F; Colonna, G. Unfolding pathway of myoglobin: molecular properties of intermediate forms. *Arch. Biochem. Biophys.,* 1986, 244, 459–469.

[47] Ishikawa, K; Takenaga, K; Akimoto, M; Koshikawa, N; Yamaguchi, A; Imanishi, H; Nakada, K; Honma, Y; Hayashi, J. ROS–generating mitochondrial DNA mutations can regulate tumor cell metastasis. *Science,* 2008, 320, 661–664.

[48] Iwazaki, T; Obinata, R; Sugimoto, W; Takasu, K. High oxygen-reduction activity of silk-derived activated carbon. *Electrochem. Commun.*, 2009, 11, 376–378.

[49] Jafri, RI; Rajalakshmi, N; Ramaprabhu, S. Nitrogen doped graphene nanoplatelets as catalyst support for oxygen reduction reaction in proton exchange membrane fuel cell. *J. Mater. Chem.,* 2010, 20, 7114–7117.

[50] Jay, D; Hitomi, H; Griendling, KK. Oxidative stress and diabetic cardio–vascular complications. *Free Radic. Biol. Med.,* 2006, 40, 183–192.

[51] Kaul, N; Forman, HJ. Activation of NF–κB by the respiratory burst of macrophages. *Free Radic. Biol. Med.,* 1996, 21, 401–405.

[52] Kelemen, R; Afeworki, M; Gorbaty, ML; Kwiatek, PJ; Solum, MS; Hu, JZ; Pugmire, RJ. XPS and ^{15}N NMR study of nitrogen forms in carbonaceous solids. *Energy Fuel,* 2002, 16, 1507–1515.

[53] Khachatryan, L; Vejerano, E; Lomnicki, S; Dellinger, B. Environmentally persistent free radicals (EPFRs). 1. Generation of reactive oxygen species in aqueous solutions. *Environ. Sci. Technol.,* 2011, 45, 8559–8566.

[54] Kim, G; Lee, YEK; Xu, H; Philbert, MA; Kopelman, R. Nanoencapsulation method for high selectivity sensing of hydrogen peroxide inside live cells. *Anal. Chem.,* 2010, 82, 2165–2169.

[55] Kim, G; Lee. YK; Xu, H; Phibert, MA; Kopelman, R. Nanoencapsulation method for high selectivity sensing of hydrogen peroxide inside live cells. *Anal. Chem.,* 2010, 82, 2165–2169.

[56] Kojima, H; Nakatsubo, N; Kikuchi, K; Kawahara, S; Kirino, Y. Detection and imaging of nitric oxide with novel fluorescent indicators: diaminofluoresceins. *Anal. Chem.,* 1998, 70, 2446–2453.

[57] Kundu, S; Nagaiah, TC; Xia, W; Wang, Y; Dommele, SV; Bitter, JH; Santa, M; Grundmeier, G; Bron, M; Schuhmann, W; Muhler, M. Electrocatalytic activity and stability of nitrogen-containing carbon nanotubes in the oxygen reduction reaction. *J. Phys. Chem. C,* 2009, 113, 14302–14310.

[58] Lardinois, OM; Mestdagh, MM; Rouxhet, PG; Peroxidatic degradation of azide by catalase and irreversible enzyme inactivation. *Biochim. Biophys. Acta,* 1996, 1295, 222–238.

[59] Li, B; Cao, H; Shao, J; Qu, M. Enhanced anode performances of the Fe_3O_4-Carbon-rGO three dimensional composite in lithium ion batteries. *Chem. Commun.,* 2011, 47, 10374–10376.

[60] Li, BX; Zhang, ZJ; Zhao, LX. Flow injection technique for biochemical analysis with chemiluminescence detection in acidic media. *Anal. Chim. Acta,* 2001, 445, 161–167.

[61] Li, CY; Zhang, H; Wu, P; Gong, ZN; Xu, GL, Cai, CX. Electrochemical detection of extracellular hydrogen peroxide released from RAW 264.7 murine macrophage cells based on horseradish peroxidase–hydroxyapatite nanohybrids. *Analyst,* 2011, 136, 1116–1123.

[62] Li, HM; Li, QL; Wang, X; Xu, KH; Chen, ZZ; Gong, XC; Liu, X; Tong, LL; Tang, B. Simultaneous determination of superoxide and hydrogen peroxide in macropgage RAW 264.7 cell extracts by microchip electrophoresis with laser–induced fluorescence detection. *Anal. Chem.,* 2009, 81, 2193–2198.

[63] Li, JZ; Dasgupta, PK. Measurement of gaseous hydrogen peroxide with a liquid core waveguide chemiluminescence detector. *Anal. Chim. Acta,* 2001, 442, 63–70.

[64] Li, X; Wang, H; Robinson, JT; Sanchez, H; Diankov, G; Dai, H. Simultaneous nitrogen doping and reduction of graphene oxide. *J. Am. Chem. Soc.,* 2009, 131, 15939–15944.

[65] Lin, MT; Beal, MF. Mitochondrial dysfunction and oxidative stress in neurodegenerative diseases. *Nature,* 2006, 443, 787–795.

[66] Liu, R; Wu, D; Feng, X; Müllen, K. Nitrogen-doped ordered mesoporous graphitic arrays with high electrocatalytic activity for oxygen reduction. *Angew. Chem., Int. Ed.,* 2010, 49, 2565–2569.

[67] Liu, Y; Dong, X; Chen, P. Biological and chemical sensors based on graphene materials. *Chem. Soc. Rev.,* 2012, 41, 2283–2307.

[68] Long, D; Li, W; Ling, L; Miyawaki, J; Mochida, I; Yoon, SH. Preparation of nitrogen-doped graphene sheets by a combined chemical and hydrothermal reduction of graphene oxide. *Langmuir,* 2010, 26, 16096–16102.

[69] Lu, C; Song, GQ; Lin, JM. Reactive oxygen species and their chemiluminescence detection methods. *Trends in Anal. Chem.,* 2006, 25, 985–995.

[70] Luo, D; Zhang, G; Liu, J; Sun, X. pH-Sensitive highly dispersed reduced graphene oxide solution using lysozyme via an in situ reduction method. *J. Phys. Chem. C,* 2011, 115, 11327–11335.

[71] Luo, W; Abbas, ME; Zhu, L; Deng, K; Tang, H. Rapid quantitative determination of hydrogen peroxide by oxidation decolorization of methyl orange using a fenton reaction system. *Anal. Chim. Acta,* 2008, 629, 1–5.

[72] Luo, YP; Liu, HQ; Q. Rui; Tian, Y. Detection of extracellular H_2O_2 released from human liver cancer cells based on TiO_2 nanoneedles with enhanced electron transfer of cytochrome c. *Anal. Chem.,* 2009, 81, 3035–3041.

[73] Luo, Z; Lim, S; Tian, Z; Shang, J; Lai, L; MacDonald, B; Fu, C; Shen, Z; Yu, T; Lin, J. Pyridinic N doped graphene: synthesis, electronic structure, and electrocatalytic property. *J. Mater. Chem.,* 2011, 21, 8038–8044.

[74] Lv, YF; Yin, YJ; Wu P; Cai, CX. Direct electrochemistry and bioelectrocatalysis of myoglobin at a carbon nanotube–modified electrode. *Acta Phys.-Chim. Sin.,* 2007, 23, 5–11.

[75] Maeda, H; Fukuyasu, Y; Yoshida, S; Fukuda, M; Saeki, K; Matsuno, H; Yamauchi, Y; Yoshida, K; Hirata, K; Miyamoto, K. Fluorescent probes for hydrogen peroxide based on a non–oxidative mechanism. *Angew. Chem.,Int. Ed.,* 2004, 43, 2389–2391.

[76] Matsubara, C; Kudo, K; Kawashita, T; Takamura, K. Spectrophotometric determination of hydrogen peroxide with titanium 2–((5–bromopyridyl)azo)–5– (N–pro–pyl–N–sulfopropylamino) phenol reagent and its application to the determination of serum glucose using glucose oxidase. *Anal. Chem.,* 1985, 57, 1107–1109.

[77] Matter, PH; Zhang, L; Ozkan, US. The role of nanostructure in nitrogen-containing carbon catalysts for the oxygen reduction reaction. *J. Catal*., 2006, 239, 83–96.

[78] Meng, L, Wu, P, Chen, GX; Cai, CX; Sun, YM; Yuan, ZH. Low potential detection of glutamate based on the electrocatalytic oxidation of NADH at thionine/single–walled carbon nanotubes composite modified electrode. *Biosens. Bioelectron.,* 2009, 24, 1751–1756.

[79] Meng, L; Yang, LG; Zhou, B; Cai, CX. Cerium phosphate nanotubes: synthesis, characterization and biosensing. *Nanotechnology,* 2009, 20, 035502.

[80] Miao, SD; Leeman, HG; Feyter, SD; Schoonheydt, RA. Three–component Langmuir–Blodgett films consisting of surfactant, clay mineral, and lysozyme: construction and characterization. *Chem. Eur. J.,* 2010, 16, 2461–2469.

[81] Miao, SD; Liu, ZM; Zhang, ZF; Han, BX; Miao, ZJ; Ding, KL; An, GM. An ionic liquid–assisted immobilization of Rh on attapulgite and its application in cyclohexene hydrogenation. *J. Phys. Chem. C,* 2007, 111, 2185–2190.

[82] Miller, EW; Tulyathan, O; Isacoff, EY; Chang, CJ. Chemistry and biology of reactive oxygen species in signaling or stress responses. *Nat. Chem. Biol.,* 2007, 3, 263–267.

[83] Mosmann, T. Rapid colorimetric assay for cellular growth and survival: application to proliferation and cytotoxicity assays. *J. Immunol. Methods,* 1983, 65, 55–63.

[84] Mroz, P; Bhaumik, J; Dogutan, DK; Aly, Z; Kamal, Z; Khalid, L; Kee, HL; Bocian, DF; Holten, D; Lindsey, JS; Hamblin, MR. Imidazole metalloporphyrins as photosensitizers for photodynamic therapy: role of molecular charge, central metal and hydroxyl radical production. *Cancer Lett.,* 2009, 282, 63–76.

[85] Muller, FL; Lustgarten, MS; Jang, Y; Richardson, A; Van Remmen, H. Trends in oxidative aging theories. *Free Radic. Biol. Med.,* 2007, 43, 477–503.

[86] Nakajima, T; Matsuo, Y. A case of angiocentric immunoproliferative lesions in an ATL-virus carrier. *Carbon,* 1994, 32, 469–475.

[87] Natalello, A; Ami, D; Brocca, S; Lotti, M; Doglia, SM. Secondary structure, conformational stability and glycosylation of a recombinant Candida rugosa lipase studied by Fourier–transform infrared spectroscopy. *Biochem. J.,* 2005, 385, 511–517.

[88] Neaman, A; Singer, A. The effects of palygorskite on chemical and physico–chemical properties of soils: a review. *Soil Sci. Soc. Am. J.,* 2000, 64, 427–436.

[89] Negre–Salvayre, A; Augé, N; Duval, C; Robbesyn, F; Thiers, JC; Nazzal, D; Benoist, H; Salvayre, R. Detection of intracellular reactive oxygen species in cultured cells using fluorescent probes. *Methods Enzymol.,* 2002, 352, 62–71.

[90] Negre–Salvayre, A; Auge□ N; Duval, C; Robbesyn, F; Thiers, JC; Nazzal, D; Benoist, H; Salvayre, R. Detection of intracellular reactive oxygen species in cultured cells using fluorescent probes. *Methods Enzymol.,* 2002, 352, 62–71.

[91] Niwa, K; Furukawa, M; Niki, K. Ir reflectance studies of electron transfer promoters for cytochrome c on a gold electrode. *J. Electroanal. Chem.,* 1988, 245, 275–285.

[92] Nozaki, O; Kawamoto, H. Reactivation of horseradish peroxidase with imidazole for continuous determination of hydrogen peroxide using a microflow injection–chemiluminescence detection system. *Luminescence,* 2003, 18, 203–206.

[93] Nozaki, O; Kawamoto, H. Reactivation of inactivated horseradish peroxidase with ethyleneurea and allantoin for determination of hydrogen peroxide by micro–flow injection horseradish peroxidase–catalyzed chemiluminescence. *Anal. Chim. Acta,* 2003, 495, 233–238.

[94] Panchakarla, LS; Subrahmanyam, KS; Saha, SK; Govindaraj, A; Krishnamurthy, HR; Waghmare, UV; Rao, CN. Synthesis, structure, and properties of boron- and nitrogen-doped graphene. *Adv. Mater.,* 2009, 21, 4726–4730.

[95] Park, PJ; HEO, SJ; Park, EJ; Kim, SK; Byun, HG; Jeon, BT; Jeon, YJ. Reactive oxygen scavenging effect of enzymatic extracts from sargassum thunbergii. *J. Agric. Food Chem.,* 2005, 53, 6666–6672.

[96] Pels, JR; Kapteijn, F; Moulijn, JA; Zhu, Q; Thomas, KM. Evolution of nitrogen functionalities in carbonaceous materials during pyrolysis. *Carbon,* 1995, 33, 1641–1653.

[97] Peng, Y; Jiang, DL; Su, L; Zhang, L; Yan, M; Du, JJ; Lu, YF; Liu, YN; Zhou, FM. Mixed monolayers of ferrocenylalkanethiol and encapsulated horseradish peroxidase for sensitive and durable electrochemical detection of hydrogen peroxide. *Anal.Chem.,* 2009, 81, 9985–9992.

[98] Rossi, DJ; Jamieson, CH; Weissman, IL. Stems cells and the pathways to aging and cancer. *Cell,* 2008, 132, 681–696.

[99] Rouahi, M; Championb, E; Gallet, O; Jada, A; Anselme, K. Physico–chemical characteristics and protein adsorption potential of hydroxyapatite particles: influence on in vitro biocompatibility of ceramics after sintering. *Colloids Surf. B: Biointerfaces,* 2006, 47, 10–19.

[100] Setsukinai, K; Urano, Y; Kakinuma, K; Majima, HJ; Nagano, T. Development of novel fluorescence probes that can reliably detect reactive oxygen species and distinguish specific species. *J.Biol. Chem.,* 2003, 278, 3170–3175.

[101] Shao, Y; Zhang, S; Engelhard, MH; Li, G; Shao, G; Wang, Y; Liu, J; Aksay, IA; Lin, YH. Nitrogen-doped graphene and its electrochemical applications. *J. Mater. Chem.,* 2010, 20, 7491–7496.

[102] Sharpless, NE; Depinho, RA. How stem cells age and why this makes us grow old. *Nat. Rev. Mol. Cell. Biol.,* 2007, 8, 703–713.

[103] Song, Y; Chen, Y; Feng, L; Ren, J; Qu, X. Selective and quantitative cancer cell detection using target-directed functionalized graphene and its synergetic peroxidase-like activity. *Chem. Commun.,* 2011, 47, 4436–4438.

[104] Stankovich, S; Dikin, DA; Piner, RD; Kohlhaas, KA; Kleinhammes, A; Jia, YY; Wu, Y; Nguyen, ST; Ruoff, RS. Synthesis of graphene-based nanosheets via chemical reduction of exfoliated graphite oxide. *Carbon,* 2007, 45, 1558–1565.

[105] Szymczyk, NH; Kerr, BA; Freeman, TA; Adams, CS; Steinbeck, MJ. Involvement of hydrogen peroxide in the differentiation and apoptosis of preosteoclastic cells exposed to arsenite. *Biochem. Pharmacol.,* 2006, 72, 761–769.

[106] Tang, J; Wang, B; Wu, Z; Han, X; Dong, S; Wang, E. Lipid membrane immobilized horseradish peroxidase biosensor for amperometric determination of hydrogen peroxide. *Biosens. Bioelectron.,* 2003, 18, 867–872.

[107] Tang, L; Feng, H; Cheng, J; Li, J. Uniform and rich-wrinkled electrophoretic deposited graphene film: a robust electrochemical platform for TNT sensing. *Chem. Commun.,* 2010, 46, 5882–5884.

[108] Tanner, PA; Wong, AY. Spectrophotometric determination of hydrogen peroxide in rainwater. *Anal. Chim. Acta,* 1998, 370, 279–287.

[109] Terrones, M; Martin, O; González, M; Pozuelo, J; Serrano, B; Cabanelas, JC; Vega-Diaz, SM; Baselga, J. Interphases in graphene polymer-based nanocomposites: achievements and challenges. *Adv. Mater.,* 2011, 23, 5302–5310.

[110] Travis, ER; Wightman, RM. Spatio–temporal resolution of exocytosis from individual cells. *Annu. Rev. Biophys. Biomol. Strucy.,* 1998, 27, 77–103.

[111] Usachov, D; Vilkov, O; Grüneis, A; Haberer, D; Fedorov, A; Adamchuk, VK; Preobrajenski, AB; Dudin, P; Barinov, A; Oehzelt, M; Laubschat, C; Vyalikh, DV. Nitrogen-doped graphene: efficient growth, structure, and electronic properties. *Nano Lett.,* 2011, 11, 5401–5407.

[112] Wallace, GG; Chen, J; Li, G; Moulton, SE; Razal, JM. Nanostructured carbon electrodes. *J. Mater. Chem.,* 2010, 20, 3553–3562.

[113] Wang, JH; Li, LM; Huang, WH; Cheng, JK. Development and evaluation of a rotary cell for capillary electrophoresis chemiluminescence detection. *Anal. Chem.,* 2010, 82, 5380–5383.

[114] Wang, Q; Xu, W; Wu, P; Zhang, H; Cai, CX; Zhao, B. New insights into the effects of thermal treatment on the catalytic activity and conformational structure of glucose

oxidase studied by electrochemistry, IR spectroscopy, and theoretical calculation. *J. Phys. Chem. B,* 2010, 114, 12754–12764.

[115] Wang, X; Li, X; Zhang, L; Yoon, Y; Weber, PK; Wang, H; Guo, J; Dai, H. N-doping of graphene through electrothermal reactions with ammonia. *Science,* 2009, 324, 768–771.

[116] Wei, D; Liu, Y; Wang, Y; Zhang, H; Huang, L; Yu, G. Synthesis of N-doped graphene by chemical vapor deposition and its electrical properties. *Nano Lett.,* 2009, 9, 1752–1758.

[117] Winterbourn, CC; Hampton, MB; Livesey, JH; Kettle, AJ. Modeling the reactions of superoxide and myeloperoxidase in the neutrophil phagosome: implications for microbial killing. *J. Biol. Chem.,* 2006, 281, 39860–39869.

[118] Wolfbeis, OS; Du□rkop, A; Wu, M; Lin, Z. A europium–ion based luminescent sensing probe for hydrogen peroxide. *Angew. Chem., Int. Ed.,* 2002, 41, 4495–4498.

[119] Wolff, S. Ferrous ion oxidation in presence of ferric ion indicator xylenol orange for measurement of hydroperoxides. *Methods Enzymol.,* 1994, 233, 182–189.

[120] Wu, G; Li, D; Dai, C; Wang, D. Well-dispersed high-loading pt nanoparticles supported by shell-core nanostructured carbon for methanol electrooxidation. *Langmuir,* 2008, 24, 3566–3575.

[121] Wu, J; Xie, L; Li, Y; Wang, H; Ouyang, Y; Guo, J; Dai, H. Controlled chlorine plasma reaction for noninvasive graphene doping. *J. Am. Chem. Soc.,* 2011, 133, 19668–19671.

[122] Wu, P; Cai, ZW; Chen, J; Zhang, H; Cai, CX. Electrochemical measurement of the flux of hydrogen peroxide releasing from RAW 264.7 macrophage cells based on enzyme–attapulgite clay nanohybrids. *Biosens. Bioelectron.,* 2011, 26, 4012–4017.

[123] Wu, P; Cai, ZW; Gao, Y; Zhang, H; Cai, CX. Enhancing the electrochemical reduction of hydrogen peroxide based on nitrogen–doped graphene for measurement of its releasing process from living cells. *Chem. Commun.,* 2011, 47, 11327–11329.

[124] Wu, P; Qian, YD; Du, P; Zhang, H; Cai, CX. Facile synthesis of nitrogen-doped graphene for measuring the releasing process of hydrogen peroxide from living cells. *J. Mater. Chem.,* 2012, 22, 6402–6412.

[125] Wu, XM; Hu, YJ; Jin, J; Zhou, NL; Wu, P; Zhang, H; Cai, CX. Electrochemical approach for detection of extracellular oxygen released from erythrocytes based on graphene fFilm integrated with laccase and 2,2-azino-bis(3-ethylbenzothiazoline -6-sulfonic acid). *Anal. Chem.,* 2010, 82, 3588–3596.

[126] Xhattopadhyay, K; Mazumdar, S. Structural and conformational stability of horseradish peroxidase: effect of temperature and pH. *Biochemistry,* 2000, 39, 263–270.

[127] Xu, J; White, Y; Li, P; He, C; Han, YF. Hydroxyapatite foam as a catalyst for formaldehyde combustion at room temperature. *J. Am. Chem. Soc.,* 2010, 132, 13172–13173.

[128] Yang, HM; Tang, AD; Jing, OY; Li, M; Mann, S. From natural attapulgite to mesoporous materials: methodology, characterization and structural evolution. *J. Phys. Chem. C,* 2010, 114, 2390–2398.

[129] Yang, HM; Yang, AD; Ouyang, J; Li, M; Mann, S. From natural attapulgite to mesoporous materials: methodology, characterization and structural evolution. *J. Phys. Chem. B,* 2010, 114, 2390–2398.

[130] Yang, S; Feng, X; Wang, X; Müllen, K. Graphene-based carbon nitride nanosheets as efficient metal-free electrocatalysts for oxygen reduction reactions. *Angew. Chem., Int. Ed.,* 2011, 50, 5339–5343.

[131] Yang, XX; Liu, L; Wu, MH; Wang, WL; Bai, XD; Wang, E. Wet-chemistry-assisted nanotube-substitution reaction for high-efficiency and bulk-quantity synthesis of boron- and nitrogen-codoped single-walled carbon nanotubes. *J. Am. Chem..Soc.,* 2011, 133, 13216–13219.

[132] Yao, T; Suzuki, S; Nishino, H; Nakahara, T. Handling low frequency drifting background noise in data from a swept–potential electrochemical detector. *Electroanalysis,*1995, 7, 1114–1117.

[133] Ye, FG; Huang, Y; Xu, Q; Shi, M; Zhao, SL. Quantification of taurine and amino acids in mice single fibrosarcoma cell by microchip electrophoresis coupled with chemiluminescence detection. *Electrophoresis,* 2010, 31, 1630–1636.

[134] Yin, YJ; Lv, YF; Wu, P; Du, P; Shi, YM; Cai, CX. Immobilization and direct electrochemistry of cytochrome c at a single-walled carbon nanotube-modified electrode. *J. Solid State Electrochem.,* 2007, 11, 390–397.

[135] Yin, YJ; Zhang, H; Wu, P; Zhou B; Cai, CX. Iron phosphate nanostructures synthesized by microwave method and their applications in biosensing. *Nanotechnology,* 2010, 21, 425504.

[136] Zhang, C; Fu, L; Liu, N; Liu, M; Wang, Y; Liu, Z. Synthesis of nitrogen-doped graphene using embedded carbon and nitrogen sources. *Adv. Mater.,* 2011, 23, 1020–1024.

[137] Zhang, JP; Wang, Q; Wang, AQ. Synthesis and characterization of chitosan-g-poly (acrylicacid)/attapulgite superabsorbent composites. *Carbohydr. Polym.*, 2007, 68, 367–374.

[138] Zhang, L; Xia, Z. Mechanisms of oxygen reduction reaction on nitrogen-doped graphene for fuel cells. *J. Phys. Chem. C,* 2011, 115, 11170–11176.

[139] Zhang, Y; Ali, SF; Dervishi, E; Xu, Y; Li, Z; Casciano, D; Biris, AS. Cytotoxicity effects of graphene and single-wall carbon nanotubes in neural phaeochromocytoma-derived PC12 cells. *ACS Nano,* 2010, 4, 3181–3186.

[140] Zhao, L; He, R; Rim, KT; Schiros, T; Kim, KS; Zhou, H; Gutiérrez, C; Chochalingam, SP; Arguello, CJ; Pálová, L; Nordlund, D; Hybertsen, MS; Reichman, DR; Heinz, TF; Kim, P; Pinczuk, A; Flynn, GW; Pasupathy, AN. Visualizing individual nitrogen dopants in monolayer graphene. *Science,* 2011, 333, 999–1003.

[141] Zilkha, E; Koshy, A; Obrenovitch, TP; Bennetto, HP; Symon, L. Amperometric biosensors for on-line monitoring of extracellular glucose and glutamate in the brain. *Anal. Lett.,* 1994, 27,453–473.

In: Hydrogen Peroxide
Editors: Gilberto Aguilar and Raphael A. Guzman
ISBN: 978-1-62257-414-8

Chapter 3

HYDROGEN PEROXIDE: ROLE IN INFLAMMATION

Stefania Marzocco, Giuseppe Bianco and Giuseppina Autore
Department of Pharmaceutical and Biomedical Sciences, University of Salerno, Fisciano, SA, Italy

ABSTRACT

Hydrogen peroxide (H_2O_2) is an highly reactive compound produced by cells after secondary reactions as result of the leakage of electrons from the electron transport chain in mitochondria or as a consequence of other enzymatic reactions.

H_2O_2 is able to oxidize and thereby to induce damage in numerous cellular constituents. Over the last decade it has become accepted that H_2O_2 serves important biological roles outside of its capacity to cause oxidative damage. The main factor which determines H_2O_2 effects is its own level associated to the activation of different redox-sensitive transcription factors and to distinct biological responses. Once accumulates in cells, H_2O_2 can, for example, modify proteins and the underlining mechanisms involved attracting great interest at the moment. One of the main proteins target for H_2O_2 is the thiol group of the amino acid cysteine in protein regulating many cell responses (e.g. phosphatases and transcription factors).

Inflammation is a crucial function of the innate immune system that protects host tissues against dangerous insults that are detrimental to tissue homeostasis, including wound damage and pathogen invasion. During the course of inflammatory processes, H_2O_2 may be generated in large amounts. This occurrence may be due to the activation of mast cells, macrophages, eosinophils, and neutrophils, which generate superoxide radical ($O_2\cdot^-$), predominantly via NADPH oxidase (NOX). Superoxide dismutase (SOD) rapidly converts $O_2\cdot^-$ into H_2O_2 which can easily penetrate the membranes of surrounding cells. There is no known receptor for H_2O_2, but potentially H_2O_2 gradients might be "read" via intracellular sensors such as phosphatases that subsequently regulate the chemotactic response. Alternatively, there may be other extracellular (or intracellular) substrates that H_2O_2 modifies in some way to convert them into chemotactic factors or other modulators of cell motility. Several observations suggest that H_2O_2 acts as a common second messenger following cellular exposure to agents that induce inflammatory response. Interactions between H_2O_2 and inflammatory mediator (e.g. NF-κB) might be a component of the intracellular signalling process that leads to inflammation. Treatment with H_2O_2 directly activates NF-κB in some cells, and over expression of SOD, enhances the tumor necrosis factor-α induced activation of NF-κB. In this context H_2O_2 represent a messenger able to alter redox homeostasis contributing, at several levels, to inflammatory-related disease. Despite H_2O_2 is not an inherently reactive compound it can be also converted into highly reactive and deleterious products killing cells. This is a non-specific process direct to pathogens but it can induce damage also to the human body.

Here we describe the role of H_2O_2 in cellular signalling focusing our attention on H_2O_2 role in inflammation. Moreover we also examine the H_2O_2 signalling in some cardiovascular, renal and pulmonary inflammatory-based disease.

Hydrogen Peroxide as a Signal Molecule

Cell signalling is one of the most important aspects of modern biological sciences. It is clear that for their survivor cells must be able to respond to the environment and change their activity accordingly. The observation that nitric oxide (NO) was taking part in cell signalling opened the door to ask the question about the involvement of redox in general. In fact, also other compounds that come under the umbrella of reactive oxygen species (ROS) as hydrogen peroxide (H_2O_2), superoxide anions (O_2^-), hydroxyl radicals ($OH\cdot$) and peroxynitrite ($ONOO^-$) are involved in redox-mediated signal transduction [1]. A growing body of evidence suggests that ROS levels manipulation in and around cells represents a key signal in various physio-pathological pathways, ranging from the response to growth factor stimulation to the generation of the inflammatory response. Accumulating evidences also suggest that dysregulation of oxidant signalling may cause or accelerate a host of pathological conditions.

Among ROS much attention is addressed to H_2O_2. Despite its short biological half-life, H_2O_2 is indicated as an important signalling molecule and over the last decade it has become accepted that it has important biological roles outside of its capacity to cause oxidative damage, even though many questions remain unanswered about how it functions. Although produced in small amounts under homeostatic circumstances acting as intra-cellular signalling molecules or extra-cellular (paracrine) signalling, H_2O_2 production may increase in response to cell stresses (e.g. inflammation). H_2O_2 is generated both as a by-product of mitochondrial respiration, and other metabolic processes, or intentionally by NADPH oxidases. The mitochondrial electron transport chain produces H_2O_2 by O_2; this mechanism could occur when the incomplete reduction of O_2 during respiration produces O_2^-, which dismutates to H_2O_2. Mitochondria can clearly be important contributors to ROS generation under some conditions, particularly when their electron transport proteins are mutated [2-4] or exposed to certain toxins [5,6].The discovery of a family of "professional" ROS generating enzymes, the NADPH oxidase (NOX) and dual oxidase (DUOX) enzymes, in a wide variety of tissues [7-9] resulted in an ongoing reassessment of cellular sources of ROS. The incomplete reduction of O_2 during respiration can also induce H_2O_2 generation by these enzymes which key function appears to be H_2O_2, or O_2^-, synthesis (Figure 1).

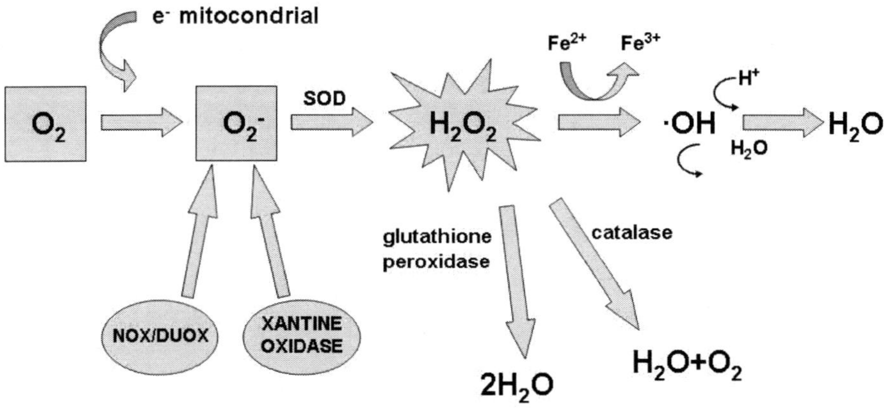

Figure 1. Generation of reactive oxygen species (ROS). Molecular oxygen (O_2) react with an impaired electron (e^-) to form superoxide anion (O_2^-). O_2^- develops also by NOXes/DUOXes and xantine oxidase enzymatic activity. O_2^- is thus converted to hydrogen peroxide (H_2O_2) by superoxide dismutase (SOD). H_2O_2 is then converted to the highly reactive hydroxyl radical (OH·); alternatively, it can be alsodetoxified via either gluthathione peroxidase or catalase to H_2O and O_2.

NOX and the related DUOX family members have a wide tissue distribution. The first discovered NOX, NOX2, was discovered in phagocytic cells where it plays a central role in microbial killing: professional phagocytes, as neutrophils and macrophages, produce large amounts of O2·$^-$, which leads to downstream production of other ROS, including H_2O_2, hypohalides, OH·, and singlet oxygen ($^1O_2$). Under a variety of conditions, ROS generation and/or accumulation is not limited to the phagosome but also occurs in the extracellular milieu. DUOX is so named because contain not only an NADPH-oxidase domain but also a domain homologous to heme-containing peroxidases such as myeloperoxidase and lactoperoxidase [10-12]. DUOX are expressed in thyroid but also in lung and gastrointestinal tract epithelium. Collectively, the highly reactive products of NOXes and DUOXes contribute are involved in different kinds of functions as host defense, mainly killing invading microbes (Table 1).

The identification of related enzyme complexes in non-phagocytic cells suggested on adaptation of the phagocyte NOX in a signalling role as the regulation and resolution of the inflammatory response [13-15]. Cellular responses to various growth factors, hormones and inflammatory mediators, including platelet-derived growth factor (PDGF), epidermal growth factor (EGF), insuline-like growth factor (IGF), insulin, angiotensin II (ANGII), interleukin (IL)-1β, *tumor necrosis factor* (TNF) and Fas, utilize redox transformations in their signaling cascades [16-21]. In all these pathways H_2O_2 is widely accepted to be a second messenger.

Usually signals need to be perceived by cells, either when they arrives from the outside or when they are produced intra-cellularly. One unresolved issue in redox signalling is how specificity is achieved with highly diffusible agents such as H_2O_2. It is hard to conceive that there are receptors for ROS and that H_2O_2 is too small molecule to take part in a ligand-receptor binding event for example. Recent evidence suggests that H_2O_2 might preferentially enter into the cell through specific plasma membrane aquaporin channels, although it had been generally assumed that, once generated, H_2O_2 could simply diffuse back across the plasma membrane, H_2O_2 mediated signalling is tightly related to the control of its accumulation to discrete cells areas [22, 23]. This mechanism has been studied through the antioxidant peroxiredoxin 1 (Prx1), which usually degrades H_2O_2. The presence of growth factor on its receptor inactivated Prx1. The inactivation of this peroxide scavenger only occurs in the region surrounding the stimulated growth factor inducing a local accumulation in this area of H_2O_2 that can exert its effects on protein transcription factor or cell-signalling-mediating proteins [24].

Table 1. NOX and DUOX enzymes isoforms schematization reporting their activating stimulus and related functions mediated

NOX isoforms	Stimulus	Function(s)
NOX1	H_2O_2	Proliferation
	Thrombine	
	Angiotensin II	Hypertrophy
	PDGF	Migration
	FGF	
	TNF-α	Necrosis
	LPS-Flagelline	Host defense
NOX2	Microorganism	Host defense
	Cytokines	
	ATP	Apoptosis
	TNF-α	Necrosis
	VEGF	Angiogenesis
NOX3	Cisplatin	Ototoxicity
NOX4	IGF-1	Migration
	PDGF	
	Angiotensin II	Survivor
	TGF-β	
	Insulin	Differentiation
NOX5	IL-4	Gene expression
	PDGF	Proliferation
DUOX1/DUOX2	n.d.	Hormones iodination Defense and inflammation

The site or compartments where an oxidant is generated is an important determinant of how the cell will respond to oxidative stress. Exogenously added oxidants must either act on the surface or traverse the cellular membrane. If cells generate their own oxidants extracellularly or in compartments such as mitochondria or endosomes, membrane permeability could be also an obstacle. Even for H_2O_2, which is readily diffusible, there is a concentration differential across the membrane that depends on the consumption rates on the two sides [25]. Thus it is hypothesised that H_2O_2 generated in the mitochondrial matrix should therefore have different targets to that extra-cellularly derived [26] (Figure 2).

Intracellular H_2O_2 is due to mitochondrial activiy or is derived from extracellular production of H_2O_2 channeled back into the cell through specific plasma membrane aquaporins; once in cell H_2O_2 can performs proteins modification on transcription factors activating inflammatory gene transcription.

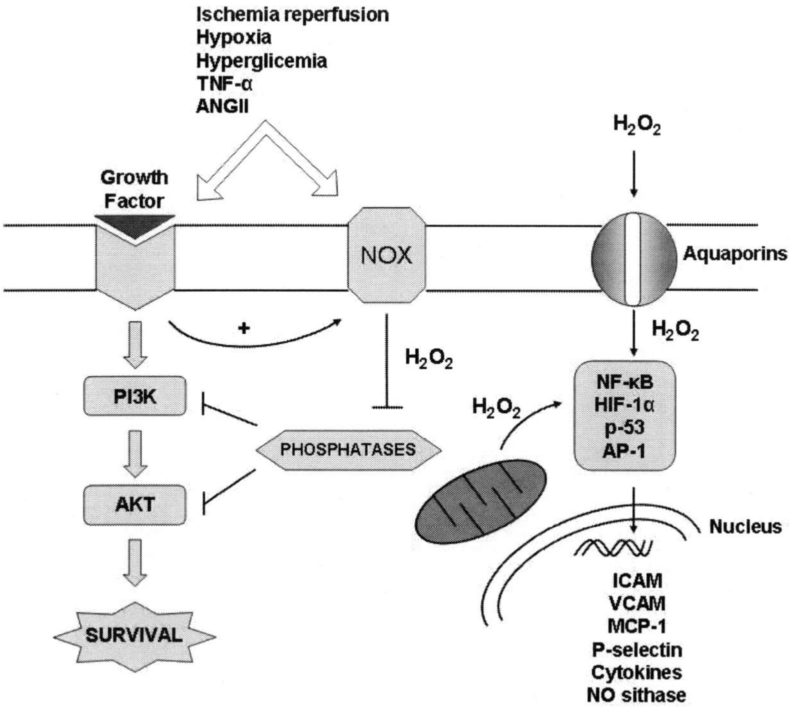

Figure 2. Signalling pathway by H_2O_2 during inflammation is mainly related NOXes/DUOXes stimulation by growth factor and pathological condition. Ligand binding to its receptor stimulates H_2O_2 production then H_2O_2 produced can activates cell survival by phosphatases inhibition.

Once accumulates in cells, H_2O_2 could interact chemically modifying proteins and the underlining mechanisms involved attracting great interest at the moment. Usually aminoacid modifications would potentially be associated with conformational changes in the protein with a resulting activity change, either increasing it or decreasing it. One of the main proteins target for H_2O_2 is the thiol group of the amino acid cysteine protein regulating cell response (e.g. phosphatases and transcription factors). Much like phosphorylation on serine or threonine residues, also modification of the thiol group can in turn modify enzymatic activity, for which there is no doubt about its importance in cell signalling (Figure 3).

Phosporilation of aminoacidic residues by kinases occur in intracellular signal transduction cascades, usually after the stimulation of a receptor. Phosphatases, deputated to kinases dephosphorylation, may be inactivated in response to increased intracellular H_2O_2 and their inactivation usually leads to enhanced activity of the reciprocal kinase. Several kinases including those related to prosurvival, proapoptotic or proinflammatory process are known to be redox regulated [27]. Also transcription factors, co-activators and repressors have been identified as elements redox-controlled. Among these Specificity Protein 1 (Sp1), the glucocorticoid receptor, Early growth response protein 1 (Egr1), p53, Nuclear Factor kappa B (NF-κB), Nuclear factor erythroid-derived 2 (Nrf2), Hypoxia Inducible Factor-1α (HIF-1α) and Nuclear Respiratory Factor-1 (NRF-1). Among these NF-κB is one of the most important transcription factors H_2O_2 responsive. NF-κB is normally bounded in the cytoplasm to IκB in its inactive state. Stimuli that activate NF-κB induce the proteosomal degradation of IκB, allowing NF-κB to translocate into the nucleus and bind to κB motifs in the promoter region of many inflammatory genes, including TNF-α and inducible nitric oxide synthase (iNOS). Several studies reported that H_2O_2 clearly modulates NF-κB function which role is very important during inflammatory states [28].

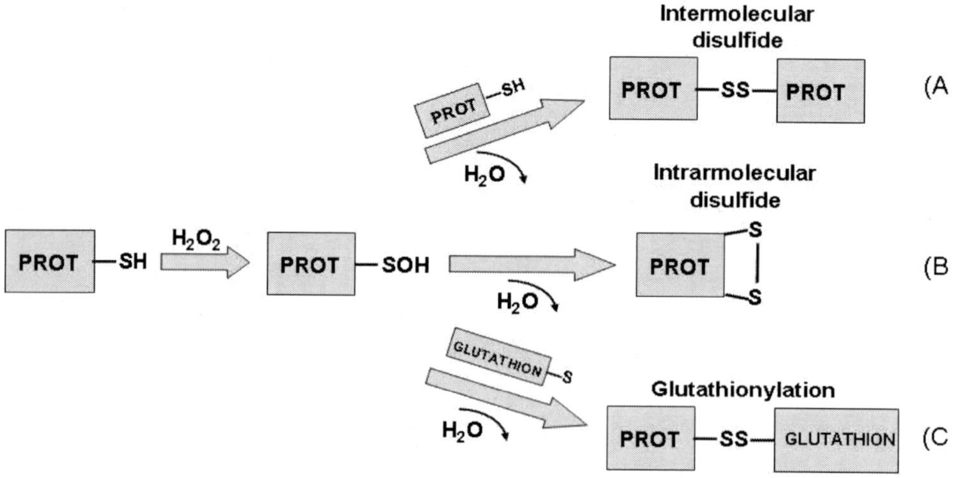

Figure 3. Protein thiol modifications and reductive pathways during proteic oxidant stress by H_2O_2. These modifications are able to activate H_2O_2 biological response. All paths (A, B, C) represent the protein thiol modifications occurring during exposure to H_2O_2 starting with the formation of sulfenic acid group. The reaction proceeds forming intermolecular disulfide when sulfenic group reacts with another thiol group of other proteins (A), whereas the disulfide is intramolecular when the bound occurs by two thiol groups on the same protein (B). Protein sulfenic group could also bounds glutathion forming glutathionylated protein (C).

In eukaryotes, as in bacteria, H_2O_2 signalling pathways increase the expression of antioxidant and phase 2 detoxification genes as a vital part of adaptive responses to oxidative stress. The main factor which determines H_2O_2 effect is its own level, associated to the activation of different redox-sensitive transcription factors and to distinct biological responses. Low H_2O_2 concentrations mediate fine-tune responses to growth factors and cytokines in order to control biological responses (e.g. cell growth, division, differentiation, migration and apoptosis). Moreover H_2O_2 at low level is associated to low oxidative stress and primarily induces Nrf2, a transcription factor implicated in the transactivation of gene coding for antioxidant enzymes [29]. An intermediate amount of H_2O_2 triggers an inflammatory response through the activation of nuclear factor-κB (NF-κB) and AP-1. Finally high level of H_2O_2 induces perturbation of the pore mitochondrial transmembrane potential and disruption of the electron transfer that mainly results in apoptosis or necrosis [30].

Hydrogen Peroxide in Inflammation

Inflammation is considered as the immune response of tissues due to bodily injury. Clinical characteristics of inflammation include pain, heat, swelling and redness at the site of the injury. The inflammatory process may also include loss of function of the involved tissues. This type of acute inflammation is normally a localized and protective response following trauma or infection.

At the site of infection or injury, mast cells, platelets, nerve endings, endothelial cells and other resident cells release signalling molecules and chemoattractants able to recruit leukocytes into the affected area. Neutrophils, a type of granulocyte, are the first leukocytes to appear at the injured site. These cells phagocytose (engulf) and kill invading microorganisms through the release of non-specific toxins, such as H_2O_2 and other ROS that kill pathogens as well as adjacent cells, sick and healthy alike. Neutrophils, and also macrophages, release cytokines as interleukin IL-1, IL-6, TNF-α, gamma interferon (INF-γ) and others. Such pro-inflammatory cytokines in turn induce the liver to synthesize various acute phase reactant proteins and also induce systemic inflammatory responses (e.g., fever and leukocytosis arise in the number of white blood cells).

However, if the agent responsible of inflammation persists for a prolonged period of time, the inflammary process becomes chronic. Chronic inflammation can result from a viral or microbial infection, environmental antigen (e.g., pollen), autoimmune reaction, or persistent activation of inflammatory molecules and it is primarily mediated by monocytes and long-lived macrophages; monocytes mature into macrophages once they leave the bloodstream and enter into the tissues. Macrophages engulf and digest microorganisms, foreign invaders and senescent cells. They are able to release mediators, as IL-1, TNF-α and prostaglandins, that amplify the pro-inflammatory response. At later stages, other cells, including lymphocytes, invade the affected tissues: T lymphocytes kill virus-infected cells and B lymphocytes produce antibodies that specifically target the invading microorganisms for destruction. Macrophages and leukocytes release ROS and proteases that destroy the source of inflammation; however damage to the body's own tissues often results. Tissue damage is an hallmark of chronic inflammation as well as the repair of the damaged tissue by replacement with cells of the same type or with fibrous connective tissue.

Granulocytes and macrophages, which are the "professional" phagocytes, respond to a variety of stimuli with aggregation, activation of movement, secretion, endocytosis, activation of the respiration and mediators production. Through their response macrophages exert a key role in the defence against invading organisms and tumor cells. One of the most impressive response is the activation of the respiration with a concomitant production of an enormous amount of intermediate products of O_2 reduction such as $O2^{\cdot -}$, $OH^{\cdot}$, H_2O_2 and $^1O_2$. The activation of the respiration is called "respiratory burst" and is due to the activation of a membrane bound NOX, which is dormant in resting state, induced by perturbation of the plasma membrane during phagocytosis or following interaction between the cell surface and a number of environmental stimuli.

Among ROS, H_2O_2 plays a relevant role in host defences, in the evolution of the inflammatory process and in various cytotoxic effects responsible for tissue damage. Emerging reports suggest that H_2O_2 may enhance cellular activation by suppressing the activity of anti-inflammatory enzymes such as histone deacetylases [31] and mitogen-activated protein kinase phosphatases (MKPs) [32]. Moreover H_2O_2 can suppress the negative regulatory functions of IκB-α, prolonging NF-κB nuclear localization and pro-inflammatory transcriptional responses in activated cells.

While in physiological condition grow factors are mainly involved in H_2O_2 signalling during inflammation also pathological condition are able to activate H_2O_2 production.

Thus evidences are continuously growing that H_2O_2 exerts amplificatory and modulatory functions both on the normal cell physiology and on the inflammatory process involving several tissue and organ as vasculature, kidney and lung.

HYDROGEN PEROXIDE IN VASCULAR INFLAMMATION

All vascular cell types (endothelial cells, vascular smooth muscle cells and adventitial fibroblasts) produce ROS primarily via cell membrane-associated NOX.

Among these H_2O_2, O_2^- and NO are important signalling molecules participating in growth, apoptosis and migration of vascular smooth muscle cells, moreover they participate to the modulation of extracellular matrix and endothelial function [33].

Until recently, major emphasis was placed on the production of O_2^- in the vasculature as a result of its ability to directly attenuate the biological activity of endothelium derived NO. O_2^- has a relatively short biologic half life, and thus is unlikely to mediate effects distant from where it is produced thus limiting its role as an important paracrine hormone in vascular biology. The charge on the O_2^- makes it cell-membrane impermeable; it generally does not cross membranes except possibly via ion channels. In contrast, H_2O_2 has a longer biologic lifespan than O_2^- and it is readily diffusible across lipid bilayers and is capable to modulate both contractile and growth-promoting pathways with more extensive effects [34].

Under physiological conditions H_2O_2 is produced in a controlled manner at low concentrations and play function as signalling molecules regulating vascular smooth muscle cell (VSMC) contraction/relaxation and growth [35] to maintain vascular integrity, in regulating cell apoptosis [36], induction of inflammatory proteins [37], endothelium–leukocyte interaction and vascular remodelling [38, 39]. During stress condition the activation of growth factor receptors is able to activate vascular NOX to produce H_2O_2 which, in turn, can activate signalling cascade associated to growth factor themselves. Receptor tyrosine kinases such as those for EGF, PDGF, fibroblast growth factors (FGF), vascular endothelial growth factor (VEGF) [40] and non-receptor tyrosine kinases such as Janus kinase 2 (JAK2) [41], Src, [42], and focal adhesion kinase (FAK) respond to H_2O_2. Growth factor activation often lie upstream of MAPK and H_2O_2 potently activates extracellular-signal-regulated kinase1/2 (ERK1/2), p38MAPK, c-Jun N-terminal kinase (JNK), and ERK5 in both vascular endothelial and smooth muscle cells [32, 42-45]. Redox-sensitive transcriptional factors including NF-kB, activator protein 1 (AP-1), and HIF-1 are often activated via MAPK to modulate changes in gene expression and cellular function [46, 36]. ROS stimulated NF-kB and AP-1 induce also the expression of pro-inflammatory genes, including monocyte chemotactic protein-1 (MCP-1), adhesion molecules, and interleukins [47], able to play a role in vascular inflammation associated with hypertension and atherosclerosis. Increased activation of vascular NF-kB and AP-1 and associated inflammatory and mitogenic responses, attributed in part to oxidative excess(REF), have been demonstrated in hypertensive rats [48]. Phosphorylation-dependent postranslational regulation of proteins also occurs in response to H_2O_2. For example, it has been shown that H_2O_2 induces phosphatidylinositol (PI) 3-kinase/Akt-dependent phosphorylation of endothelial nitric oxide synthase (eNOS), leading to a compensatory, transient increase in NO production, [49, 50] which may act as an intermediate step for vascular long-term detrimental consequences [51].

The main NOXes involved in vascular H_2O_2 production are: NOX1, NOX 2, NOX 4 and NOX 5. Their signalling cascades is triggered by stresses, hormones and vasoactive agents, even if also cytokines affect NOX isoforms, enphatizing the involvement of H_2O_2 in vasculature inflammation [52]. H_2O_2 controlled production is guaranteed by constitutive oxidase activities and endogenous scavenger systems, including catalase and glutathione peroxidases, that thus maintain an homeostatic balance in favour of normal wall thickness.

However upon stimulation of the oxidases by mechanical strech and/or by vasoactive hormones, increased production of peroxide and/or the release of cytokines participate in a feed-forward activation of vascular NOX isoforms leading to abnormally high levels of growth-promoting H_2O_2. Infact, certain vascular diseases such as atherosclerosis, hypertension, restenosis as well as risk factors for coronary artery disease (e.g. hyperlipidemia, diabetes and cigarette smoking) can disrupt the homeostatic redox state. During this process, the adventitial NOX is expected to produce large amounts of H_2O_2, that eventually tips the balance in favour of the inflammatory response. H_2O_2 overproduction leads to events as endothelial dysfunction, increased contractility, VSMC growth and apoptosis, monocyte migration, lipid peroxidation and increased deposition of extracellular matrix proteins mainly contributing to vascular damage in cardiovascular disease [53, 54]. Moreover emerging evidence has demonstrated that uniquely, H_2O_2 is able to amplify its own production in vascular cells, and this phenomenon likely contributes to its long-lasting pathological effects [55].

Although atherogenesis it was formerly considered a bland lipid storage disease, substantial advances in basic and experimental science have illuminated the role of inflammation in this disease. Compelling evidence for the importance of inflammation in atherosclerosis at both the basic and clinical level has evolved in parallel. Atherosclerosis is a multifactorial, multistep disease that involves chronic inflammation at every stage, from initiation to progression and, eventually, plaque rupture [56]. In atherosclerosis the normal homeostatic functions of the endothelium are altered promoting an inflammatory response [57, 58]. Accumulating data indicate that insights gained from the link between inflammation and atherosclerosis can yield predictive and prognostic information of considerable clinical utility [56]. H_2O_2 increases eNOS expression and activity and directly oxidizes protein kinase GIα, promoting vasodilatation [59]. H_2O_2 may then be converted into reactive nitrogen and reactive chlorine, which attack both low-density lipoprotein (LDL) and high-density lipoprotein (HDL), contributing to plaque formation [60, 61]. ROS, as H_2O_2, also act on metalloproteinases alterating vascular structure in hypertension by increasing deposition of extracellular matrix proteins, such as collagen and fibronectin, leading to modification of extracellular structures homeostasis. In particular H_2O_2 and O_2^- influence activity of vascular matrix metalloproteinase-2 e 9 (MMP-2 and MMP-9), which promote degradation of basement membrane and elastin [62] and induces the development and complications in atherosclerosis.

Hypertension follows closely behind lipids on the classical risk factors for atherosclerosis. Increasing evidence supports the view that, like atherosclerosis itself, inflammation may participate in hypertension providing a pathophysiological link between these two diseases. Despite extensive research, the etiology of hypertension remains elusive. Low-grade inflammation is recognized as a key player in the pathogenesis of hypertension. Inflammation participates in many processes that contribute to the development of elevated blood pressure. In the vasculature, for example, inflammation increases the proliferation of smooth muscle cells and participates in vascular remodeling [63]. Angiotensin II (ANGII), in addition to its vasoconstrictor properties, can instigate intimal inflammation. ANGII is the major bioactive peptide of the renin-angiotensin-aldosteron system (RAAS). It is produced systemically and locally within the vascular wall and is a potent vasoactive peptide that also stimulates vascular smooth muscle cell growth, inflammation and fibrosis through myriad signaling pathways [64-66]. Accordingly, ANGII plays an important physiological role in maintaining vascular tone by regulating immediate vasoconstriction. ANGII has also a pathophysiological role in cardiovascular diseases such as hypertension, atherosclerosis and heart failure, conditions associated with endothelial dysfunctions, vascular hyper-reactivity and structural remodeling. ANGII and the activation of its type I receptor (AT1R) induces a cascade of intra-cellular signalling events that ultimately leads to vascular regulation of NOX1, NOX2, NOX4, and NOX5 are expressed in vascular tissue and are probably the major sources of ROS in the cardiovascular system [52]. Vascular NOX, especially NOX4, are constitutively active but also are regulated by many vasoactive hormones like ANGII, endothelin-1 (ET-1), aldosterone, growth factors (PDGF, transforming growth factor-β, TGF-β), mechanical stimuli (shear stress and stretch), and cardiovascular risk factors [67-69], to produce H_2O_2.

ANGII is a potent inducer of vascular H_2O_2 generation and NOX activation. The mechanism of action of ANGII involves protein kinase C (PKC), phospholipase D (PLD), c-Src, and receptor tyrosine kinases [70, 67, 71]. Mechanisms linking ANGII/AT1 to NOX and upstream signalling in vascular cells have not been fully elucidated but PDGF, TGF-β, TNF-α and thrombin are able to activate NOX in vascular smooth muscle cells (VSMCs) [72-75].

H_2O_2 is generally considered as a vasodilator, which promotes endothelium-dependent relaxation and reduces blood pressure in mice [76, 77]. Mechanisms of H_2O_2-mediated vasodilation are complex but probably involve regulation of K^+ and Ca^{2+} channels through direct actions and indirect actions via second messengers [77]. These redox-regulated Ca^{2+} processes may be more important both in stress responses than in receptor-mediated signalling by growth factors or cytokines and may play a role in altered vascular contractility during hypertension. Despite this role of H_2O_2 in modulating vascular tone is quite complex, as confirmed by studies showing conflicting results on vasomotor tone; H_2O_2 can exert either a contractile or a relaxant response depending on the cellular and enzymatic source of H_2O_2, the intracellular H_2O_2 localization, the vascular bed and the contractile state [78, 79]. Mechanisms contributing to H_2O_2-induced vasoconstriction, include: (i) an increase in intracellular $[Ca^{2+}]i$ through regulation of L-type Ca^{2+} channels [80, 81], (ii) generation of arachidonic acid metabolites with vasoconstrictor activity, and (iii) a direct Ca^{2+}-independent tonic effect on the vascular smooth muscle contractile apparatus.

Vascular over-expression of catalase in mice, which converts H_2O_2 to H_2O, reduced the pressor response to vasoconstrictor agents and decreased blood pressure, suggesting the importance of endogenous H_2O_2 as a vasoconstrictor and regulator of blood pressure. H_2O_2 interacts with phospholipase C (PLC), PKC, and phosphoinositide 3-kinase (PI3K), which may contribute to molecular mechanisms underlying H_2O_2-induced vasoconstriction.

Cardiac myocytes modulate coronary vascular tone through H_2O_2, purinergic components (adenosine and adenosine diphosphate, ADP) and ANGII, especially in ischemic conditions, also associated to an inflammatory process. H_2O_2 released from cardiac myocytes induced vasodilatory effects and ANGII released from cardiac myocytes exhibited vasoconstrictor effects in the coronary circulation in response to oxygen supply [82]. In vascular cells H_2O_2

also regulate cell survival inducing, at high concentrations (>100 mmol/l), apoptosis and anoikis (cell detachment and shedding), whereas at lower concentrations they stimulate growth and differentiation [83-85].

Oxidative stress may have a critical role in the development of certain vascular disease states. However, the failures of classical antioxidant therapies demonstrate that they are insufficient to inhibit oxidative stress. Simply ignoring the importance of ROS, as H_2O_2, will never lead to the achievement of clinical benefits. Indeed, further research into selective molecular inhibitors of vascular ROS generation, particularly those that interfere with NADPH oxidase activation, are warranted. In this regard the therapeutic potential in cardiovascular disease of some drugs is just related to their ability to interact with NOX. Activators of peroxisome proliferator-activated receptors (PPARs), statins and antihypertensive drugs, such as β-blockers, Ca^{2+} channel blockers, ACE inhibitors, and AT1R blockers, downregulate expression of oxidase subunits and decrease NOX activity [86-89, 48, 90, 91]. Thus the selective inhibition of dysfunctional NOX homologs appears to be the most reasonable approach at present with the potential to be far more efficient than non-selective scavenging of ROS by the administration of antioxidants.

Hydrogen Peroxide in Renal Inflammation

Cardiovascular disease and chronic kidney disease (CKD) share several common antecedents, including increased blood pressure, dyslipidemia and diabetes. Moreover similarities between atherosclerosis and glomerulosclerosis, two vassel inflammatory-based disease, have been proposed by various investigators; both atherosclerosis and glomerulosclerosis lead to influx of monocytes, production of lipid-laden macrophages, increased presence of cholesterol and cholesterol esters, proliferation of contractile cells, and matrix expansion resulting in fibrosis [92, 93].

Inflammation of the kidney is usually called nephritis. In particular the most important inflammatory kidney pathology are glomerulonephritis, also called glomerular disease, that happens when the internal kidney structures (glomeruli) become inflammed. Membranoproliferative glomerulonephritis is another glomerular disease that occurs mostly in children and teenagers and interstitial nephritis when appears inflammation of the kidney cells that are not part of the fluid-collecting units. Over the time these inflammatory conditions lead to acute renal failure or chronic kidney disease. ROS play an important role in inflammatory-related renal pathology with the primary involvement of H_2O_2 that interact with RAAS and cooperate with myeloperoxidase (MPO) to induce inflammatory renal injury.

Intensive research in the past 20 years has provided convincing evidences that ANGII, the primary effector of RAAS is a key contributor in the progression of renal failure by stimulating growth, inflammation and fibrosis of the kidney [94]. Also aldosterone, usually involved in kidney function, also induces ROS formation by increasing expression of the p47phox and p67phox subunits of NOX [95, 96]. In fact ROS formation in the heart induced by ANGII infusion was prevented by the mineralocorticoid receptor antagonist spironolactone and depended on the presence of NOX2 [97].

At molecular level recent evidences, using ischemic reperfusion model, suggest that ANGII stimulates also mitochondrial ROS generation through the opening of mitochondrial

K^+/ATP channels, leading to redox sensitive activation of MAPK [98]. Renal inflammatory injury is related also with ROS-activated MMP-2 and 9, which degradates extracellular matrix components. ANGII-induced ROS are potential regulators of adhesion molecule expression, that together with NF-κB-mediated transcriptional activation of chemokines and contribute to renal inflammation [99].

Another key factor in H_2O_2-mediated renal inflammation is the heme protein MPO, a catalyst for (lipo)protein oxidation present in mononuclear cells. The ability of MPO to generate hypochlorous acid/hypochlorite from H_2O_2 in the presence of chloride ions is a unique and defining activity for this enzyme. The MPO- H_2O_2 -chloride system leads to a variety of chlorinated protein and lipid adducts that in turn may cause dysfunctions of cells in different compartments of the kidney. MPO contribute to renal injury trough inflammation and enhances the development of renal fibrosis and other inflammatory related renal pathologies (e.g. glomerulonephritis, diabetic nephropathy, hypertension, acute renal failure, progression of renal disease) [99].

In diabetic nephropathy the role of glucose in oxidative stress-induced kidney injury has been extensively documented. Chronic hyperglycemia contributes to diabetic complications through the formation of advanced glycation end products (AGEs), which are irreversibly formed biochemical end products of non enzymatic glycation. In vitro experiments have shown that synthesis of acrolein, 2-hydroxypropanal, and glycoaldehyde required chloride, H_2O_2 and MPO, respectively, thus a role for H_2O_2 in diabetic nephropaty is postulated.

In the kidney, monocytes represent the major source of MPO in lipid-associated glomerulopathy a process very similar to lipid-associated development of atherosclerosis; the release of ROS as H_2O_2 by monocytes evidently can lead to oxidative modification of lipoproteins, which consequently perpetuate the process of disease.

In membranous glomerulonephritis, glomerular epithelial cells may generate H_2O_2 by expressing NOX in analogy to neutrophilic granulocytes during the respiratory burst reaction. Evidence for a role of the MPO- H_2O_2 -chloride system in the synthesis of reactive oxygen species in damaged glomerular basement membranes from patients with membranous glomerulonephritis [100] was based on immunohistochemistry of HOCl-modified proteins/epitopes co-localizing with immune deposits in affected glomeruli.

Thus it is evident that the MPO- H_2O_2-chloride system plays a relevant role not only in oxidant mediated antimicrobial defence by granulocytes but also in the progression of degenerative and inflammatory mediated kidney diseases.

Hydrogen Peroxide in Lung Inflammation

Because of their direct exposure to ambient air, lung cells are subjected to an enhanced oxidant stress by environmental irritants and pollutants including oxidants such as cigarette smoke, ozone and free radical–generating environmental carcinogens. Moreover the respiratory tract is the target organ for inorganic dusts (asbestos, silica, and coal) and particulate matter 2.5 (PM2.5) contributing to asthma, chronic obstructive pulmonary disease (COPD), fibrosis, and constrictive bronchiolitis [101].

A typical component in most lung disorders and lung infections is inflammation and activation of inflammatory cells with consequent free radical generation. In addition, several therapies, most importantly chemotherapy and radiation, lead to free radical-induced tissue damage where again the lung is the most commonly involved organ. Thus, lung represents a unique tissue exposed not only directly to higher oxygen tensions and environmental oxidants but also to oxidants produced by a variety of lung diseases and in the course of their therapies [102].

ROS as H_2O_2 and $O_2\cdot^-$, as well MPO are produced by inflammatory cells exposed to environmental toxins in the lower respiratory tract and contribute to lung inflammation and injury. During this process etiologic agents are phagocytosed by alveolar macrophages or bronchial epithelial cells which release chemotactic factors that recruit inflammatory cells to the lung. Chemotactic factors attract and activate neutrophils, eosinophils, mast cells and lymphocytes that further activate macrophages to release oxidant as H_2O_2 [103]. Oxidants from macrophages and bronchial epithelial cells cause inflammation leading to injury and fibrosis [104]. These events are particularly important to the upper airways leading to asthma and constrictive bronchiolitis. In the lower respiratory tract exogenous oxidants can activate macrophages to release endogenous oxidants, thus attracting other inflammatory cells and injuring the delicate walls of the peripheral respiratory bronchioles and alveolar sacs [105]. Typical sources of ROS in inflammate lung tissue include generation *in situ* by the activation of NOX on cell membranes and inducible NOS activation in the cells cytosolic compartments.

H_2O_2 is produced in lung tissue by bronchial epithelial cells and regulate physiological homeostasis remodelling of extracellular matrix, apoptosis, mitochondrial respiration, cell proliferation, maintenance of surfactant and the antiprotease screen, alveolar repair and immune modulation [104, 105]. H_2O_2 also represent an important role in the pathogenesis of obstructive lung diseases (e.g. asthma and COPD), parenchymal lung diseases (e.g., idiopathic pulmonary fibrosis and lung granulomatous diseases) and lung malignancies.

Asthma is associated with lung chronic inflammation and an increased oxidant burden. Inflammatory cells of the airways of patients with asthma are activated and they generate increased amounts of free radicals. Exhaled air of patients with asthma contains elevated levels of H_2O_2, NO and organic volatile compounds, indicating an H_2O_2 overproduction in this condition [106, 107].

Increased oxidative stress or reduced antioxidant resources seem to be associated also with COPD, an heterogenous syndrome characterised by irreversible progressive airflow limitation [108, 110]. A common feature of lung inflammation in COPD is the activation of epithelial cells and resident macrophages as well as the recruitment and activation of neutrophilis [110] which generate ROS via NOX. ROS can trigger the peroxidation of polyunsaturated fatty acids in biological tissue resulting in the transformation of the fatty acid into lipid hydroperoxides which play an important role in the decline of lung function in COPD [111]. Several indicators of oxidative stress, such as H_2O_2 exhalation, lipid peroxidation products and degraded proteins, are indeed elevated in COPD patients, as a result on imbalance of redox system in COPD patients [112].

H_2O_2 in exhaled breath directly reflects oxidant generation in the lung. It has been shown that also smokers have higher levels of exhaled H_2O_2 than ex-smokers with COPD or non-smokers [113, 114]. During acute exacerbations of COPD these H_2O_2-levels are even higher. The source of the enhanced exhalation of H_2O_2 is unknown, but it has been suggested to partly originate from the increased release of O_2^- by smokers alveolar macrophages [115, 116]. Additionally, the combination xanthine/ xanthine oxidase, capable of generating O_2^- and

H_2O_2, is increased in the bronchoalveolar lavage and plasma of COPD patients and smokers when compared with healthy subjects and nonsmokers respectively [117, 118].

Thus the assessment of airway inflammation by non-invasive methods, such as H_2O_2 levels increase, could provide a signal to start anti-inflammatory treatment before the onset of symptoms and the impairment of lung function. It could also be useful in the follow-up of patients with lung disease and for guiding drug treatment. Lung inflammatory disorders mentioned above, including asthma, COPD, adult respiratory distress syndrome and idiopathic pulmonary fibrosis involve the recruitment of immune and inflammatory cells in the lung. These cells are activated and produce mediators of inflammation including oxidants and cytokines [119].

It has been studied the molecular mechanism of oxidant mediated proinflammatory response and it has been suggested that oxidant generating systems and pro-inflammatory mediators influence histone acetylation/phosphorylation via a mechanism dependent on the activation of the MAPK pathway [120-122]. In fact both TNF-α and H_2O_2 alter histone acetylation/deacetylation, and the activation of NF-κB and AP-1, leading to the release of the pro-inflammatory cytokine IL-8 in human alveolar epithelial cells [123]. Moreover oxidants generate airway inflammation by activating NF-κB in macrophages leading to the production of proinflammatory cytokines IL-1β, TNF-α, IL-8, and IL-6 [124]. Then oxidants are a defence mechanism against pathogens which adds to the amount of oxidative stress. Moreover also cigarette smoking contributes to oxidative stress directly and stimulates the release of chemokines to attract inflammatory cells [125].

Further understanding of the effects and roles of ROS in basic cellular functions as amplification of pro-inflammatory and immunological responses, signalling pathways, activation of transcription factors (NF-κB and AP-1), chromatin modelling (histone acetylation and deacetylation), and gene expression will provide important information regarding basic pathological processes contributing to chronic inflammatory diseases [123].

Hydrogen Peroxide in other Inflammatory Diseases

H_2O_2 generation by NOXes also contribute to tissue damage in other inflammatory pathologies such as intestinal bowel disease (IBD) and joint diseases.

At bowel level potential role for NOX enzymes in Crohn's disease and ulcerative colitis have been discussed for quite some time, but most evidence has been rather indirect. Recently, stronger arguments were provided by the observation that in a genome-wide screen, there was a strong genetic association between a polymorphism of the neutrophil cytosol factor (NCF) 4 gene (coding for the p40phox subunit of NOX2) and Crohn's disease. At this point, however, it is not clear whether gain of function or loss in NOX2 function might be associated with Crohn's disease. As some patients with chronic granulomatous disease (hereditary NOX2 deficiency) develop an inflammatory colitis, the latter possibility should not be excluded. Thus, at this point, it is impossible to predict whether NOX-based therapeutics might be beneficial for inflammatory bowel disease.

Regard musculoskeletal diseases ROS-related the question of whether various forms of arthritis are due to either increased or decreased production of ROS as H_2O_2, is open. Although the exact causes of rheumatoid arthritis (RA) are unknown, involvement of ROS is suspected. Several studies suggest a beneficial effect of antioxidants such as vitamin E in RA. At foci of inflammation, activation of T cells and macrophages leads to a large increase in oxygen consumption, whose corollary is increased release of ROS.

Both increased and decreased NOX2 activity has been implied in arthritis, and, accordingly, NOX inhibitors and NOX activators have been suggested to improve experimental arthritis [126-127].

REFERENCES

[1] Ullrich, V; Kissner, R. Redox signaling: bioinorganic chemistry at its best. *J Inorg Biochem*, 2006 100(12), 2079-86.

[2] Ishii, N; Ishii, T; Hartman, PS. The role of the electron transport SDHC gene on lifespan and cancer. *Mitochondrion*, 2007 7, 24-28.

[3] Ishii, T; Yasuda, K; Akatsuka, A; Hino, O; Hartman, PS; Ishii, N. A mutation in the SDHC gene of complex II increases oxidative stress, resulting in apoptosis and tumorigenesis. *Cancer Res*, 2005 65, 203-209.

[4] Wallace, DC. Mitochondrial diseases in man and mouse. *Science*, 1999 283, 1482-1488.

[5] Castello, PR; Drechsel, DA; Patel, M. Mitochondria are a major source of paraquat-induced reactive oxygen species production in the brain., *J Biol Chem*, 2007 282, 14186-14193.

[6] Senft, AP; Dalton, TP; Nebert, DW; Genter, MB; Hutchinson, RJ; Shertzer, HG Dioxin increases reactive oxygen production in mouse liver mitochondria. *Toxicol Appl Pharmacol*, 2002 178, 15-21.

[7] Bedard, K; Krause, KH. The NOX family of ROS generating NADPH oxidases: physiology and pathophysiology. *Physiol Rev*, 2007 87, 245–313.

[8] Lambeth, JD. NOX enzymes and the biology of reactive oxygen. *Nat Rev Immunol*, 2004 4, 181-189.

[9] Lambeth, JD; Kawahara, T; Diebold, B. Regulation of Nox and Duox enzymatic activity and expression. *Free Radic Biol Med*, 2007 43, 319-331.

[10] Lambeth, JD; Cheng, G; Arnold, RS; Edens, WE. Novel homologs of gp91phox. *TIBS*, 2000 25, 459-461.

[11] De Deken, X; Wang, D; Many, MC; Costagliola, S; Libert, F; Vassart, G; Dumont, JE; Miot, F. Cloning of two human thyroid cDNAs encoding new members of the NADPH oxidase family. *Journal of Biological Chemistry*, 2000 275, 23227-23233.

[12] Dupuy, C; Ohayon, R; Valent, A ; Noe-Hudson, M; Dee, D; Virion, A.. Purification of a novel flavoprotein involved in the thyroid NADPH oxidase. *Journal of Biological Chemistry*, 1999 274, 37265–37269.

[13] Hampton, MB; Keenan, JI; Vissers, MCM; Winterbourn, CC. Oxidant mediated phosphatidylserine exposure and macrophage uptake of activated neutrophils: possible impairment in chronic granulomatous disease. *Journal Leukoc Biol*, 2002 71, 775-781.

[14] Romani, L; Fallarino, F; De Luca, A; Montagnoli, C; D'Angelo, C; Zelante, T; Vacca, C; Bistoni, F; Fioretti, MC; Grohmann, U; Segal, BH; Puccetti, P. Defective tryptophan catabolism underlies inflammation in mouse chronic granulomatous disease. *Nature*, 2008 451, 211-215.

[15] Brown, JR; Goldblatt, D; Buddle, J; Morton, L; Thrasher, AJ. Diminished production of anti-inflammatory mediators during neutrophil apoptosis and macrophage phagocytosis in chronic granulomatous disease (CGD). *J. Leukoc. Biol.*, 2003 73, 591-599.

[16] Lee, SR; Kwon, KS; Kim, SR; Rhee, SG. Reversible inactivation of proteintyrosine phosphatase 1B in A431 cells stimulated with epidermal growth factor. *J. Biol. Chem.*, 1998 273, 15366-15372.

[17] Sundaresan, M; Yu, ZX; Ferrans, VJ; Irani, K; Finkel, T. Requirement for generation of H_2O_2 for platelet-derived growth factor signal transduction. *Science.* 1994 270, 296-299.

[18] Zafari, AM; Ushio-Fukai, M; Akers, M; Yin, Q; Shah, A; Harrison, DG; Taylor, WR; Griendling, KK. Role of NADH/NADPH oxidase-derived H_2O_2 in angiotensin II-induced vascular hypertrophy. *Hypertension*, 1998 32, 488-495.

[19] Suzuki, Y; Ono, Y; Hirabayashi, Y.. Rapid and specific reactive oxygen species generation via NADPH oxidase activation during Fas-mediated apoptosis. *FEBS Lett.*, 1998 425, 209-212.

[20] Chiarugi, P; Fiaschi, T; Taddei, ML; Talini, D; Giannoni, E; Raugei, G; Ramponi, G. Two vicinal cysteines confer a peculiar redox regulation to low molecular weight protein tyrosine phosphatase in response to platelet-derived growth factor receptor stimulation. *J. Biol. Chem.*, 2001 276, 33478-33487.

[21] Mahadev, K; Wu, X; Zilbering, A; Zhu, L; Lawrence, JT; Goldstein, BJ;. Hydrogen peroxide generated during cellular insulin stimulation is integral to activation of the distal insulin signaling cascade in 3T3-L1 adipocytes. *J. Biol. Chem.*, 2001 276, 48662-48669.

[22] Bienert GP; Møller AL; Kristiansen KA; Schulz A; Møller IM; Schjoerring JK; Jahn TP.. Specific aquaporins facilitate the diffusion of hydrogen peroxide across membranes. *J. Biol. Chem.*, 2007 282, 1183-1192.

[23] Miller EW; Dickinson BC; Chang CJ. Aquaporin-3 mediates hydrogen peroxide uptake to regulate downstream intracellular signaling. *Proc. Natl. Acad. Sci.*, 2010 107, 15681-15686.

[24] Niethammer P; Grabher C; Look AT; Mitchison TJ. A tissue-scale gradient of hydrogen peroxide mediates rapid wound detection in zebrafish. Nature, 2009 459, 996-999.

[25] Antunes, F; Cadenas, E. Estimation of H2O2 gradients across biomembranes. *FEBS Lett.*, 2000 475, 121-126.

[26] Chakraborti T; Das S; Mondal M; Roychoudhury S; Chakraborti S. Oxidant, mitochondria and calcium: an overview. *Cell Signal.*, 1999 112, 77-85.

[27] Cho SH; Lee CH; Ahn Y; Kim H; Kim H; Ahn CY; Yang KS; Lee SR. Redox regulation of PTEN and protein tyrosine phosphatases in H(2)O(2) mediated cell signaling. *FEBS Lett.*, 2004 560, 7-13.

[28] Gloire G; Legrand-Poels S; Piette J; NF-kappaB activation by reactive oxygen species: fifteen years later. *Biochem Pharmacol.*, 2006 72, 1493-1505.

[29] Ungvari Z; Bailey-Downs L; Sosnowska D; Gautam T; Koncz P; Losonczy G; Ballabh P; de Cabo R; Sonntag WE; Csiszar A. Vascular oxidative stress in aging: a homeostatic failure due to dysregulation of NRF2-mediated antioxidant response. *Am J Physiol Heart Circ Physiol.*, 2011 301, 363-372.

[30] Klamt F; Zdanov S; Levine RL; Pariser A; Zhang Y; Zhang B; Yu LR; Veenstra TD; Shacter E; Oxidant-induced apoptosis is mediated by oxidation of the actin-regulatory protein cofilin. *Nat Cell Biol.*, 2009 11, 1241-1246.

[31] Ito, K; Hanazawa, T; Tomita, K; Barnes, PJ; and Adcock, IM. Oxidative stress reduces histone deacetylase 2 activity and enhances IL-8 gene expression: role of tyrosine nitration. *Biochem. Biophys. Res. Commun.*, 2004 315, 240-245.

[32] Sundaresan M; Yu ZX; Ferrans VJ; Irani K; Finkel T; Requirement for generation of H_2O_2 for platelet-derived growth factor signal transduction. *Science.*, 1995 270, 296-299.

[33] Arora S; Vaishya R; Dabla PK; Singh B. NAD(P)H oxidases in coronary artery disease. *Adv Clin Chem.*, 2010 50, 65-86.

[34] Ardanaz N; Pagano PJ. Hydrogen peroxide as a paracrine vascular mediator: regulation and signaling leading to dysfunction. *Exp Biol Med*, 2006 231, 237-251.

[35] Touyz RM; Schiffrin EL Ang II-stimulated superoxide production is mediated via phospholipase D in human vascular smooth muscle cells. *Hypertension*, 1999 34, 976-98.

[36] Taniyama Y; Griendling KK. Reactive oxygen species in the vasculature: molecular and cellular mechanisms. *Hypertension*, 2003 42, 1075-1081.

[37] Griendling KK; Sorescu D; Lassegue B; Ushio-Fukai M. Modulation of protein kinase activity and gene expression by reactive oxygen species and their role in vascular physiology and pathophysiology. *Arterioscler Thromb Vasc Biol.*, 2000 20, 2175-2183.

[38] Lessner SM; Galis ZS. Matrix metalloproteinases and vascular endothelium-mononuclear cell close encounters. *Trends Cardiovasc Med.*, 2004 14,105-111.

[39] Pasterkamp G; Galis ZS; de Kleijn DP. Expansive arterial remodeling: location, location, location. *Arterioscler Thromb Vasc Biol.*, 2004 24, 650-657.

[40] Aslan M; Ozben T. Oxidants in receptor tyrosine kinase signal transduction pathways. *Antioxid Redox Signal.*, 2003 5, 781-788.

[41] Simon AR; Rai U, Fanburg BL; Cochran BH. Activation of the JAK-STAT pathway by reactive oxygen species. *Am J Physiol.*, 1998 275, 1640-1652.

[42] Abe J; Takahashi M; Ishida M; Lee JD; Berk BC. c-Src is required for oxidative stress-mediated activation of big mitogen-activated protein kinase 1. *J Biol Chem.*, 1997 272, 20389-20394.

[43] Ushio-Fukai M; Alexander RW; Akers M; Griendling KK. p38 Mitogenactivated protein kinase is a critical component of the redox-sensitive signaling pathways activated by angiotensin II. Role in vascular smooth muscle cell hypertrophy. *J Biol Chem.*, 1998 273, 15022-15029.

[44] Yoshizumi M; Abe J; Haendeler J; Huang Q; Berk BC. Src and Cas mediate JNK activation but not ERK1/2 and p38 kinases by reactive oxygen species. *J Biol Chem.*, 2000 275, 11706-11712.

[45] Cai H; Li Z; Davis ME; Kanner W; Harrison DG; Dudley SC Jr. Akt dependent phosphorylation of serine 1179 and mitogen-activated protein kinase kinase/extracellular signal-regulated kinase 1/2 cooperatively mediate activation of the endothelial nitric-oxide synthase by hydrogen peroxide. *Mol Pharmacol.*, 2003 63, 325-331.

[46] Li JM; Shah AM. Endothelial cell superoxide generation: regulation and relevance for cardiovascular pathophysiology. *Am J Physiol Regul Integr Comp Physiol.*, 2004 287, 1014-1030.

[47] Brigelius-Flohé R; Banning A; Kny M; Böl GF. Redox events in interleukin-1 signaling. *Arch Biochem Biophys.*, 2004 423, 66-73.

[48] Taddei S; Virdis A; Ghiadoni L; Magagna A; Pasini AF; Garbin U; Cominacini L; Salvetti A. Effect of calcium antagonist or beta blockade treatment on nitric oxide-dependent vasodilation and oxidative stress in essential hypertensive patients. *J Hypertens*, 2001 19, 1379-1386.

[49] Cai H; Davis ME; Drummond GR; Harrison DG. Induction of endothelial NO synthase by hydrogen peroxide via a Ca(2+)/calmodulin-dependent protein kinase II/janus kinase 2-dependent pathway. *Arterioscler Thromb Vasc Biol.*, 2001 21, 1571-1576.

[50] Thomas SR; Chen K; Keaney JF Jr. Hydrogen peroxide activates endothelial nitric-oxide synthase through coordinated phosphorylation and dephosphorylation via a

phosphoinositide 3-kinase-dependent signalling pathway. *J Biol Chem.*, 2002 277, 6017-6024.

[51] Cai H; Li Z; Dikalov S; Holland SM; Hwang J; Jo H; Dudley SC Jr; Harrison DG. NAD(P)H oxidase-derived hydrogen peroxide mediates endothelial nitric oxide production in response to angiotensin II. *J Biol Chem.*, 2002 277, 48311-48317.

[52] Lassègue B; Griendling KK; NADPH Oxidases: Functions and Pathologies in the Vasculature. *Arterioscler Thromb Vasc Biol.*, 2010 30, 653-661.

[53] Rao GN; Berk BC. Active oxygen species stimulate vascular smooth muscle cell growth and proto-oncogene expression. *Circ Res.*, 1992 70, 593-599.

[54] Harrison DG. Cellular and molecular mechanisms of endothelial cell dysfunction. *J Clin Invest.*, 1997, 2153-2157.

[55] Cai H. Vascular Disease NAD(P)H Oxidase-Dependent Self-Propagation of Hydrogen Peroxide and Vascular Disease. *Circulation Research*, 2005 96, 818-822.

[56] Libby P; Ridker PM; Maseri A; Inflammation and Atherosclerosis. *Circulation*, 2002 105, 1135-1143.

[57] Prescott SM; McIntyre TM; Zimmerman GA; Stafforini DM. Molecular events in acute inflammation. *Arterioscler Thromb Vasc Biol.*, 2002 22, 727-733.

[58] Ross R. Atherosclerosis: an inflammatory disease. *N Engl J Med.*, 1999 340, 115-126.

[59] Takac I; Schröder K; Brandes RP. The Nox family of NADPH oxidases: friend or foe of the vascular system? *Curr Hypertens Rep.*, 2012 14, 70-78.

[60] Lyle AN; Griendling KK. Modulation of vascular smooth muscle signaling by reactive oxygen species. *Physiology* (Bethesda)., 2006 21, 269-80.

[61] Nicholls SJ; Hazen SL. The role of myeloperoxidase in the pathogenesis of coronary artery disease. *Jpn J Infect Dis.*, 2004 57, 21-22.

[62] Rajagopalan S; Meng XP; Ramasamy S; Harrison DG; Galis ZS. Reactive oxygen species produced by macrophage-derived foam cells regulate the activity of vascular matrix metalloproteinases in vitro. Implications for atherosclerotic plaque stability. *J Clin Invest.*, 1996 98, 2572-2579.

[63] Leibowitz A; Schiffrin EL. Immune mechanisms in hypertension. *Curr Hypertens Rep.*, 2011 13, 465-72.

[64] Touyz RM; Schiffrin EL. Signal transduction mechanisms mediating the physiological and pathophysiological actions of angiotensin II in vascular smooth muscle cells. *Pharmacol Rev.*, 2000 52, 639-72.

[65] Mehta PK; Griendling KK. Angiotensin II cell signalling: physiological and pathological effects in the cardiovascular system. *Am J Physiol Cell Physiol.*, 2007 292, 82-97.

[66] Lemarié CA; Schiffrin EL. The angiotensin II type 2 receptor in cardiovascular disease. *J Renin Angiotensin Aldosterone Syst.*, 2010 11, 19-31.

[67] Lassegue B; Clempus RE: Vascular NAD(P)H oxidases: specific features, expression, and regulation. *Am J Physiol Regul Integr Comp Physiol.*, 2003 285, 277-297.

[68] Li JM; Shah AM: Intracellular localization and preassembly of the NAD(P)H oxidase complex in cultured endothelial cells. *J Biol Chem.*, 2002 277, 19952-19960.

[69] Orosz Z; Csiszar A; Labinskyy N; Smith K; Kaminski PM; Ferdinandy P; Wolin MS; Rivera A; Ungvari Z. Cigarette smoke-induced proinflammatory alterations in the endothelial phenotype: role of NAD(P)H oxidase activation. *Am J Physiol Heart Circ Physiol.*, 2007 292, 130-139.

[70] Touyz RM; Chen X; Tabet F; Yao G; He G; Quinn MT; Pagano PJ; Schiffrin EL. Expression of a functionally active gp91phox-containing neutrophil-type NAD(P)H oxidase in smooth muscle cells from human resistance arteries: regulation by angiotensin II. *Circ Res.*, 2002 90, 1205-1213.

[71] Touyz RM; Schiffrin EL. Increased generation of superoxide by angiotensin II in smooth muscle cells from resistance arteries of hypertensive patients: role of phospholipase D-dependent NAD (P)H oxidase-sensitive pathways. *J Hypertens.*, 2001 19, 1245-1254.
[72] Marumo T; Schini-Kerth VB; Fisslthaler B; Busse R. Platelet-derived growth factor-stimulated superoxide anion production modulates activation of transcription factor NFkappaB and expression of monocyte chemoattractant protein 1 in human aortic smooth muscle cells. *Circulation*, 1997 96, 2361-2367.
[73] De Keulenaer GW; Alexander RW; Ushio-Fukai M; Ishizaka N; Griendling KK Tumour necrosis factor alpha activates a p22phox-based NADH oxidase in vascular smooth muscle. *Biochem J.,* 1998 329:653-657.
[74] Gorlach A; Diebold I; Schini-Kerth VB; Berchner-Pfannschmidt U; Roth U; Brandes RP; Kietzmann T; Busse R. Thrombin activates the hypoxia-inducible factor-1 signaling pathway in vascular smooth muscle cells: role of the p22(phox)-containing NADPH oxidase. *Circ Res.*, 2001 89, 47-54.
[75] Brandes RP; Miller FJ; Beer S; Haendeler J; Hoffmann J; Ha T; Holland SM; Gorlach A; Busse R. The vascular NADPH oxidase subunit p47phox is involved in redox-mediated gene expression. *Free Radic Biol Med.,* 2002 32, 1116-1122.
[76] Ray R; Murdoch CE; Wang M; Santos CX; Zhang M; Alom-Ruiz S; Anilkumar N; Ouattara A; Cave AC; Walker SJ; Grieve DJ; Charles RL; Eaton P; Brewer AC; Shah AM. Endothelial Nox4 NADPH oxidase enhances vasodilatation and reduces blood pressure in vivo. *Arterioscler Thromb Vasc Biol.*, 2011 31, 1368-76.
[77] Nguyen Dinh Cat A; Touyz RM. Cell Signaling of Angiotensin II on Vascular Tone: Novel Mechanisms. *Curr Hypertens Rep.*, 2011 13, 122-128.
[78] Suvorava T; Lauer N; Kumpf S; Jacob R; Meyer W; Kojda G. Endogenous vascular hydrogen peroxide regulates arteriolar tension in vivo. *Circulation*, 2005 112, 2487–2495.
[79] Ardanaz N; Pagano PJ; Hydrogen peroxide as a paracrine vascular mediator: regulation and signaling leading to dysfunction. *Exp Biol Med.*, 2006 231, 237-51.
[80] Touyz RM. Reactive oxygen species as mediators of calcium signaling by angiotensin II: implications in vascular physiology and pathophysiology. *Antioxid Redox Signal.*, 2005 7, 1302-1314.
[81] Amberg GC; Eraley S; Glapa SA. Local regulation of arterial L-type calcium channels by reactive oxygen species. *Circ Res.*, 2010 107, 1002-1010.
[82] Kaneshiro T; Saitoh S; Machii H; Yamaguchi O; Ishibashi T; Maruyama Y; Takeishi Y. Metabolic regulation of coronary vascular tone: role of hydrogen peroxide, purinergic components, and angiotensin. *Eur J Pharmacol.*, 2010 645, 127-34.
[83] Deshpande NN; Sorescu D; Seshiah P; Ushio-Fukai M; Akers M; Yin Q; Griendling KK. Mechanism of hydrogen peroxide- induced cell cycle arrest in vascular smooth muscle. *Antioxid Redox Signal.*, 2002 4, 845-854.
[84] Li AE; Ito H; Rovira II; Kim KS; Takeda K; Yu ZY; Ferrans VJ; Finkel T. A role for reactive oxygen species in endothelial cell anoikis. Circ Res., 1999 85, 304-310.
[85] Li J; Li W; Su J: Liu W; Altura BT; Altura BM. Hydrogen peroxide induces apoptosis in cerebral vascular smooth muscle cells: possible relation to neurodegenerative diseases and strokes. *Brain Res Bull.*, 2003 62, 101-106.
[86] Dandona P; Karne R; Ghanim H; Hamouda W; Aljada A; Magsino CH. Carvedilol inhibits reactive oxygen species generation by leukocytes and oxidative damage to amino acids. *Circulation*, 2000 101, 122-124.
[87] Dhalla SN; Temsah RM; Netticadan T. Role of oxidative stress in cardiovascular diseases. *J Hypertens.*, 2000 18, 655-673.

[88] Mantle D; Patel VB; Why HJ; Ahmed S; Rahman I; MacNee W; Wassif WS; Richardson PJ; Preedy VR. Effects of lisinopril and amlodipine on antioxidant status in experimental hypertension. *Clin Chim Acta.*, 2000 299, 1-10.
[89] Ohtahara A; Hisatome I; Yamamoto Y; Furuse M; Sonoyama K; Furuse Y. The release of the substrate for xanthine oxidase in hypertensive patients was suppressed by angiotensin converting enzyme inhibitors and a1-blockers. *J Hypertens.*, 2001 19, 575-582.
[90] Diep QN; Amiri F; Touyz RM; Cohn JS; Endemann D; Neves MF; Schiffrin EL. PPARalpha activator effects on Ang IIinduced vascular oxidative stress and inflammation. *Hypertension*, 2002 40, 866-871.
[91] Wassmann S; Laufs U; Muller K; Konkol C; Ahlbory K; Baumer AT; Linz W; Bohm M; Nickenig G. Cellular antioxidant effects of atorvastatin in vitro and in vivo. *Arterioscler Thromb Vasc Biol.*, 2002 22, 300-305.
[92] Diamond JR. Analogous pathobiologic mechanisms in glomerulosclerosis and atherosclerosis. *Kidney Int Suppl.,* 1991 31, 29-34.
[93] Keane WF; Kasiske BL; O'Donnell MP. Lipids and progressive glomerulosclerosis. A model analogous to atherosclerosis. *Am J Nephrol.*, 1988 8, 261-271.
[94] Wolf G; Butzmann U; Wenzel UO; The reninangiotensin system and progression of renal disease: From hemodynamics to cell biology. *Nephron Physiol.*, 2003 93, 3-13.
[95] Miyata K; Rahmann M; Shokoji T; Nagai Y; Zhang GX; Sun GP; Kimura S; Yukimura T; Kiyomoto H; Kohno M; Abe Y; Nishiyama A. Aldosterone stimulates reactive oxygen species production through activation of NADPH oxidase in rat mesangial cells. *J Am Soc Nephrol.*, 2005 16, 2906-2912.
[96] Sun Y; Ahokas RA; Bhattacharya SK; Gerling IC; Carbone LD; Weber KT. Oxidative stress in aldosteronism. *Cardiovasc Res.*, 2006 71, 300-309.
[97] Johar S; Cave AC; Narayanapanicker A; Grieve DJ; Shah AM. Aldosterone mediates angiotensin II-induced interstitial cardiac fibrosis via a Nox2-containing NADPH oxidase. *FASEB J.,* 2006 20, 1546-1548.
[98] Kimura S; Zhang GX; Nishiyama A; Shokoji T; Yao L; Fan YY; Rahman M; Suzuki T; Maeta H; Abe Y. Role of NAD(P)H oxidase- and mitochondria-derived reactive oxygen species in cardioprotection of ischemic reperfusion injury by angiotensin II. *Hypertension*, 2005 45, 860-866.
[99] Sachse A; Wolf G. Angiotensin II–Induced Reactive Oxygen Species and the Kidney. *Am Soc Nephrol.*, 2007 18:, 2439-2446.
[100] Gröne HJ; Gröne EF; Malle E. Immunohistochemical detection of hypochlorite-modified proteins in glomeruli of human membranous glomerulonephritis. *Lab Invest.*, 2004 82, 5-14.
[101] Iwagaki A; Choe N; Li Y; Hemenway DR; Kagan E. Asbestos inhalation induces tyrosine nitration associated with extracellular signal-regulated kinase 1/2 activation in the rat lung. *Am J Respir Cell Mol Biol.*, 2003 28, 51-60.
[102] Kinnula VL; Crapo JD. Superoxide ismutases in the lung and human lung diseases. *Am J Respir Crit Care Med.,* 2003 167, 1600-1619.
[103] Rom WN. Role of Oxidants in Interstitial Lung Diseases: Pneumoconioses, Constrictive Bronchiolitis, and Chronic Tropical Pulmonary Eosinophilia. *Mediators Inflamm.*, 2011:407657. Epub 2011 Oct 30.
[104] J.Wang; S.Wang, R; Manzer, G McConville; R. J. Mason. Ozone induces oxidative stress in rat alveolar type II and type I-like cells. *Free Radic Biol Med.*, 2006 40, 1914-1928.
[105] Alessandrini F; Beck-Speier I; Krappmann D; Weichenmeier I; Takenaka S; Karg E; Kloo B; Schulz H; Jakob T; Mempel M; Behrendt H. Role of oxidative stress in

ultrafine particle-induced exacerbation of allergic lung inflammation. *Am J Respir Crit Care Med.*, 2009 179, 984-991.
[106] Corradi M; Montuschi P; Donnelly LE; Pesci A; Kharitonov SA; Barnes PJ. Increased nitrosothiols in exhaled breath condensate in inflammatory airway diseases. *Am J Respir Crit Care,* 2001 163, 854-858.
[107] Kharitonov SA; Barnes PJ. Exhaled markers of pulmonary disease. *Am J Respir Crit Care*, 2001 163, 1693-1722.
[108] Agusti AG. Systemic effects of chronic obstructive pulmonary disease. *Novartis Found Symp.*, 2001 234, 242-254.
[109] Rennard SI. Overview of causes of COPD. New understanding of pathogenesis and mechanisms can guide future therapy. *Postgrad Med.*, 2002 111, 28-30, 33-4, 37-8.
[110] MacNee W. Oxidative stress and lung inflammation in airways disease. *Eur J Pharmacol.*, 2001 429, 195-207.
[111] Tsukagoshi H; Shimizu Y; Iwamae S; Hisada T; Ishizuka T; Iizuka K; Dobashi K; Mori M. Evidence of oxidative stress in asthma and COPD: potential inhibitory effect of theophylline. *Respir Med.*, 2000 94, 584-588.
[112] Boots AW; Haenen GRMM; Bast A; Oxidant metabolism in chronic obstructive pulmonary disease. *Eur Respir J.,* 2003 22, 14-27.
[113] Nowak D; Antczak A; Krol M; Pietras T; Shariati B; Bialasiewicz P; Jeczkowski K; Kula P. Increased content of hydrogen peroxide in the expired breath of cigarette smokers. *Eur Respir J.*, 1996 9, 652-657.
[114] Macnee W; Rahman I. Oxidants and antioxidants as therapeutic targets in chronic obstructive pulmonary disease. *Am J Respir Crit Care Med.*, 1999 160, 58-65.
[115] Dekhuijzen PN; Aben KK; Dekker I; Aarts LP; Wielders PL; van Herwaarden CL; Bast A. Increased exhalation of hydrogen peroxide in patients with stable and unstable chronic obstructive pulmonary disease. *Am J Respir Crit Care Med.*, 1996 154, 813-816.
[116] Hoidal JR; Fox RB; LeMarbe PA; Perri R; Repine JE. Altered oxidative metabolic responses in vitro of alveolar macrophages from asymptomatic cigarette smokers. *Am Rev Respir Dis.*, 1981 123, 85-89.
[117] Heunks LM; Vina J; van Herwaarden CL; Folgering HT; Gimeno A; Dekhuijzen PN. Xanthine oxidase is involved in exercise-induced oxidative stress in chronic obstructive pulmonary disease. *Am J Physiol.*, 1999 277, 1697-1704.
[118] Pinamonti S; Muzzoli M; Chicca MC; Papi A; Ravenna F; Fabbri LM; Ciaccia A. Xanthine oxidase activity in bronchoalveolar lavage fluid from patients with chronic obstructive pulmonary disease. *Free Radic Biol Med.*, 1996 21, 147-155.
[119] Brennan FM; Maini RN; Feldemann M. Cytokine expression in chronic inflammatory disease.*Br Med Bull.*, 1995 51, 368-84.
[120] Tikoo K; Lau SS; Monks TJ; Histone H3 phosphorylation is coupled to poly-(ADP-ribosylation) during reactive oxygen species-induces cell death in renal proximal tubular epithelial cells. *Mol. Pharmacol.*, 2001 60, 394-402.
[121] Miyata Y; Towatari M; Maeda T; Ozawa Y; Saito H. Histone acetylation induces by granulocyte colony-stimulating factor in a MAP kinase-dependent manner. *Biochem. Biophys. Res. Commun.,* 2001 283, 655-660.
[122] Böhm L; Schneeweiss FA; Sharan RN; Feinendegen LE. Influence of histone acetylation on the modification of cytoplasmic and nuclear proteins by ADP-ribosylation in response to free radicals. Biochim. *Biophys. Acta*, 1997 1334, 149-154.
[123] Rahman I. Oxidative stress, transcription factors and chromatin remodelling in lung inflammation. *Biochem Pharmacol.*, 2002 64, 935-942.

[124] YJanssen YM; Barchowsky A; Treadwell M; Driscoll KE; Mossman BT. Asbestos induces nuclear factor κB (NF-κB) DNA-binding activity and NF-κB-dependent gene expression in tracheal epithelial cells. *Proc Natl Acad Sci.*,1995 92, 8458-8462.

[125] Flouris AD; Metsios GS; Carrillo AE; Jamurtas AZ; Gourgoulianis K; Kiropoulos T; Tzatzarakis MN; Tsatsakis AM; Koutedakis Y. Acute and short-term effects of secondhand smoke on lung function and cytokine production. *Am J Respir Crit Care Med.*, 2009 179, 1029-1033.

[126] Hougee S; Hartog A; Sanders A; Graus YM; Hoijer MA; Garssen J; van den Berg WB; van Beuningen HM; Smit HF. Oral administration of the NADPH-oxidase inhibitor apocynin partially restores diminished cartilage proteoglycan synthesis and reduces inflammation in mice. *Eur J Pharmacol.*, 2006 531, 264-269.

[127] Hultqvist M; Olofsson P; Gelderman KA; Holmberg J; Holmdahl R A new arthritis therapy with oxidative burst inducers. *PLoS Med.*, 2006 3, 1625-1636.

In: Hydrogen Peroxide
Editors: Gilberto Aguilar and Raphael A. Guzman
ISBN: 978-1-62257-414-8

Chapter 4

HYDROGEN PEROXIDE DEPLETES PHOSPHATIDYLINOSITOL-3-PHOSPHATE FROM ENDOSOMES IN A P38 MAPK-DEPENDENT MANNER AND PERTURBS ENDOCYTIC TRANSPORT AND SIGNAL TRANSDUCTION IN PATHOGENIC CELLS

Masayuki Murata and Fumi Kano
Department of Life Sciences, Graduate School of Arts and Sciences,
The University of Tokyo, Tokyo, Japan

ABSTRACT

Hydrogen peroxide (H_2O_2) has gained much attention not only as a stressor that oxidizes components of the cell but also as signaling molecule in physiological processes. On the minus side, elevation of reactive oxygen species (ROS) including H_2O_2 causes oxidative stress, and is reported to be involved in many diseases, such as diabetes, atherosclerosis, Parkinson's disease, and Alzheimer's disease. On the plus side, H_2O_2 is produced when cells respond to various stimuli, and mediates downstream signals. Recently, we found that H_2O_2 treatment depleted HeLa cell endosomes of phosphatidylinositol-3-phosphate (PI3P). To quantify and visualize the intracellular localization of PI3P we have developed a novel analytical method that employs a semi-intact cell system, a permeabilized cell that enables the addition of various exogenous molecules into the cells, and protein probe for PI3P, GST-2xFYVE protein. Using this method, we found that H_2O_2 decreased the amount of PI3P in endosomes in a manner dependent upon p38 MAPK, a kinase that is activated during oxidative stress or pathogenic conditions. In addition, H_2O_2 treatment delayed vesicular transport in a broad range of endocytic pathways: the lysosomal degradation of EGFR, the retrograde transport of cholera toxin from the plasma membrane to the *trans*-Golgi network (TGN), and the uptake and recycling of transferrin.

The physiological role of PI3P in endosomes has received a lot of attention with respect to endocytosis and signal transduction via endosomes. For example, PI3P in the endosomal membrane appears to act as a microdomain for the recruitment of various proteins that contain a PI3P-binding (FYVE) domain. Interestingly, FYVE-containing proteins are involved not only in the regulation of endosomal processing but also in signal transduction via endosomes.

In this chapter, we demonstrate the effect of H_2O_2 on early endocytic pathways and discuss the utility of H_2O_2 as a useful oxidative stressor to investigate various intracellular events under pathological conditions.

1. Introduction

Oxidative stress occurs when the concentration of oxidants such as reactive oxygen species (ROS) becomes too high to allow their removal by antioxidants. ROS and free radicals are known to damage various components of cells, including proteins, lipids, and DNA. As a consequence, oxidative stress is reported to be involved in many diseases, including diabetes, atherosclerosis, Parkinson's disease, and Alzheimer's disease. ROS also have an important role as signaling molecules in the regulation of various biological processes, such as enhanced growth triggered by growth factors, tumorigenesis, host defense, phagocytosis and autophagy [1-4]. Hydrogen peroxide (H_2O_2) is a relatively stable ROS, and it can cross the cell membrane, which makes it suitable for use as a signaling molecule. For example, the production of intracellular H_2O_2 was increased upon treatment of A431 cells with epidermal growth factor (EGF) [5,6], and was dependent upon the tyrosine kinase activity of EGF receptor [6]. Furthermore, H_2O_2 production was necessary for EGF-mediated down regulation of E-cadherin via p38 MAPK activation in SKOV3 cells [7]. One potential target of H_2O_2 is protein tyrosine phosphatases (PTPs), which inhibit growth factor receptor tyrosine kinase signaling. H_2O_2 treatment leads to oxidation of cysteine in the active site of PTPs, effectively blocking their enzymatic activity [8-10]. The inhibition of phosphatase activity may also enhance tyrosine phosphorylation-dependent signaling [11,12]. Thus, H_2O_2 acts as both a mediator of cell cytotoxicity and a signal transduction molecule in living cells. In this chapter, we investigated the effects of H_2O_2 as one of oxidative stressor on the endocytic vesicular traffic in mammalian cells.

Endocytosis is the process by which molecules outside a cell are internalized into the cell by the formation of vesicles from the plasma membrane. In general, the molecules are first loaded into vesicles in a clathrin-dependent or -independent manner. After internalization, the vesicles are trafficked to early endosomes, in a process that depends on the small GTPase Rab5 and its effector EEA1 [13,14]. Given that there are various endocytic pathways, early endosomes serve as a sorting station that determines the final destination of internalized ligands and/or receptors. For example, upon binding of EGF, EGFR is transported rapidly in a Rab7-dependent manner to lysosomes where it is degraded, which downregulates the EGF signal [15]. In contrast, transferrin (Tf) is recycled from early endosomes back to the plasma membrane both directly, by a process that requires Rab4 [16], and via recycling endosomes, which requires Rab11 [17]. Shiga toxin and cholera toxin, which are the bacterial exotoxins produced by *Shigella dysenteriae* and *Vibrio cholerae*, respectively, undergo retrograde transport from endosomes to the *trans*-Golgi network (TGN), via a pathway that is dependent on sorting nexin (SNX) 1, 2, or 8 [18-20]. The molecular mechanisms of the sorting processes in early endosomes have been studied and characterized in detail. Recently, however, the role of early endosomes as signaling organelles has started to garner attention. Endosomes are ideally suited to act as a specialized signaling platform because they have a distinct biochemical composition that promotes the selective recruitment of various scaffold proteins and signaling mediators. In this way, signaling by growth factor receptors can be modulated by extending or shortening the sojourn time of the receptors in endosomes ([21], see below).

One of the characteristic features of early endosomes is their high content of phosphatidylinositol-3-phosphate (PI3P). The high concentration of PI3P enables the early endosome to act as a platform where proteins that contain a PI3P-binding domain (FYVE domain) are recruited and concentrated [22-24]. Interestingly, proteins that contain a FYVE domain are involved not only in the regulation of endosomal processing but also in signal transduction via endosomes. For example, the FYVE domain proteins EEA1 and Hrs mediate endosomal fusion and endocytic sorting, respectively [25-29]. Another FYVE domain protein, SMAD anchor for receptor activation (SARA), is localized to early endosomes to transduct the signal from the transforming growth factor TGFβ receptor to the mediator SMAD efficiently [30,31]. The PI3P content in early endosomes also affected the signal transduction through regulating the sorting process of receptors. Petiot et al. [32] showed that interfering with PI3P using the PI3 kinase inhibitor, wortmannin, or overexpression of GFP-tagged tandem FYVE protein, trapped EGF receptors in early/recycling endosomes, and inhibited transport of the receptors to late endosomes. In addition, Zoncu et al. [33] have demonstrated that endosomes associated with APPL1 (APPL-endosomes) are the first station for early endocytic traffic and that APPL-endosomes mature to canonical EEA1-positive endosomes as PI3P accumulates. APPL1 was initially identified as an Akt-binding protein, and was shown to be involved in various signaling pathways through binding to receptors for these pathways [34-36]. These authors showed that the amount of PI3P in early endosomes affects signaling through APPL-endosomes and depletion of PI3P from early endosomes results in expansion of the lifetime of APPL-endosomes and in enhanced signaling by growth factor receptors via Akt and MAPK [34,35]. Therefore, a decrease in the amount of PI3P in endosomes would perturb the association of FYVE domain proteins with early endosomes, and as a consequence lead to the inhibition of a wide range of endocytotic trafficking pathways and signal transduction via endosomes.

The effect of oxidative stress on endocytosis has been studied extensively to date, but the results of these studies are controversial and depend on the type of ligand and cell used [37-41]. Many factors that regulate H_2O_2-dependent vesicular transport in cells have been elucidated so far. Among the factors that are involved in endocytosis under oxidative stress, a particular role has been shown for p38 MAPK, a kinase that is activated under stress conditions. It has been demonstrated that the activation of p38 MAPK by UV irradiation or incubation with TNF-alpha [42] or anisomycin [43] induces the activation and ligand-independent internalization of EGFR. In addition, Cavalli et al. [41] have shown that p38, which is activated by H_2O_2, stimulates the formation of the Rab5/Rab-GDI complex both *in vivo* and *in vitro*, and accelerates the endocytosis of HRP. Given that p38 MAPK has an emerging role in biological processes that are involved in different pathologies [44], it is important to understand how p38 acts as a stress-activated regulator of endocytosis.

In the work described herein, we have developed a novel analytical method to quantify the amount of PI3P in endosomes using a GST-2xFYVE protein probe and a semi-intact cell system [45]. Using this method, we found that treatment of HeLa cells with H_2O_2 decreased the amount of PI3P in endosomes in a p38 MAPK-dependent manner. We also examined the effect of H_2O_2 treatment on various endocytic pathways: the lysosomal degradation of EGFR, the retrograde transport of cholera toxin from the plasma membrane to the TGN, and the uptake and recycling of Tf. Treatment with H_2O_2 delayed endocytic transport in general, especially with respect to post-early endosome transport. The retrograde transport of cholera toxin was especially dependent on the amount of PI3P in endosomes.

2. H_2O_2 Treatment Decreases Endosomal PI3P Content

Recently, the physiological role of PI3P in endosomes has received a lot of attention with respect to endocytosis and signal transduction via endosomes [33,46,47]. This is because PI3P in the endosomal membrane seems to act as a microdomain for the recruitment of various proteins that contain a PI3P-binding (FYVE) domain. In this way, PI3P is involved in endosomal processing and in the transduction of extracellular signals from EGF and TGFβ to the nucleus via endosomes. So, at first, to investigate the effect of the oxidative stressor H_2O_2 on the intracellular localization of PI3P in HeLa cells, we expressed a recombinant protein that consisted of a tandem repeat of the FYVE peptide sequence, a domain that binds specifically to PI3P, tagged with YFP (YFP-2xFYVE) in HeLa cells. As described by others [48], YFP fluorescence was observed as punctate structures throughout the cytoplasm (Figure 1A, control). Immunofluorescence analysis revealed that the dots of YFP fluorescence were colocalized with EEA1, a marker of the early endosome, but not with p230, a marker of the TGN. However, as has been reported by other researchers [49], we observed that EEA1-positive endosomes in cells that expressed YFP-2xFYVE were dilated compared with intrinsic ones (data not shown). Despite this perturbation of endosomal structures by overexpression of the fusion protein, YFP-2xFYVE dissociated from endosomes and diffused throughout the cytoplasm after treatment with wortmannin, which is an inhibitor of PI3 kinase (Figure 1A, wortmannin). This finding indicated that YFP-2xFYVE recognized PI3P even in the transfected cells.

Next, we examined the localization of YFP-2xFYVE in H_2O_2-treated or untreated HeLa cells. As shown in Figure 1A (H_2O_2) and 1B (5 min), the fluorescence of YFP-2xFYVE that was associated with the endosomes became blurred and dispersed throughout the cytoplasm even after treatment with 500 μM H_2O_2 for only 5~10 min. After incubation with H_2O_2 for 15 or 30 min, the YFP-2xFYVE fluorescence had disappeared from endosomes and diffused more into the cytoplasm (Figure 1B, 15 and 30 min). These results suggested that treatment with H_2O_2 reduced the amount of PI3P in endosomes.

To examine directly the reduction in the amount of PI3P associated with endosomes in H_2O_2-treated cells, we performed lipid blot analysis using GST-tagged 2xFYVE protein (GST-2xFYVE) and PI3P Mass Strip Kit, which is a commercially available kit for the detection and quantification of specific lipids in cell lysates (Figure 2A).

As shown in Figures 2B, C and D, treatment with H_2O_2 decreased the PI3P content to 48.12 ± 3.64%, compared to 100% in control cells. Interestingly, after the H_2O_2 had been washed out, the PI3P content recovered rapidly to 98.56 ± 0.61% and 91.94 ± 2.55% after incubation for 10 and 30 min in medium, respectively. The fact that the YFP-2xFYVE probe was targeted specifically to endosomes, and not to other organelles, implied that PI3P was restricted to endosomes. Therefore, these results indicated that treatment with H_2O_2 reduced the amount of PI3P in endosomes.

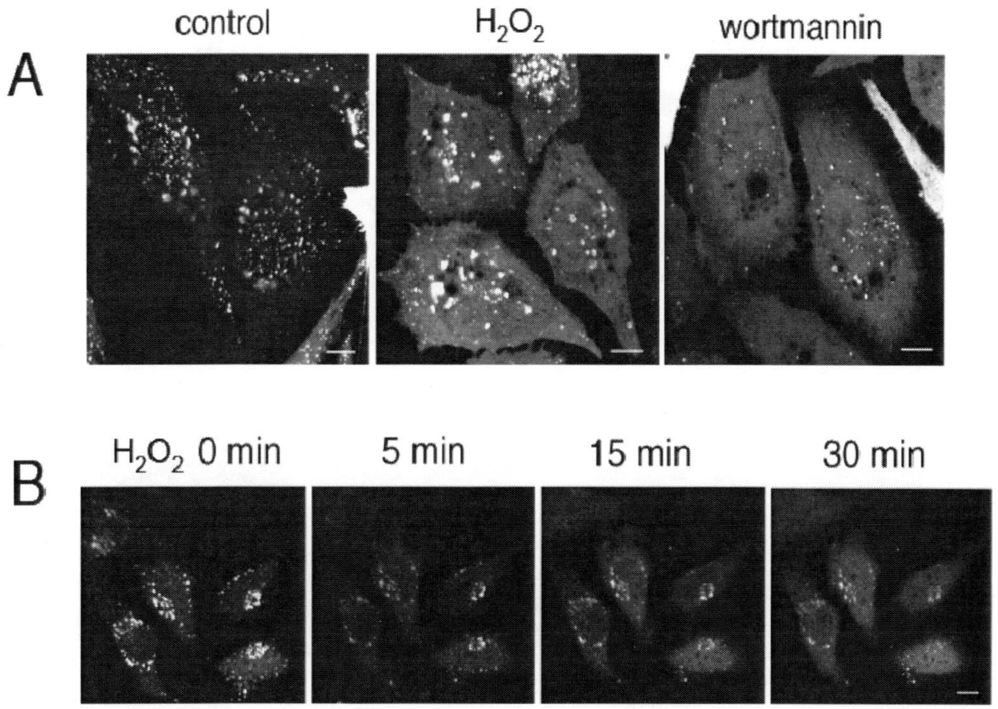

Figure 1. Treatment with H_2O_2 dissociated YFP-2xFYVE from endosomes in HeLa cells. A. HeLa cells that were expressing YFP-2xFYVE were incubated with or without 500 μM H_2O_2 at 37°C for 10 min, or with 1 μM wortmannin at 37°C for 1 hr. The cells were fixed and observed by confocal microscopy. Bar = 10 μm. B. HeLa cells that were expressing YFP-2xFYVE were treated with 500 μM H_2O_2 at 37°C for 0, 5, 15, and 30 min. Dissociation of YFP-2xFYVE from endosomes into the cytosol was observed. Bar = 10 μm.

Next, we performed the thin layer chromatography (TLC) method to detect the difference in lipid contents between cells, which had or had not been treated with H_2O_2. Total lipids were extracted from H_2O_2-treated or untreated HeLa cells, and then separated using TLC. For the majority of lipids, the content did not change between the H_2O_2-treated and untreated cells. These results suggested that H_2O_2 treatment did not perturb the intracellular lipids severely. To investigate the effect of H_2O_2 treatment on phosphoinositide levels further, we analyzed the acidic lipids by TLC and observed no significant changes of ratio in phosphatidylinositol contents by H_2O_2 treatment, although we observed that phosphatidic acid (PA) appeared to be decreased by H_2O_2 treatment. One possible reason for the decreased amount of PA by H_2O_2 treatment is the inactivation of diacylglycerol kinase, an enzyme that metabolizes diacylglycerol to PA, as reported by Atsumi et al. [50]. Since both diacylglycerol and PA act as lipid signal messengers, these results suggest that diverse signal transduction pathways could be perturbed by H_2O_2 treatment.

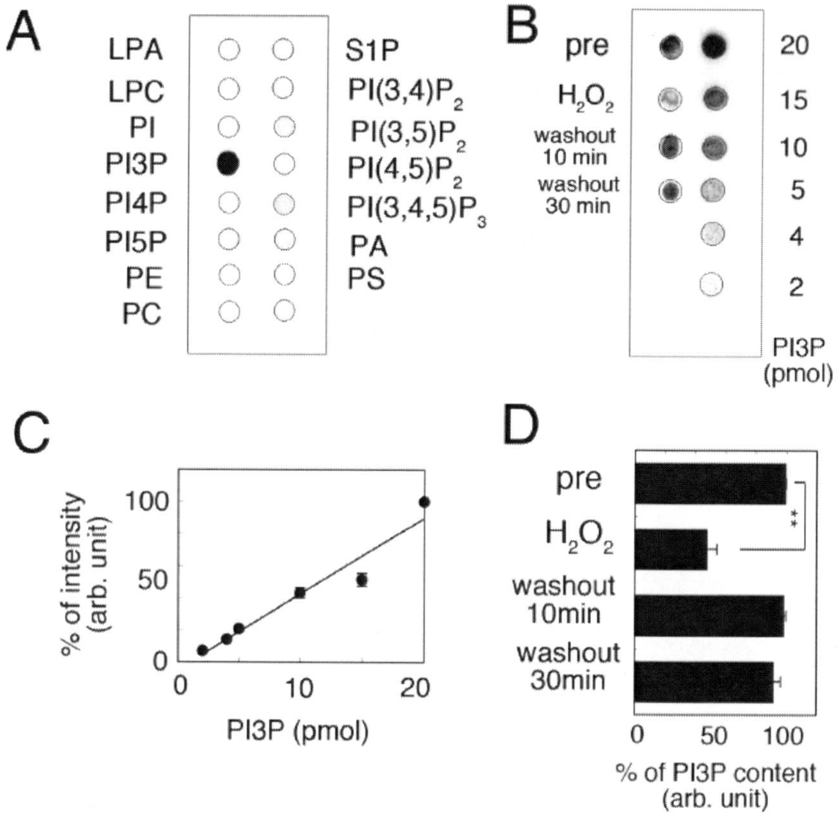

Figure 2. H_2O_2 treatment decreased the intracellular amount of PI3P. A. Lipid blot of GST-2xFYVE. A nitrocellulose membrane onto which a variety of lipids had been spotted (PIP Strip) was incubated with GST-2xFYVE recombinant protein. GST-2xFYVE specifically recognized PI3P. B. HeLa cells were incubated with or without 500 μM H_2O_2 for 10 min (pre and H_2O_2). After the treatment, the cells were washed and incubated with medium at 37°C for 10 and 30 min (washout 10 min and 30 min). Lipids were extracted from the cells and lipid blot analysis for PI3P was performed. Right lane of the figure showed the spotted PI3P for the standard of the lipid blot assay. C. Standard curve of PI3P derived by lipid blot analysis. We quantified the intensity of the dots of serially diluted PI3P shown in B and assigned the intensity of 20 pmol as 100%. The means and standard deviations are shown in the graph. N = 4. The line indicates the linear regression line, $y = 4.7237x - 4.6584$, and the R^2 value was 0.943. D. Using the standard curve for each blot, we quantified the intracellular PI3P in each sample and assigned the content of PI3P in control cells as 100% to allow comparison between experimental samples. The experiments were performed in triplicate, and the means and standard deviations are shown in the graph. ** $P < 0.01$.

3. A Semi-Intact Cell Assay Allows Quantification of the Amount of Endogenous PI3P and Its Localization in Endosomes

As demonstrated above, the expression in living cells of GFP-tagged proteins that contain the FYVE domain (the YFP-2xFYVE probe used in this chapter) is used widely to visualize and analyze the dynamic behavior of intracellular PI3P. Although this is a convenient and powerful method to elucidate the dynamics of PI3P in living cells, a concern remains about whether overexpression of FYVE domain proteins in living cells might activate the fusion of endosomes and result in the enlargement of endosomal structures (Figure 1). This enlargement could affect the localization of endosomal proteins and perturb the amount of PI3P in the endosomal membrane, which in turn could disrupt the function of the endosome.

To eliminate the artificial effect of overexpression of YFP-2xFYVE on the estimation of the amount of PI3P in endosomes, we have developed a new analytical method using recombinant GST-2xFYVE protein and a semi-intact cell system (Figure 3).

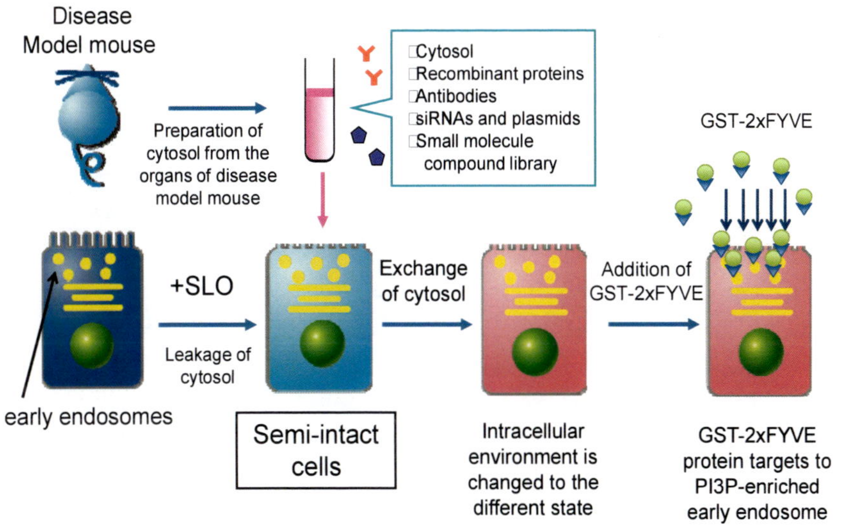

Figure 3. Scheme of GST-FYVE targeting assay using the semi-intact cell system.

Semi-intact cells are cells whose plasma membrane has been permeabilized with streptolysin O (SLO), a streptococcal pore-forming toxin. SLO binds to cholesterol in plasma membranes at 4˚C. At warmer temperatures (> 32˚C), SLO assembles to form amphiphilic oligomers, leading to the generation of small, stable transmembrane pores [51]. SLO-induced pores are approximately 30 nm in diameter, which allows large molecules such as immunoglobulin G (150 kDa) to enter into the cells. After permeabilization, almost ~80% of the cytosol flows out through the pores into the medium. Despite this loss of cytosol, the relative intracellular configuration of the cytoskeleton and organelles or between independent organelles can be maintained because damage to the membranes of the intracellular organelles caused by entry of SLO into the semi-intact cells can be minimized by washing away any excess SLO at 4°C before pore formation is initiated. By exchanging cytoplasmic proteins with exogenously added proteins, antibodies or cytosol that has been prepared from cells at distinct stages of the cell cycle or differentiation, or from disease states, we can modulate the intracellular environment and reconstitute various physiological phenomena in semi-intact cells. Using the semi-intact cell system, we have established many types of assay for cell cycle-dependent changes in organelle morphology and membrane trafficking [52, 53, 54]. In this study, we developed the semi-quantitative assay to measure the intracellular PI3P using the semi-intact cell system.

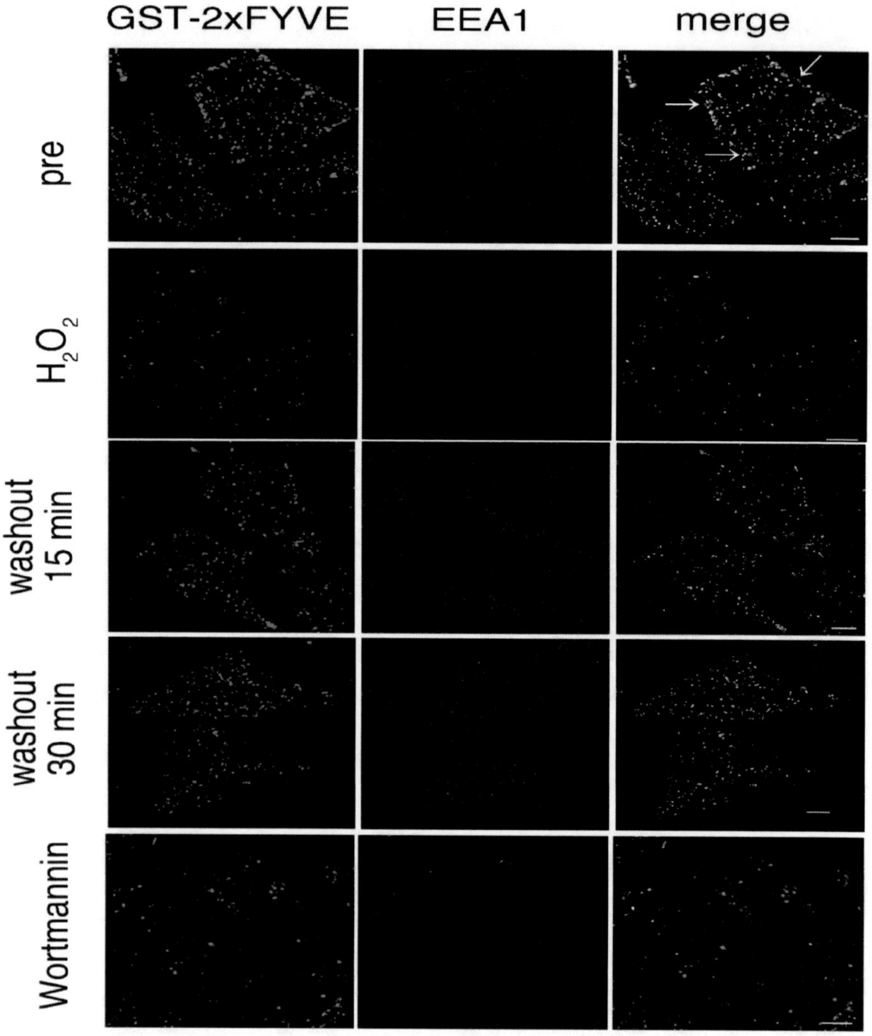

Figure 4. The GST-2xFYVE targeting assay in semi-intact cells to estimate the content of intracellular PI3P. HeLa cells were incubated with or without 500 μM H_2O_2 for 10 min (pre and H_2O_2). After the treatment, the cells were washed and incubated with medium at 37°C for 15 or 30 min (washout 15 min and 30 min). In the samples labeled wortmannin, HeLa cells were incubated with 1 μM wortmannin for 2 hr. The cells were then permeabilized with streptolysin O and subjected to the GST-2xFYVE targeting assay. Then the cells were fixed and immunostained using antibodies against GST and EEA1. Arrows indicate the GST-FYVE-positive, EEA1-negative structures. Bar = 10 μm.

HeLa cells were incubated with/without 500 μM H_2O_2 for 10 min. The cells were washed and incubated with medium at 37°C for 15 or 30 min. After permeabilization with SLO, we added 5 μg of GST-tagged tandem FYVE recombinant protein (GST-2xFYVE), together with an ATP-regenerating system but no cytosol, to semi-intact HeLa cells. Incubation without cytosol is crucial to halt the progression of physiological events within the endosomes. After incubation with GST-2xFYVE at 32°C for 15 min, the semi-intact cells were fixed and subjected to immunofluorescence analysis using antibodies against GST and EEA1.

We detected many GST-2xFYVE-positive punctate membranous structures throughout the cytoplasm of the semi-intact cells, which were stained in part with the early endosomal marker, EEA1 (Figure 4, pre). It is noteworthy that early endosomes appeared to have retained their small punctate structure in this semi-intact cell assay, compared with the endosomes seen in cells that overexpressed YFP-2xFYVE (Figure 1). We determined that treatment with H_2O_2 for 10 min was sufficient to induce the dispersal of GST-2xFYVE and EEA1 throughout the cytoplasm (Figure 4, H_2O_2), which suggested that these molecules dissociated from endosomes into the cytosol very rapidly upon the addition of H_2O_2. Interestingly, the H_2O_2-induced dissociation of GST-2xFYVE and EEA1 from endosomes was reversible. After the H_2O_2 had been washed out, EEA1 was recruited back rapidly to early endosomes and GST-2xFYVE reappeared in EEA1-positive endosomes (Figure 4, washout 15 min and 30 min). In addition, we observed that, upon treatment with wortmannin, GST-2xFYVE was dispersed throughout cytoplasm (Figure 4, wortmannin). The results of the lipid blot analysis showed that the intracellular levels of PI3P recovered rapidly after H_2O_2 had been washed out (Figure 2B and D), but they did not reveal the location of the PI3P within the cell. These results, together with the ones obtained by lipid blot analysis and shown in Figure 2, suggested that the semi-intact cell assay could be suitable to detect the endogenous intracellular localization of PI3P and to estimate the amount of intrinsic PI3P in cells.

4. Dissociation and Reassociation of Endogenous FYVE Domain Proteins, EEA1 and SARA, upon H_2O_2 Treatment Was Dependent on PI3P Content in Endosomes

We also examined the localization and/or membrane association of endogenous FYVE domain proteins, EEA1 and SARA, biochemically. Membrane fractions were collected from HeLa cells that had or had not been treated with H_2O_2 and were subjected to western blotting with antibodies against EEA1 and SARA (Figure 5A, EEA1 and SARA).

As shown in Figure 5, in H_2O_2-treated cells, the amount of membrane-associated both proteins decreased compared to that in untreated cells. After H_2O_2 had been washed out, both proteins reassociated with the membrane within 15 or 30 min (Figure 5B, EEA1 and SARA). We confirmed that the total amount of EEA1 and SARA in the cells did not change. Collectively, the results obtained with the semi-intact cell system show that treatment with H_2O_2 reduces the amount of PI3P in endosomes and results in the rapid dissociation of both GST-2xFYVE and endogenous FYVE domain proteins from endosomal membranes. The observation that the content of PI3P appeared to be restored shortly after the H_2O_2 had been washed out suggests that turnover of PI3P in the cells is so fast that most of the PI3P is replaced within 10 minutes, and are consistent with the report by Bergeland et al. [55], who investigated 2xFYVE turnover on single endosomes using fluorescence recovery after photobleaching (FRAP) and demonstrated a very rapid exchange of PI3P with a half life of less than a second.

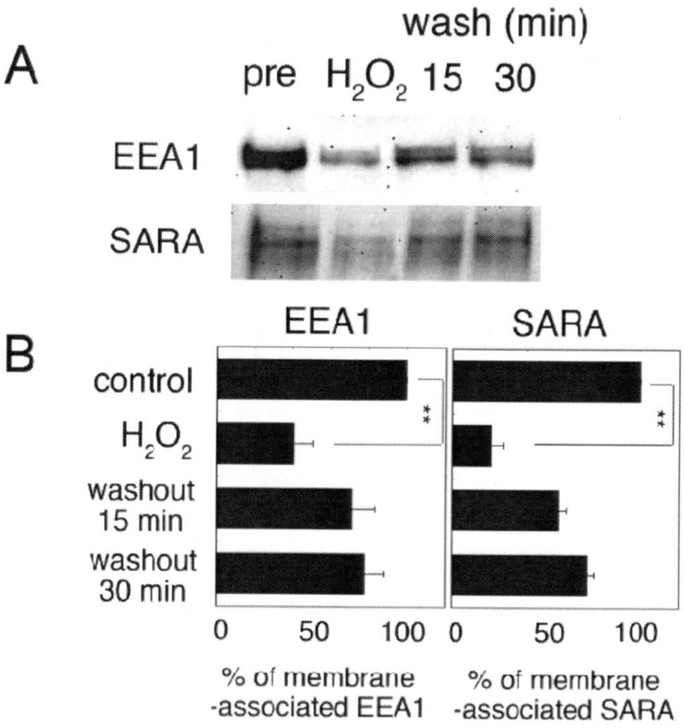

Figure 5. The FYVE domain proteins EEA1 and SARA were dissociated from membranes by H_2O_2 treatment. A. HeLa cells were incubated with or without 500 µM H_2O_2 for 10 min (H_2O_2 and pre). After the H_2O_2 had been washed out, the cells were further incubated with medium at 37°C for 15 or 30 min. The membrane fraction was collected and subjected to western blotting using antibodies against EEA1 and SARA. B. Quantification of membrane-associated EEA1 and SARA. The intensities of the bands that corresponded to membrane-bound EEA1 and SARA shown in A were quantified and values were expressed as the mean ± S.E.M. (N = 5 for EEA1 and 3 for SARA). We assigned the band intensity of membrane-bound EEA1 or SARA in control cells as 100%. $^{**}P$ values were < 0.01.

5. A Range of Endocytic Pathways Were Delayed by H_2O_2 Treatment

EEA1 is reported to play a critical role in endosomal processing, especially endosomal fusion [56, 57]. Given that we found that EEA1 was dissociated from early endosomes by treatment with H_2O_2, we examined the effect of H_2O_2 on membrane trafficking, especially endocytic pathways (Figure 6).

Firstly, we examined the clathrin-dependent endocytosis and subsequent degradation of EGFR in lysosomes after the addition of EGF to H_2O_2-treated and untreated cells (Figure 6, dotted line). In the untreated cells, EGFR was degraded to approximately 60% of the original level within 5 min after the addition of EGF (Figure 7A and B, -H_2O_2).

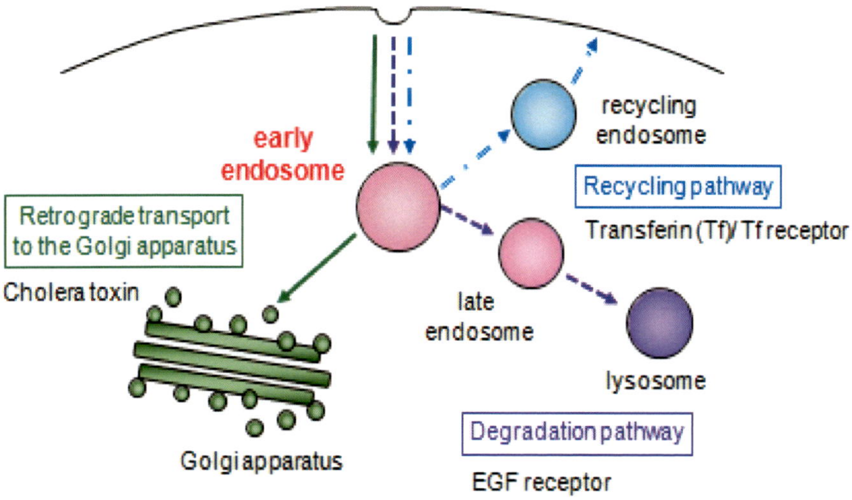

Figure 6. Various endocytic pathways via early endosomes.

In contrast, the degradation of EGFR was substantially inhibited in the H_2O_2-treated cells and only approximately 10% of the EGFR was degraded 5 min after the addition of EGF (Figure 7A and B, +H_2O_2). After incubation for 60 min, H_2O_2 no longer showed an inhibitory effect and the amount of EGFR in both treated and untreated cells had decreased to a similar level, approximately 30% of the initial amount (Figure 7A and B). The different kinetics of EGFR degradation in the H_2O_2-treated and untreated cells indicated that an early step in the endocytosis of EGFR was perturbed in the treated cells.

Secondly, we examined the endocytosis of CtxB in H_2O_2-treated and untreated HeLa cells. CtxB is internalized and trafficked to endosomes in a clathrin-independent (caveolae-dependent) manner, and then is transported to the ER via the Golgi apparatus (Figure 6, solid line). We estimated the retrograde transport of endocytosed CtxB from endosomes to the Golgi using morphometric analysis. The results showed that the amount of CtxB that bound to the plasma membrane was similar in H_2O_2-treated and untreated cells (96.17 ± 2.71% in H_2O_2-treated cells, when the amount in control cells was assigned as 100 ± 5.76%, by flow cytometry). The rate of internalization and trafficking to endosomes in H_2O_2-treated cells appeared to be indistinguishable from that in untreated cells (Figure 8A, 5 min). However, as shown in Figure 8A and B, the accumulation of CtxB at the Golgi was delayed by treatment with H_2O_2, but after the cells had been incubated for 60 min, the same amount of CtxB had accumulated at the Golgi in both H_2O_2-treated and untreated cells. These results suggest that the Golgi apparatus is a rate-limiting step for the post-Golgi transport of CtxB and an early step of endocytosis, probably after internalization, is perturbed in H_2O_2-treated cells, but the total amount of CtxB that is internalized or transported is not affected by the treatment.

Thirdly, we investigated the effect of H_2O_2 treatment on the uptake and subsequent recycling of Tf (Figure 6, dashed line). We could not detect any significant difference in the fluorescence intensity of TRITC-Tf that was associated with the plasma membrane between H_2O_2-treated and untreated cells (Figure 9A, 0 min and B).

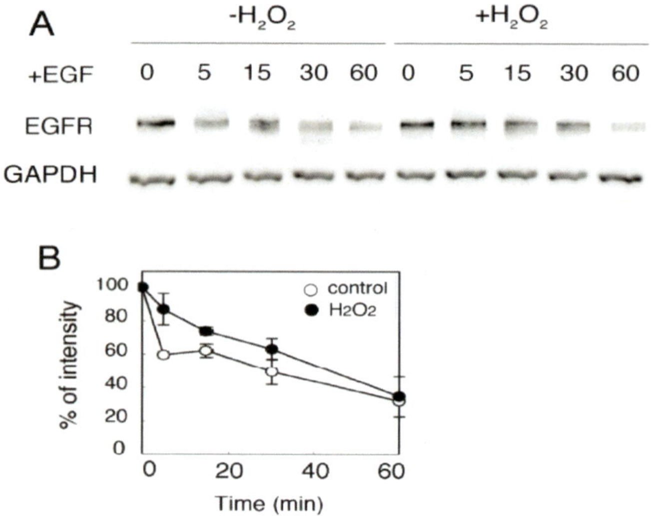

Figure 7. H_2O_2 treatment delayed the endocytic trafficking of EGFR. A. H_2O_2-treated (+H_2O_2) or untreated (-H_2O_2) HeLa cells were incubated with 10 ng/ml EGF at 37°C for 0, 5, 15, 30, and 60 min. EGFR and GAPDH in the cell lysates were detected by western blotting. B. The band intensities of EGFR were quantified, and we assigned the value of intensity [EGFR]/intensity [GAPDH] at the 0 min time point as 100%. The means and standard deviations are plotted in the graph. Open circles represent control cells, and closed circles represent H_2O_2-treated cells.

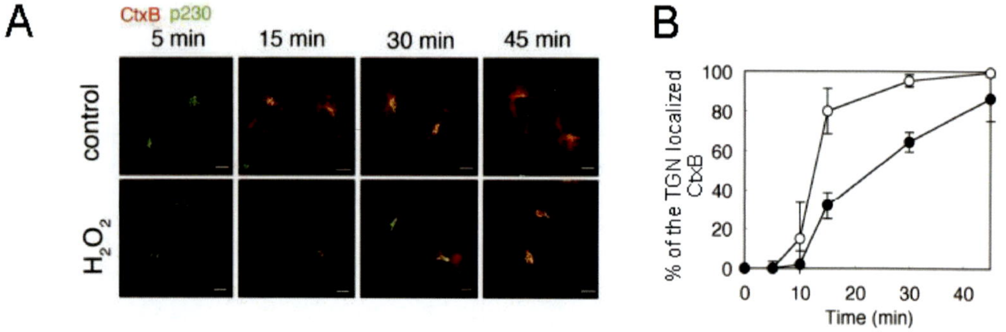

Figure 8. H_2O_2 treatment delayed the endocytic trafficking of cholera toxin. A. H_2O_2-treated (H_2O_2) or untreated (control) HeLa cells were incubated with TRITC-labeled cholera toxin B subunit (CtxB, red) on ice for 30 min. After the H_2O_2 had been washed out, the cells were incubated with medium at 37°C for 5, 15, 30, and 45 min. The cells were fixed and stained with anti-p230 antibody (green). Bar = 10 µm. B. The number of cells in which CtxB was localized to the Golgi was counted, and the means and standard deviations of the percentages are plotted in the graph. Open circles represent control cells, and closed circles represent H_2O_2-treated cells.

This indicated that the ability of TRITC-Tf to bind to the receptor was not influenced by treatment with H_2O_2. TRITC-Tf was internalized into the cells and localized to punctate structures after incubation for 5 min in both H_2O_2-treated and untreated cells (Figure 9A, 5 min). In untreated cells, the fluorescent signal of TRITC-Tf decreased to 68.7 ± 10.2% and 41.3 ± 7.9% of the original level after 10 min and 15 min, respectively. However, in H_2O_2-treated cells, the decline in the fluorescence of TRITC-Tf was delayed and it had only decreased to 94.1 ± 16.6% of the original level after 10 min, and 68.7 ± 17.6% after 15 min

(Figure 9C). After 30 min, both H_2O_2-treated and untreated cells contained less than 30% of the original amount of TRITC-Tf (Figure 9C). These results indicated that H_2O_2 treatment might inhibit mainly the recycling of TRITC-Tf /Tf receptor, rather than its internalization.

To quantify the rate of Tf recycling more precisely, we measured the total amount of Alexa488-Tf that remained in each cell after different chase times using flow cytometry. We confirmed that the amount of Alexa488 conjugated-Tf (Alexa488-Tf) that bound to the plasma membrane was almost identical in H_2O_2-treated and untreated cells. After a 2 min or 5 min chase, the difference in the fluorescence intensity of Alexa488-Tf between H_2O_2-treated cells and untreated cells was the same as that observed for a chase of 0 min, which indicated that the internalization of Alexa488-Tf occurred at a similar rate in both types of cell.

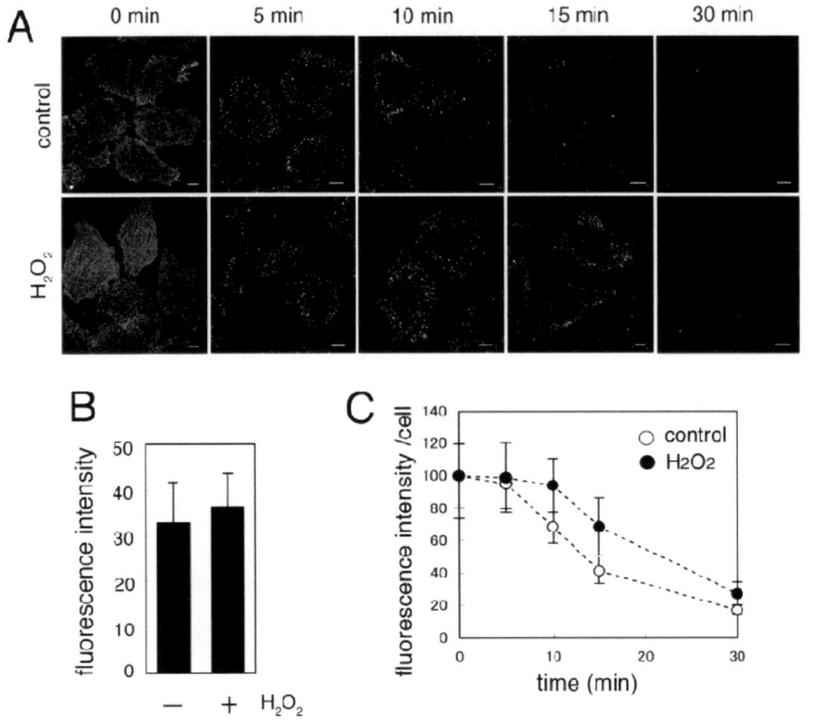

Figure 9. H_2O_2 treatment delayed the recycling of transferrin. A. H_2O_2-treated (H_2O_2) or untreated (control) HeLa cells were incubated with 100 μg/ml TRITC-labelled transferrin (TRITC-Tf) on ice for 30 min. The cells were washed and chased at 37°C for 0, 5, 10, 15, and 30 min. The cells were fixed and observed by confocal microscopy. Bar = 10 μm. B. The fluorescence intensity of TRITC-Tf bound to the plasma membrane was quantified morphometrically in control (-) and H_2O_2-treated (+) cells. C. H_2O_2-treated (•) or untreated (○) HeLa cells were incubated with TRITC-Tf on ice for 30 min. After the cells had been incubated with medium at 37°C for the indicated times, they were fixed and the fluorescence intensity of Tf that remained in the cells was quantified.

After a 15 min chase time, the Alexa488-Tf in untreated cells had decreased to approximately 60% of the initial value, which suggested that the internalized Alexa488-Tf had been recycled from endosomes to the plasma membrane/medium. In contrast, in the H_2O_2-treated cells, almost all of the Alexa488-Tf remained in the cells after 15 min. Again, these results suggest that H_2O_2 treatment perturbs mainly the recycling step of the endocytic pathway of Tf.

Taken together, our results demonstrated that H_2O_2 affects the early steps of endocytosis, in particular post-endosomal pathways, including: 1] the transport of EGFR from endosomes to the lysosome; 2] the transport of CtxB from endosomes to the Golgi apparatus; and 3] the recycling of Tf from endosomes to the plasma membrane. A reduction in the amount of PI3P in endosomes might be involved in the perturbation of post-endosomal trafficking during endocytosis. In contrast, H_2O_2 treatment appeared to have no effect on the normal exocytic pathway, as exemplified by the transport of VSVG-ts045 (data not shown).

6. H_2O_2-Induced Depletion of PI3P from Endosomes and the Dissociation of FYVE Domain-Containing Proteins Were Dependent on p38 MAPK

Cavalli et al. [41] reported that p38 MAPK stimulates the formation of the Rab5:Rab-GDI complex, which regulates early endocytic membrane traffic. They showed that, upon treatment with H_2O_2, EEA1 is distributed throughout the cytosol in a p38 MAPK-dependent manner. As a consequence, we hypothesized that p38 MAPK plays a role in the H_2O_2-induced depletion of PI3P from endosomes. To test this hypothesis, we treated HeLa cells with 30 μM SB203580, which is an inhibitor of p38 MAPK, for 50 min at 37°C and then added 500 μM H_2O_2 for 10 min. Then, we performed the GST-2xFYVE targeting assay as described above to estimate the PI3P content in the cells. Interestingly, SB203580 inhibited the H_2O_2-induced dissociation of GST-2xFYVE from endosomes (Figure 10A and B, H_2O_2 and SB203580+ H_2O_2). We also confirmed that another inhibitor of p38 MAPK, SB202190, inhibited the H_2O_2-induced dissociation of GST-2xFYVE (Figure 10B, SB202190+H_2O_2). Furthermore, a different inducer of oxidative stress, 3-aminotriazole (3-AT), which does not activate p38 MAPK, did not induce the dissociation of GST-2xFYVE from endosomes (Figure 10A and B, 3-AT). These results suggest that the reduction in the amount of PI3P that occurred upon treatment with H_2O_2 depended on the activation of p38 MAPK.

Next, we examined the effect of SB203580 treatment on H_2O_2-induced changes in the localization of EEA1 and Rab5 by immunofluorescence. As shown in Figure 10A (EEA1), EEA1 remained localized to endosomes in the presence of both SB203580 and H_2O_2. Given that EEA1 is recruited to early endosomes by binding to PI3P, these results support the hypothesis that, in the presence of SB203580, PI3P remains in endosomes in H_2O_2-treated cells. Similarly, Rab5 was localized to punctate structures throughout the cytoplasm in control cells, and dissociated from endosomes upon treatment with H_2O_2, which was inhibited by p38 MAPK inhibitors (Figure 10A, Rab5). These results indicate that EEA1 and Rab5 also dissociated from endosomes upon treatment with H_2O_2 in a p38 MAPK-dependent manner. We also performed p38 MAPK-knockdown by RNA interference (RNAi), and confirmed that the silencing of p38 MAPK restored the targeting of GST-2xFYVE and localization of EEA1 and Rab5 to endosomes (data not shown).

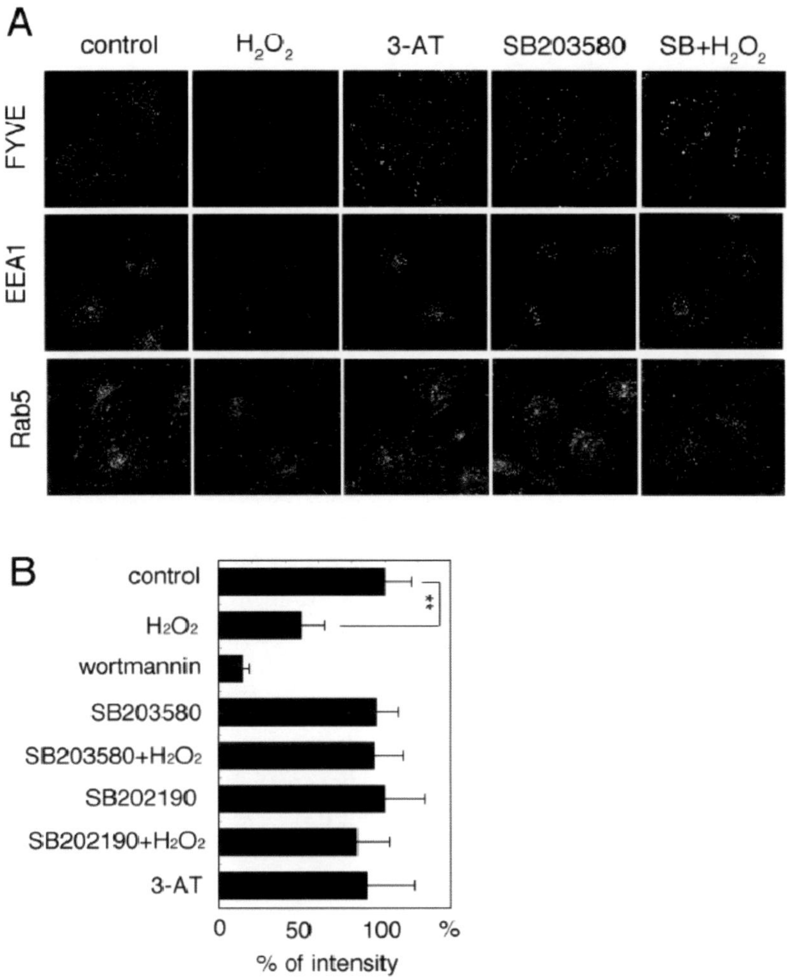

Figure 10. A p38 inhibitor prevented the H_2O_2-mediated depletion of PI3P and dissociation of FYVE domain proteins from endosomes. A. HeLa cells were incubated with (H_2O_2) or without (control) 500 μM H_2O_2 for 10 min, with 10 mM 3-aminotriazole for 60 min (3-AT), with 30 μM SB203580 for 60 min (SB203580), or with 30 μM SB203580 for 50 min and then with H_2O_2 for an additional 10 min (SB + H_2O_2). PI3P in the cells was visualized by the GST-FYVE targeting assay (FYVE). In addition, cells were treated as described above and then immunostained with antibodies against EEA1 or Rab5. B. The fluorescence intensity of cells in which GST-FYVE had been introduced to visualize intracellular PI3P was quantified. The fluorescence intensity of the control cells was assigned as 100 %. We performed these experiments in triplicate, and the means and standard deviations are plotted in the graph. ***P* value was < 0.01.

We also found that the efficiency of targeting of GST-2xFYVE was correlated with the phosphorylation of p38 MAPK. HeLa cells were incubated with or without 500 μM H_2O_2 for 10 min, permeabilized with SLO. The semi-intact cells were incubated with GST-2xFYVE at 32°C for 15 min, fixed, and subjected to immunofluorescence analysis using antibodies against GST and phosphorylated p38 MAPK. Although extensive phosphorylation of p38 MAPK and reduced targeting of GST-2xFYVE to endosomes in most cells after treatment with H_2O_2, we found that some cells exhibited weak phosphorylation of p38 MAPK together with slight association of GST-2xFYVE with endosomes, indicating the correlation between the phosphorylation of p38 MAPK and the targeting of GST-2xFYVE (data not shown).

Collectively, these results supported the previous results obtained with the p38 MAPK inhibitor, which indicated that the reduction in PI3P and dissociation of EEA1 and Rab5 upon treatment with H_2O_2 depended on the activation of p38 MAPK.

7. H_2O_2 Treatment Dissociates hVps34, a Kinase Involved in the Production of PI3P, from Endosomes

Several reports have suggested the possibility that the amount of PI3P in endosomes is regulated by the balance between the activities of PI3P kinases and phosphatases [58, 59]. Murray et al. [60] reported that the targeting of hVps34, which is a kinase involved in the production of PI3P in the endosomal membrane, to endosomes is dependent on Rab5. As a consequence, we hypothesized that the dissociation of Rab5 from endosomes upon H_2O_2 treatment results in the inhibition of hVps34 targeting, which in turn leads to a reduction in the amount of PI3P in endosomes. To test this, we estimated the amount of endogenous membrane-bound hVps34 biochemically. To do this, untreated HeLa cells or cells that had been treated with SB203580 were incubated in the presence or absence of H_2O_2 for 10 min. The cells were homogenized and the membrane fraction was isolated by centrifugation. We found that H_2O_2 treatment resulted in the dissociation of endogenous hVps34 from membranes. In addition, as in the case of Rab5, the H_2O_2-induced dissociation of hVps34 did not occur in the presence of SB203580. Collectively, these results suggested that Rab5 dissociates from endosomes in a p38-dependent manner in H_2O_2-treated cells, which results in the inhibition of the targeting of hVps34 to endosomes, although the precise localization of hVps34 remains unclear. Inhibition by H_2O_2 of the Rab5-mediated association of hVps34 with endosomes might reduce the production of PI3P on endosomes and cause EEA1 and GST-2xFYVE to dissociate from endosomal membranes in H_2O_2-treated cells.

Although the dissociation of hVPS34 from endosomal membranes could be the one of the primary reasons for the H_2O_2-induced depletion of PI3P, we could not exclude the possibility that H_2O_2 enhanced the degradation of PI3P or its conversion to different species. At this time, we have not identified the molecular species that are produced from PI3P as a result of H_2O_2 treatment. PI3P could be converted to phosphatidylinositol by myotubularin-mediated dephosphorylation [59, 61] and phosphatidylinositol-3,5-phosphate by PIKfyve-mediated phosphorylation [62]. Furthermore, Ivetac et al. [63] demonstrated that the endosomal pool of PI3P is regulated by type Ia inositol polyphosphate 4-phosphatase, which dephosphorylates $PI(3,4)P_2$ to form PI3P. However, we have not been able to evaluate the phosphatase or phosphokinase activity of these molecules in the presence of H_2O_2 to date. Therefore, it remains to be determined how PI3P is depleted in an H_2O_2-dependent manner.

8. PI3P in Endosomes Plays a Crucial Role in the Retrograde Transport of Cholera Toxin from Endosomes to the Golgi

Having found that the treatment with p38 MAPK inhibitor and knockdown of p38 MAPK by RNAi inhibited the H_2O_2-induced decrease in GST-2xFYVE at endosomes, we investigated the effect of SB203580 on the H_2O_2-induced perturbation of various endocytic pathways, in particular post-endosomal trafficking. The inhibition by SB203580 of the H_2O_2-induced reduction of PI3P in endosomes did not restore the normal transport of Tf (Figure 11A and B). The treatment with SB203580 also resulted in a delay in the degradation of EGFR. In addition, we did not observe any additive effect on the delay when the cells were treated with both H_2O_2 and SB203580 (Figure 11C and D, SB+H_2O_2). These results indicated that the H_2O_2-induced perturbation of transport of Tf and the lysosomal degradation of EGFR could be attributed to factors other than the p38-dependent depletion of PI3P.

In contrast, the retrograde transport of CtxB from the plasma membrane to the Golgi in H_2O_2-treated cells was largely affected in the presence of SB203580. As shown in Figure 12A and B, the transport of Alexa488-CtxB to the Golgi occurred normally in cells treated with either SB203580 alone or SB203580 and H_2O_2. These results suggested that the inhibition by SB203580 of the H_2O_2-induced reduction of PI3P in endosomes restored the normal transport of Alexa488-CtxB from the plasma membrane to the Golgi, in other words, the disruption of CtxB transport depended largely on the p38 MAPK-dependent depletion of PI3P from endosomes.

Furthermore, we also examined the effect of p38 MAPK knockdown on the transport of cholera toxin. We found that the accumulation of Alexa488-CtxB in the perinuclear Golgi region was inhibited in control cells by H_2O_2 treatment (Figure 12C and D, cont RNAi, +). In addition, we found that the transport of cholera toxin to the Golgi was restored in p38 MAPK knockdown cells that were treated with H_2O_2 (Figure 12C and D, p38 MAPK RNAi, +). These results indicated that the activation of p38 MAPK by H_2O_2 inhibited the transport of cholera toxin from the cell surface to the Golgi.

Thus, when the treatment with p38 inhibitor or knockdown of p38 MAPK inhibited the H_2O_2-induced depletion of PI3P from endosomes, the kinetics of CtxB transport was restored (Figure 12), but the lysosomal degradation of EGFR and the uptake and recycling of Tf remained delayed (Figure 11).

This result indicated that H_2O_2 perturbed not only the amount of PI3P in endosomes but also other factors, which were responsible for the delay in the lysosomal degradation of EGFR and in the recycling of endocytosed transferrin. Concomitantly, the results suggested that the amount of PI3P in endosomes plays a crucial role in the retrograde transport of CtxB, although it remains unclear whether it is only the amount of PI3P that is crucial for the restoration.

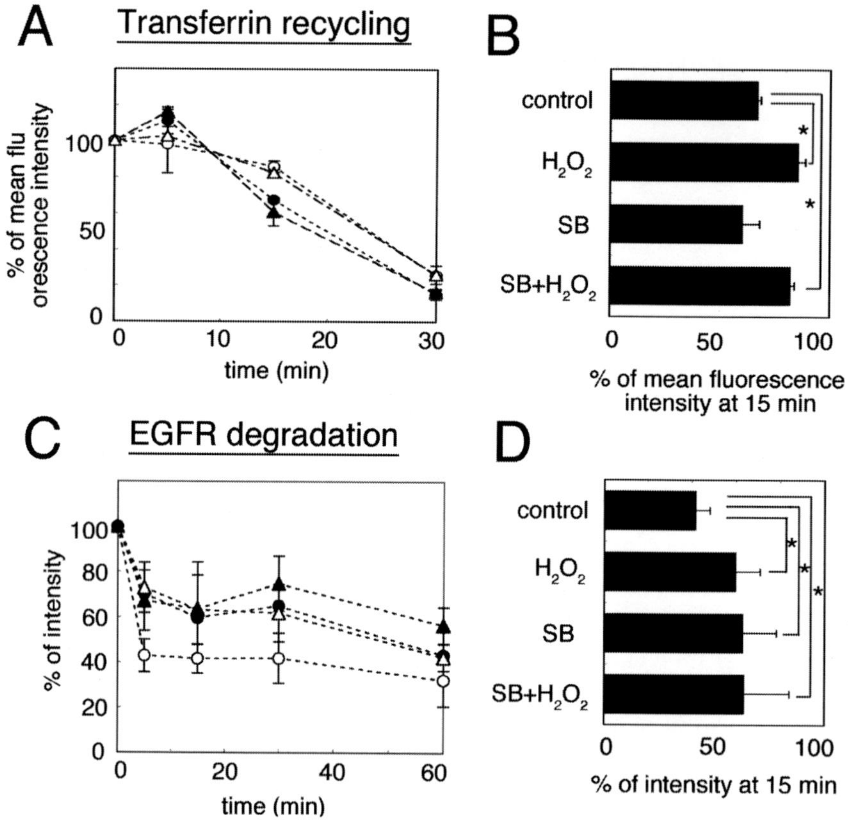

Figure 11. Effect of a p38 inhibitor on the H_2O_2-mediated inhibition of transferrin recycling and EGFR degradation. The recycling of transferrin (A, B) and degradation of EGFR (C, D) were measured in cells that had been incubated without (○, control) or with H_2O_2 (●), SB203580 (△, SB), or SB and H_2O_2 (▲, SB + H_2O_2). In B and D, the means and standard deviations for the percentage mean fluorescence intensity of transferrin at 15 min (B) or for the percentage EGFR intensity at 15 min (D) are plotted in the graphs. * *P* value was <0.01.

What are the proteins involved in the transport of CtxB whose function depends on PI3P? Sorting nexin (SNX) is one candidate. The SNX proteins are a large family of proteins that contain the PX domain, which is a phosphoinositide-binding domain, and most SNX proteins bind PI3P preferentially. Some SNX proteins are components of retromer, a complex that mediates the retrograde transport of transmembrane cargo from endosomes to the TGN [64]. Bujny et al. [19] demonstrated that SNX1 regulates the retrograde transport of Shiga toxin from early endosomes to the TGN. Skånland et al. [65] showed that the transport of ricin from endosomes to the TGN was dependent on SNX2, SNX4, and PI3P. In addition, Rojas et al. [66] demonstrated that Rab5 is required for the production of PI3K and the subsequent targeting of SNX to endosomes, whereas Rab7 is required for the targeting of Vps26. These authors also showed that both processes are essential for endosome-to-Golgi transport. Therefore, it is likely that SNX proteins are involved in the retrograde transport of CtxB, which depends on PI3P. The particular SNX proteins that are involved in the transport remain to be elucidated in the future.

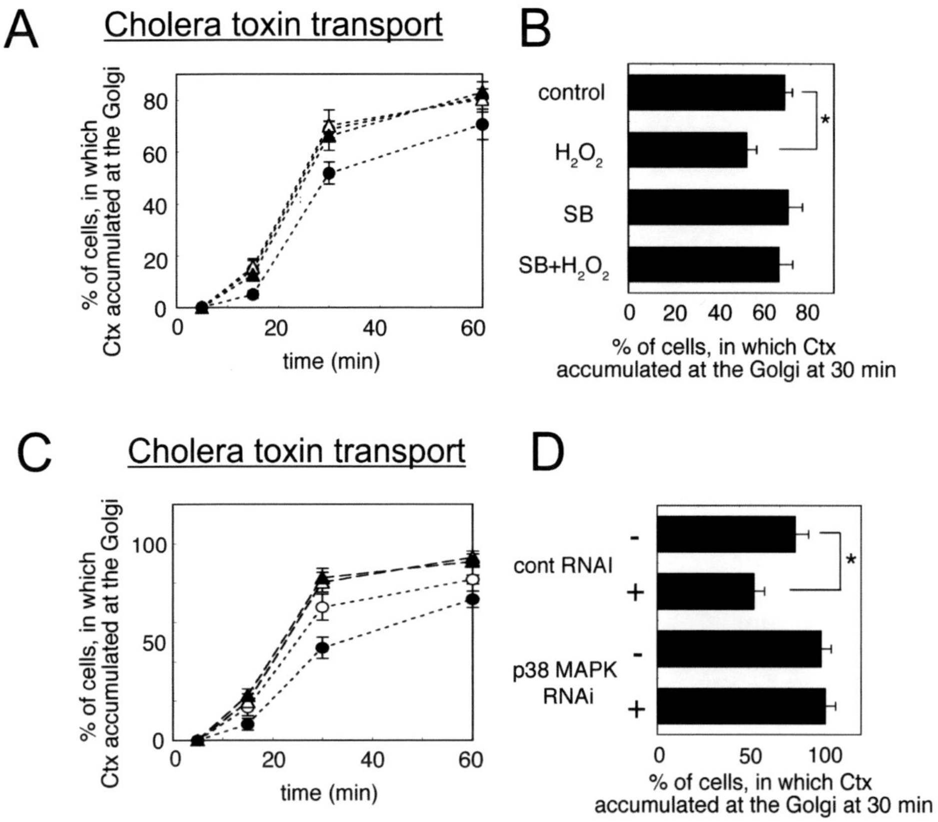

Figure 12. Effect of a p38 inhibitor and p38 knockdown on the H_2O_2-mediated inhibition of retrograde transport of cholera toxin. In A and B, the transport of cholera toxin were measured in cells that had been incubated without (○, control) or with H_2O_2 (●), SB203580 (Δ, SB,), or SB and H_2O_2 (▲, SB + H_2O_2). In C and D, the transport of cholera toxin was measured in scramble siRNA-transfected HeLa cells that were (●, cont RNAi +) or were not (○, cont RNAi -) treated with H_2O_2 or in p38 MAPK siRNA-transfected HeLa cells that were (▲, p38 MAPK RNAi +) or were not (Δ, p38 MAPK RNAi +) treated with H_2O_2. In B and D, the means and standard deviations for the percentage of cells in which cholera toxin had accumulated at the Golgi at 30 min are plotted. * P value was <0.01.

9. Mechanistic Model for H_2O_2-Mediated Perturbation of Early Endocytic Pathways

From the results that we obtained in this study, we also believe that the depletion of PI3P from endosomes is the crucial process that underlies the H_2O_2-induced dissociation of EEA1 from endosomes. Here we propose the following model for the mechanism of H_2O_2-induced EEA1 dissociation (Figure 13). Treatment with H_2O_2 activates p38 MAPK, which was reported by Cavalli et al. [41] to increase the formation of the Rab5/Rab-GDI2 complex. Given that Rab5 is reported to be involved in the recruitment of PI3 kinase (hVps34 is one candidate) to the endosomal membrane, reduction of the amount of Rab5/PI3 kinase complex due to treatment with H_2O_2 would inhibit the production of PI3P in the endosomal membrane. In fact, we found that the total amount of membrane-associated hVps34 was decreased by H_2O_2 treatment.

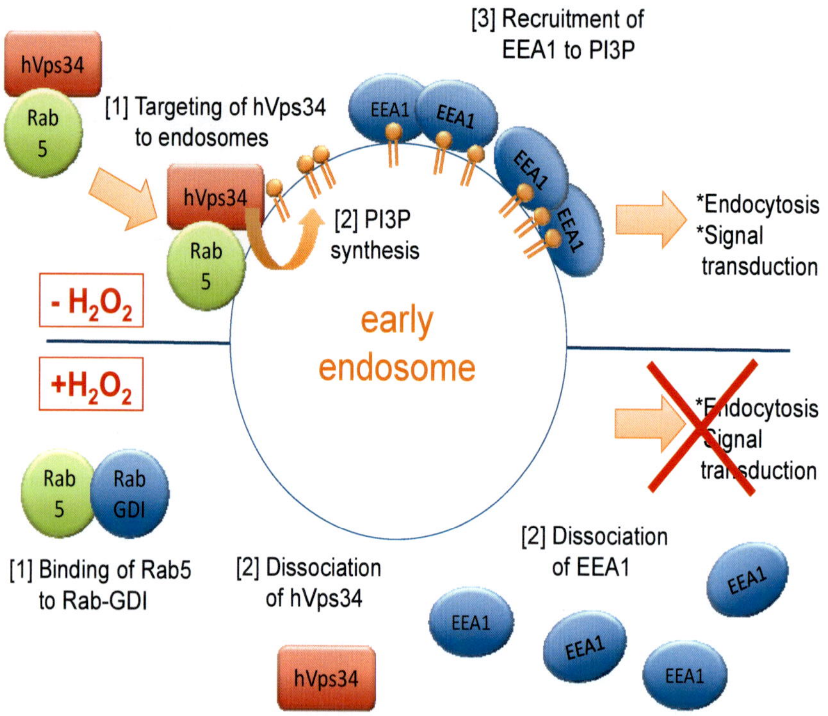

Figure 13. Schematic model of H_2O_2-induced depletion of PI3P from early endosomes.

The loss of PI3 kinase activity from endosomes in H_2O_2-treated cells might result in the dissociation of EEA1 or other FYVE domain proteins from membranes, and perturb various endocytic transport pathways that involve early endosomes. Interestingly, we found in this study that an inhibitor of p38 MAPK and knockdown of p38 MAPK prevented both the H_2O_2-induced depletion of PI3P from endosomes and the delay in the transport of CtxB from the endosome to the TGN. This suggests that the amount of PI3P in endosomes is crucial for the retrograde transport of CtxB. This p38-dependent, PI3P-sensitive transport pathway might be involved in physiologically important signal transduction under conditions of oxidative stress.

p38 MAPK is known to be activated under various stress and disease conditions [44,67]. It will be interesting and useful to investigate the correlation between the amount of PI3P in endosomes and the way in which endocytic processes or signal transduction via endosomes is perturbed in pathological cells.

As described above, by exchanging cytoplasmic proteins with exogenously added cytosol that has been prepared from organs of a disease model mouse, we can modulate the intracellular environment to the disease states and reconstitute various physiological phenomena, which are specifically perturbed under pathogenic conditions, in semi-intact cells. Therefore, the semi-intact cell assays described in this study could apply to investigate the PI3P content in endosomes and endocytic pathways under pathological states.

REFERENCES

[1] M. Sundaresan, Z.X. Yu, V.J. Ferrans, K. Irani, T. Finkel, Requirement for generation of H_2O_2 for platelet-derived growth factor signal transduction, *Science* 270 (1995) 296-299.

[2] K. Irani, Y. Xia, J.L. Zweier, S.J. Sollott, C.J. Der, E.R. Rearon, M. Sundaresan, T. Finkel, P. Goldschmidt-Clermont, Mitogenic signaling mediated by oxidants in Ras-transformed fibroblasts, *Science* 275 (1997) 1649-1652.

[3] T.P. Szatrowski, C.F. Nathan, Production of large amounts of hydrogen peroxide by human tumor cells, *Cancer Res.* 51 (1991) 794-798.

[4] J. Lee, S. Giordano, J. Zhang, Autophagy, mitochondria and oxidative stress: cross-talk and redox signalling, *Biochem J.* 441 (2012) 523-540.

[5] M. Abo, Y. Urano, K. Hanaoka, T. Terai, T. Komatsu, T. Nagano, Development of a highly sensitive fluorescence probe for hydrogen peroxide. *J. Am. Chem. Soc.* 133 (2011) 10629-10637.

[6] Y.S. Bae, S.W. Kang, M.S. Seo, I.C. Baines, E. Tekle, P.B. Chock, S.G. Rhee, Epidermal growth factor (EGF)-induced generation of hydrogen peroxide. Role in EGF receptor-mediated tyrosine phosphorylation. *J. Biol. Chem.* 272 (1997) 217-221.

[7] J.C. Cheng, C. Klausen, P.C. Leung, Hydrogen peroxide mediates EGF-induced down-regulation of E-cadherin expression via p38 MAPK and snail in human ovarian cancer cells, *Mol. Endocrinol.* 24 (2010) 1569-1580.

[8] T.C. Meng, T. Fukada, N.K. Tonks, *Reversible oxidation and inactivation of protein tyrosine phosphatases in vivo. Mol. Cell* 9 (2002) 387*–399.*

[9] J.C. Juarez, M. Manuia, M.E. Burnett, O. Betancourt, B. Boivin, D.E. Shaw, N.K. Tonks, A.P. Mazar, F. Doñate, *Superoxide dismutase 1 (SOD1) is essential for H_2O_2-mediated oxidation and inactivation of phosphatases in growth factor signaling. Proc. Natl. Acad. Sci. USA* 105 *(2008)* 7147-*7152.*

[10] P. Chiarugi, *Reactive oxygen species as mediators of cell adhesion. Ital. J. Biochem.* 52 (2003) 28-*32.*

[11] P. Chiarugi, PTPs versus PTKs: the redox side of the coin. *Free Radic. Res.* 39 (2005) 353-*364.*

[12] K. Lee, W.J. Esselman, Inhibition of PTPs by H_2O_2 regulates the activation of distinct MAPK pathways. *Free Radic. Biol. Med.* 33 (2002) 1121-*1132.*

[13] C. Bucci, R.G. Parton, I.H. Mather, H. Stunnenberg, K. Simons, B. Hoflack, M. Zerial, The small GTPase rab5 functions as a regulatory factor in the early endocytic pathway, *Cell* 70 (1992) 715– 728.

[14] S. Christoforidis, H.M. McBride, R.D. Burgoyne, M. Zerial, The Rab5 effector EEA1 is a core component of endosome docking, *Nature* 397 (1999) 621-625.

[15] B.P. Ceresa, S.J. Bahr, rab7 activity affects epidermal growth factor:epidermal growth factor receptor degradation by regulating endocytic trafficking from the late endosome, *J. Biol. Chem.* 281 (2006) 1099-1106.

[16] P. van der Sluijs, M. Hull, P. Webster, P. Mâle, B. Goud, I. Mellman, The small GTP-binding protein rab4 controls an early sorting event on the endocytic pathway, *Cell* 70 (1992) 729-740.

[17] O. Ullrich, S. Reinsch, S. Urbé, M. Zerial, R.G. Parton, Rab11 regulates recycling through the pericentriolar recycling endosome, *J. Cell Biol.* 135 (1996) 913-924.
[18] A. Utskarpen, H.H. Slagsvold, A.B. Dyve, S.S. Skånland, K. Sandvig, SNX1 and SNX2 mediate retrograde transport of Shiga toxin, *Biochem. Biophys. Res. Commun.* 358 (2007) 566-570.
[19] M.V. Bujny, V. Popoff, L. Johannes, P.J. Cullen, The retromer component sorting nexin-1 is required for efficient retrograde transport of Shiga toxin from early endosome to the trans Golgi network, *J. Cell Sci.* 120 (2007) 2010-2021.
[20] A.B. Dyve, J. Bergan, A. Utskarpen, K. Sandvig, Sorting nexin 8 regulates endosome-to-Golgi transport, *Biochem. Biophys. Res. Commun.* 390 (2009) 109-114.
[21] M. von Zastrow, A. Sorkin, Signaling on the endocytic pathway, *Curr. Opin. Cell Biol.* 19 (2007) 436-445.
[22] J.M. Gaullier, A. Simonsen, A. D'Arrigo, B. Bremnes, H. Stenmark, R. Aasland, FYVE fingers bind PtdIns(3)P, Nature 394 (1998) 432-433.
[23] C.G. Burd, S.D. Emr, Phosphatidylinositol(3)-phosphate signaling mediated by specific binding to RING FYVE domains, *Mol. Cell* 2 (1998) 157-162.
[24] V. Patki, D.C. Lawe, S. Corvera, J.V. Virbasius, A. Chawla, A functional PtdIns(3)P-binding motif, *Nature* 394 (1998) 433-434.
[25] F. Belleudi, L. Leone, M. Maggio, M.R. Torrisi, Hrs regulates the endocytic sorting of the fibroblast growth factor receptor 2b, *Exp. Cell Res.* 315 (2009) 2181-2191.
[26] S.H. Huang, L. Zhao, Z.P. Sun, X.Z. Li, Z. Geng, K.D. Zhang, M.V. Chao, Z.Y. Chen, Essential role of Hrs in endocytic recycling of full-length TrkB receptor but not its isoform TrkB.T1, *J. Biol. Chem.* 284 (2009) 15126-15136.
[27] V. Pons, P.P. Luyet, E. Morel, L. Abrami, F.G. van der Goot, R.G. Parton, J. Gruenberg, Hrs and SNX3 functions in sorting and membrane invagination within multivesicular bodies, *PLoS Biol.* 6 (2008) e214.
[28] Y. Yamashita, K. Kojima, T. Tsukahara, H. Agawa, K. Yamada, Y. Amano, N. Kurotori, N. Tanaka, K. Sugamura, T. Takeshita, Ubiquitin-independent binding of Hrs mediates endosomal sorting of the interleukin-2 receptor beta-chain, *J. Cell Sci.* 121 (2008) 1727-1738.
[29] C. Raiborg, K.G. Bache, D.J. Gillooly, I.H. Madshus, E. Stang, H. Stenmark, Hrs sorts ubiquitinated proteins into clathrin-coated microdomains of early endosomes, *Nature Cell Biol.* 4 (2002) 394-398.
[30] S. Miura, T. Takeshita, H. Asao, Y. Kimura, K. Murata, Y. Sasaki, J.I. Hanai, H. Beppu, T. Tsukazaki, J.L. Wrana, K. Miyazono, K. Sugamura, Hgs (Hrs), a FYVE domain protein, is involved in Smad signaling through cooperation with SARA, *Mol. Cell Biol.* 20 (2000) 9346-9355.
[31] F. Itoh, N. Divecha, L. Brocks, L. Oomen, H. Janssen, J. Calafat, S. Itoh, P. Dijke Pt, The FYVE domain in Smad anchor for receptor activation (SARA) is sufficient for localization of SARA in early endosomes and regulates TGF-beta/Smad signalling, *Genes Cells* 7 (2002) 321-331.
[32] A. Petiot, J. Faure, H. Stenmark, J. Gruenberg, PI3P signaling regulates receptor sorting but not transport in the endosomal pathway. *J. Cell Biol.* 162 (2003) 971-979.
[33] R. Zoncu, R.M. Perera, D.M. Balkin, M. Pirruccello, D. Toomre, P. De Camilli, A phosphoinositide switch controls the maturation and signaling properties of APPL endosomes, *Cell* 136 (2009) 1110-1121.

[34] D.C. Lin, C. Quevedo, N.E. Brewer, A. Bell, J.R. Testa, M.L. Grimes, F.D. Miller, D.R. Kaplan, APPL1 associates with TrkA and GIPC1 and is required for nerve growth factor-mediated signal transduction, *Mol. Cell Biol.* 26 (2006), 8928-8941.

[35] T. Varsano, M.Q. Dong, I. Niesman, H. Gacula, X. Lou, T. Ma, J.R. Testa, J.R. Yates 3rd, M.G. Farquhar, GIPC is recruited by APPL to peripheral TrkA endosomes and regulates TrkA trafficking and signaling, *Mol. Cell Biol.* 26 (2006), 8942-8952.

[36] M. Miaczynska, S. Christoforidis, A. Giner, A. Shevchenko, S. Uttenweiler-Joseph, B. Habermann, M. Wilm, R.G. Parton, M. Zerial, APPL proteins link Rab5 to nuclear signal transduction via an endosomal compartment, *Cell* 116 (2004) 445-456.

[37] R. Aguilar-Gaytan, J. Mas-Oliva, Oxidative stress impairs endocytosis of the scavenger receptor class A, *Biochem. Biophys. Res. Commun.* 305 (2003) 510-517.

[38] M. Bertelsen, E.E. Anggård, M.J. Carrier, Oxidative stress impairs insulin internalization in endothelial cells in vitro, *Diabetologia* 44 (2001) 605-613.

[39] R. de Wit, A. Capello, J. Boonstra, A.J. Verkleij, J.A. Post. Hydrogen peroxide inhibits epidermal growth factor receptor internalization in human fibroblasts. *Free Radic. Biol. Med.* 28 (2000) 28-38.

[40] J. Cheng, A. Vieira, Oxidative stress disrupts internalization and endocytic trafficking of transferrin in a human malignant keratinocyte line, *Cell Biochem. Biophys.* 45 (2006) 177-184.

[41] V. Cavalli, F. Vilbois, M. Corti, M.J. Marcote, K. Tamura, M. Karin, S. Arkinstall, J. Gruenberg, The stress-induced MAP kinase p38 regulates endocytic trafficking via the GDI:Rab5 complex, *Mol. Cell* 7 (2001) 421-432.

[42] Y. Zwang, Y. Yarden, p38 MAP kinase mediates stress-induced internalization of EGFR: implications for cancer chemotherapy, *EMBO J.* 25 (2006) 4195-4206.

[43] S. Vergarajauregui, A. San Miguel, R. Puertollano, Activation of p38 mitogen-activated protein kinase promotes epidermal growth factor receptor internalization, *Traffic* 7 (2006) 686-698.

[44] L.R. Coulthard, D.E. White, D.L. Jones, M.F. McDermott, S.A. Burchill, p38(MAPK): stress responses from molecular mechanisms to therapeutics, *Trends Mol. Med.* 15 (2009) 369-379.

[45] F. Kano, T. Arai, M. Matsuto, H. Hayashi, M. Sato, M. Murata, Hydrogen peroxide depletes phosphatidylinositol-3-phosphate from endosomes in a p38 MAPK-dependent manner and perturbs endocytosis, *Biochim. Biophys. Acta.* 1813 (2011) 784-801.

[46] C. Raiborg, K.G. Bache, A. Mehlum, H. Stenmark, Function of Hrs in endocytic trafficking and signalling, *Biochem. Soc. Trans.* 29 (2001) 472-475.

[47] N. Fili, V. Calleja, R. Woscholski, P.J. Parker, B. Larijani, Compartmental signal modulation: Endosomal phosphatidylinositol 3-phosphate controls endosome morphology and selective cargo sorting, *Proc. Natl. Acad. Sci. USA* 103 (2006) 15473-15478.

[48] C. Raiborg, B. Bremnes, A. Mehlum, D.J. Gillooly, A. D'Arrigo, E. Stang, H. Stenmark. FYVE and coiled-coil domains determine the specific localisation of Hrs to early endosomes, *J. Cell Sci.* 114 (2001) 2255-2263.

[49] D.J. Gillooly, I.C. Morrow, M. Lindsay, R. Gould, N.J. Bryant, J.M. Gaullier, R.G. Parton, H. Stenmark, Localization of phosphatidylinositol 3-phosphate in yeast and mammalian cells, *EMBO J.* 19 (2000) 4577-4588.

[50] H. Atsumi, M. Kitada, K. Kanasaki, D. Koya, Reversal of redox-dependent inhibition of diacylglycerol kinase by antioxidants in mesangial cells exposed to high glucose, *Mol. Med. Report* 4 (2011) 923-927.

[51] S. Bhakdi, J. Tranum-Jensen, A. Sziegoleit, Mechanism of membrane damage by streptolysin-O, *Infect. Immun.* 47 (1985) 52-60.

[52] F. Kano, K. Takenaka, A. Yamamoto, K. Nagayama, E. Nishida, M. Murata, MEK and Cdc2 kinase are sequentially required for Golgi disassembly in MDCK cells by the mitotic Xenopus extracts, *J. Cell Biol.* 149 (2000) 357-368.

[53] F. Kano, A.R. Tanaka, S. Yamauchi, H. Kondo, M. Murata, Cdc2 kinase-dependent disassembly of endoplasmic reticulum (ER) exit sites inhibits ER-to-Golgi vesicular transport during mitosis, *Mol. Biol. Cell* 15 (2004) 4289-4298.

[54] F. Kano, S. Yamauchi, Y. Yoshida, M. Watanabe-Takahashi, K. Nishikawa, N. Nakamura, M. Murata, Yip1A regulates the COPI-independent retrograde transport from the Golgi complex to the ER, *J. Cell Sci.* 122 (2009) 2218-2227.

[55] T. Bergeland, L. Haugen, O.J. Landsverk, H. Stenmark, O. Bakke, Cell-cycle-dependent binding kinetics for the early endosomal tethering factor EEA1, *EMBO Rep.* 9 (2008) 171-178.

[56] I.G. Mills, A.T. Jones, M.J. Clague, Involvement of the endosomal autoantigen EEA1 in homotypic fusion of early endosomes, *Curr. Biol.* 8 (1998) 881-884.

[57] A. Simonsen, R. Lippé, S. Christoforidis, J.M. Gaullier, A. Brech, J. Callaghan, B.H. Toh, C. Murphy, M. Zerial, H. Stenmark, EEA1 links PI(3)K function to Rab5 regulation of endosome fusion, *Nature* 394 (1998) 494-498.

[58] H.W. Shin, M. Hayashi, S. Christoforidis, S. Lacas-Gervais, S. Hoepfner, M.R. Wenk, J. Modregger, S. Uttenweiler-Joseph, M. Wilm, A. Nystuen, W.N. Frankel, M. Solimena, P. De Camilli, M. Zerial, An enzymatic cascade of Rab5 effectors regulates phosphoinositide turnover in the endocytic pathway, *J. Cell Biol.* 170 (2005) 607-618.

[59] F. Blondeau, J. Laporte, S. Bodin, G. Superti-Furga, B. Payrastre, J.L. Mandel, Myotubularin, a phosphatase deficient in myotubular myopathy, acts on phosphatidylinositol 3-kinase and phosphatidylinositol 3-phosphate pathway, *Hum. Mol. Genet.* 9 (2000) 2223–2229.

[60] J.T. Murray, C. Panaretou, H. Stenmark, M. Miaczynska, J.M. Backer, Role of Rab5 in the recruitment of hVps34/p150 to the early endosome, *Traffic* 3 (2002) 416-427.

[61] G.S. Taylor, T. Maehama, J.E. Dixon, Myotubularin, a protein tyrosine phosphatase mutated in myotubular myopathy, dephosphorylates the lipid second messenger, phosphatidylinositol 3-phosphate, *Proc. Natl. Acad. Sci. USA* 97 (2000) 8910-8915.

[62] O.C. Ikonomov, D. Sbrissa, A. Shisheva, Mammalian cell morphology and endocytic membrane homeostasis require enzymatically active phosphoinositide 5-kinase PIKfyve, *J. Biol. Chem.* 276 (2001) 26141-26147.

[63] I. Ivetac, A.D. Munday, M.V. Kisseleva, X.M. Zhang, S. Luff, T. Tiganis, J.C. Whisstock, T. Rowe, P.W. Majerus, C.A. Mitchell, The type Ialpha inositol polyphosphate 4-phosphatase generates and terminates phosphoinositide 3-kinase signals on endosomes and the plasma membrane. *Mol. Biol. Cell.* 16 (2005) 2218-2233.

[64] J.S. Bonifacino, J.H. Hurley, Retromer, *Curr. Opin. Cell Biol.* 20 (2008) 427-436.

[65] S.S. Skånland, S. Wälchli, A. Utskarpen, A. Wandinger-Ness, K. Sandvig, Phosphoinositide-regulated retrograde transport of ricin: crosstalk between hVps34 and sorting nexins, *Traffic* 8 (2007) 297-309.

[66] R. Rojas, T. van Vlijmen, G.A. Mardones, Y. Prabhu, A.L. Rojas, S. Mohammed, A.J. Heck, G. Raposo, P. van der Sluijs, J.S. Bonifacino, Regulation of retromer recruitment to endosomes by sequential action of Rab5 and Rab7, *J.s Cell Biol.* 183 (2008) 513-526.

[67] E.K. Kim, E.J. Choi, Pathological roles of MAPK signaling pathways in human diseases, *Biochim. Biophys. Acta.* 1802 (2010) 396-405.

In: Hydrogen Peroxide
Editors: Gilberto Aguilar and Raphael A. Guzman

ISBN: 978-1-62257-414-8

Chapter 5

USES OF HYDROGEN PEROXIDE IN AQUACULTURE

***Gemma Giménez Papiol*[1*] *and Ana Roque*[2]**
[1]Cawthron Institute, Nelson, New Zealand
[2]IRTA Sant Carles de la Ràpita,
Institut de Recerca i Tecnologia Agroalimentàries, Barcelona, Spain

ABSTRACT

In this chapter we revise the uses of hydrogen peroxide in aquaculture. At the beginning, hydrogen peroxide was used mainly as a treatment against ectoparasites of adult fish, such as sea lice (*Lepeophtherius salmonis*), or against the microbiota of fish eggs, mainly fungi in freshwater species.

Later, it became of interest for microbial control purposes rather than chlorine or antibiotics due to its relatively lower toxicity. Although it is one of the most powerful known oxidizers, it is easily decomposed in a water solution, particularly in the presence of organic matter, light and turbulence (all conditions easily met in aquaculture systems), leaving no toxic products. The spectrum of species and live stages has also increased; hydrogen peroxide has been tested on rotifers, crustacea, bivalves and fish, and on eggs, larvae, juveniles and adults.

Recently, issues concerning its impact on the aquaculture system (i.e. biofilters of the recirculation system), on the environment and on animal welfare of cultured species have risen.

INTRODUCTION

Aquaculture Overview

Domestication of plants and animals began about 11,000 years ago. Domestication of marine organisms started much later, about 1,000 years ago, and it has increased

* Correspondence: G. Giménez-Papiol, Cawthron Institute, 98 Halifax Street East, Private Bag 2, Nelson 7042, New Zealand. E-mail: Gemma.GimenezPapiol@cawthron.org.nz

exponentially during the last century [Duarte et al., 2007]. These authors considered the domestication of aquatic animals more successful than that of land animals, in terms of number of domesticated species and specific growth in the number of domesticated species per year, and they considered the greater diversity of domesticated aquatic taxa as the main reason for such success. Thus, the flourishing of aquaculture is a contemporary phenomenon.

The bulk of aquaculture production is destined for human consumption. In 2008, human consumption of fish accounted for 142 million tonnes, of which 52.5 million tonnes were produced by aquaculture [FAO, 2010]. It is expected that the world fish consumption will increase, and it will rely on aquaculture supply instead of capture fisheries, as the latter are already fully exploited or overexploited [FAO, 2006].

World aquaculture has had an average annual growth rate of 6.6% per year since 1970, supplying 7.8 kg of animal-food per capita in 2008 [FAO 2010]. It has increased more rapidly than capture fisheries (1.2%) and terrestrial farmed meat (2.8%) [FAO, 2006], but its growth rate is slowing due to a wide range of factors, and differs notoriously among countries [FAO 2010].

World aquaculture production is based on freshwater fishes (59.9% by production, 56.0% by value), while seawater production is focused on high-value fish and crustaceans, as well as large volumes of molluscs (Table 1).

Issues and Challenges for Aquaculture Development

Aquaculture, mainly in developed and/or the world's richer countries, must fit its development with the current social sensitivities, laws and regulations. The challenges that represent the intensification and diversification of farming practices, the advances in technological innovations in food production, as well as the continuing trade globalization, inherent to its development, must cope with the increasing demand for public health, environmental protection, biodiversity protection, animal welfare and impacts on climate change.

At international trade level, countries producing foods of animal origin and wishing to export them must satisfy animal health, public health, veterinary certification and residues requirements of the target market. Consumers are becoming more sensitive to animal welfare and environmental protection, and increasingly demand of the aquaculture industry not only a high quality and safe to eat aquaculture product, but also that it has been produced under bioethical and environmentally sustainable conditions.

Table 1. World aquaculture production of major species groups in 2008 [FAO, 2010].

	Million tonnes	Value (US $ billion)
Freshwater fishes (mainly carps)	28.8	40.5
Molluscs (mainly *Crassostrea gigas*)	13.1	13.1
Crustaceans	5.0	22.7
Diadromous fishes	3.3	13.1
Marine fishes	1.8	6.6
Other aquatic animals	0.6	2.4

The Food and Agriculture Organization (FAO) has recently identified biosecurity, climate change and standards and certificates as issues concerning aquaculture development over the next years [FAO, 2010], while animal welfare's importance is rising year by year [Conte, 2004; EFSA 2007a, 2007b, 2007c, 2008a, 2009a, 2009b]. The term "biosecurity" applies to a complex range of activities that can affect the biological compartment of the ecosystem that hosts a given aquaculture activity (e.g. farm facilities and transport containers, among others).

The two main adverse effects derived from a lack in biological control, and which are a serious threat to biodiversity, are the introduction of foreign species to new habitats and the spreading of aquatic diseases and pests. The impact of an aquaculture facility and its activities (feed release, water treatments, for instance) on the local ecosystem must also be taken into account.

Large-scale aquaculture production relies, to a greater or lesser extent, on the environment in which the facility is placed. Shellfish and marine fish, such as salmon or sea bream, are confined in cages attached to different structures like racks or ropes and, although their site location have been previously studied in order to ensure the most suitable area for the culture, the organisms are exposed to the surrounding environment.

Farmers can not take control of the environmental shifts and/or eventualities, such as harmful algal blooms, accidents of oil tankers, or other sources of contamination, they can only monitor them. Any effects of climate change on water environment (both physico-chemical and biological) will also affect large-scale aquaculture production.

On the other hand, as a human activity in which chemical compounds are broadly used (i.e. disinfectants, antibiotics, or other therapeutants), the use of alternative compounds less harmful for the environment and a strict control of water quality or its effluents are increasingly required. Requirements, such as those concerning work safety and food safety, have lead to the use of standards and certificates to label those aquaculture products that guarantee compliance to the regulations along the production chain. Finally, issues regarding animal welfare deal with the culturing and slaughtering conditions of cultured organisms, including the effects of the compounds routinely or punctually used for water treatment and disease control on their fitness.

Bottlenecks in Aquaculture Production

Beside customer and social requirements, the development of aquaculture faces different technical challenges in order to achieve an economically viable production of a given aquatic organism. There are different critical points along the production chain, from the obtaining adequate offspring from captive spawners to the quality of the end-product (i.e. nutritional composition and food safety), to which must be added the threat of diseases and other aquatic pest outbreaks at any developmental stage (egg, larvae, juveniles, adults) that can impair the growth, quality or even the survival of the cultured organism.

Interesting Characteristics of Hydrogen Peroxide for Aquaculture Purposes

Hydrogen peroxide is a natural metabolite produced by many organisms as a toxic by-product of oxygen during aerobic oxidation and by the direct action of sunlight on water. In aquatic environments, dissolved organic carbon (DOC) from naturally found organic compounds such as glycerol, benzoic acid, aniline, tryptophan, and humic acid, is excited by ultraviolet light, and the superoxide anion (O_2^-) form disproportionates and protonates into hydrogen peroxide (H_2O_2).

This reaction also releases oxygen, and is usually restricted to the water layer where the UV-light can penetrate, although hydrogen peroxide can be distributed to other layers, e.g. by convective mixing processes and turbulences, etc. Hydrogen peroxide inputs from rain or dry atmospheric deposition, or organically produced by microorganisms, are less important than that originated by UV-light reaction.

The concentration of hydrogen peroxide in freshwater is reported to range from 0.001 to 0.109 mg/L, while surface seawater concentrations range from 0.001 to 0.0136 mg/L. Higher concentrations typically occur in surface water containing high DOC. Very little hydrogen peroxide exists in deep marine or fresh water [Schmidt et al., 2006].

Hydrogen peroxide is a powerful oxidizer that, in solution, is easily decomposed, particularly in the presence of organic matter, light and turbulence, into oxygen and water. These two characteristics, the oxidizing capacity and the production of non-toxic residues which do not accumulate or give rise to complex toxic metabolites, make hydrogen peroxide a very interesting compound for different aquaculture purposes, as it can be handled without strict work safety rules. Water effluents containing hydrogen peroxide can be simply diluted, and no toxic compounds are accumulated in the cultured organisms exposed to hydrogen peroxide and will therefore have no effect on the final consumer's health.

It has been suggested as a source of oxygen for the transport of live aquatic organisms [Innes Taylor and Ross, 1988] due to its ready decomposition into water and oxygen, but it has not been widely used.

A patent [Patent CN1035413] is based on a compound that contains hydrogen peroxide, hydrogen peroxide amide carbonate, for increasing oxygen supply to fish and shrimp in rearing and transport conditions. Advantages, when compared to the current methods used, are the small amount of hydrogen peroxide amide carbonate required for obtaining large amounts of oxygen, the long active time and the low cost and simplicity of use.

The production of oxygen is also a suggested mechanism of action on ectoparasites, such as sea lice, producing a mechanical paralysis due to the formation of gas bubbles in the parasites' haemolymph that causes the organism to rise to the water surface [Treasurer and Grant, 1997].

As a reactive oxygen species (ROS) and source of other ROS such as superoxide, perhydroxyl and hydroxyl radicals, hydrogen peroxide has an array of physiological and developmental effects useful for the control of microbial and other aquatic pests as well as for the management of the physiological status of cultured organisms.

It can easily cross cell membranes and cause lipid peroxidation, modification of intracellular redox potential, destruction of protein and DNA, the activation, reversible modification and/or inactivation of enzymes, the activation, inhibition and/or modification of regulatory, metabolic and structural proteins posttranslationally as second messenger [Avendaño et al., 2006; Robbins et al., 2010; Treasurer et al., 2000].

USES OF HYDROGEN PEROXIDE IN AQUACULTURE

Treatment against Ectoparasites

The main use of hydrogen peroxide in aquaculture is as treatment against ectoparasites of adult stages of fish, mainly salmonids (Table 2). The number of effective, legal therapeutants has disminished over the last 20 years. The use of malachite green, a highly effective and once heavily used therapeutant, is no longer allowed because of concerns over teratogenicity, undesirable tissue residues, and user safety [Melendre et al., 2006]. Issues regarding user sensitization, development of resistance by the parasite, and environmental impact affect the use of formalin, acetic acid, chloramine-T, potassium permanganate, peracetic acid and other therapeutants. However, treatment efficiency of hydrogen peroxide is often inconsistent in practice, and must be tailor-made to suit specific situations.

The use of hydrogen peroxide against ectoparasites must take into account the temperature of the water and the exposure time for its effectiveness. It is less toxic at high temperatures and with high organic matter loads; both conditions favour the presence of parasites. Generally, hydrogen peroxide has variable effectiveness against parasites and the development of resistance in aquaculture facilities has been detected when it is used frequently [Treasurer et al., 2000; Bravo et al. 2010].

Table 2. Scientific articles dealing with uses of hydrogen peroxide as treatment against ectoparasites of juvenile and adult fish.

Fish species	Target organisms	Reference
Salmo salar	*Caligus rogercresseyi*	Bravo et al., 2010
Salmo salar	*Lepeophtheirus salmonis*	Bruno and Raynard, 1994
Salmo salar *Oncorhynchus tshawytscha*	*Lepeophtheirus salmonis*	Johnson et al., 1993
Seriola lalandi	*Zeuxapta seriolae*	Mansell et al., 2005
Salmo salar	*Lepeophtheirus salmonis*	McAndrew et al., 1998
Ictalurus punctatus	*Dero digitata**	Mischke et al., 2001
Polydactylus sexfilis	*Amyloodinium ocellatum*	Montgomery-Brock et al., 2001
Salmo salar	Amoeba (Amoebic gill disease)	Powell and Clark, 2004
Oncorhynchus mykiss	*Ambiphrya sp.* *Gyrodactylus sp.* *Trichodina sp.*	Rach et al., 2000
Salmo salar	*Ichthyophthitius multifiliis*	Rintamaki-Kinnunen et al., 2005
Salmo sp.	*Lepeophtheirus* sp. *Caligus* sp.	Robertson et al., 2009
Sparus aurata	*Sparicotyle chrysophrii*	Sitja-Bobadilla et al., 2006
Salmo salar	*Caligus elongatus*	Treasurer and Grant, 1997
Salmo salar	*Lepeophtheirus salmonis*	Treasurer et al. 2000

*Intermediate host in the life cycle of the parasite *Aurantiactinomyxon ictaluri*.

Adult and egg stages of crustacean parasites such as *Caligus* sp. or *Lepeophtherius* sp. are more sensitive to hydrogen peroxide, but the infestating stage is usually more resistant [Treasurer and Grant, 1997; McAndrew et al. 1998].

In experimental conditions, some authors observed that sea lice can be viable after exposure to hydrogen peroxide, and could resettle on fish, but they considered that in natural conditions this is unlikely to happen because released sea lice are swept out by water currents [McAndrew et al., 1998]. Hydrogen peroxide used against crustacean ectoparasites seems to be more effective as a preventive treatment before infestation breaks out in an aquaculture facility [Rae, 2002].

Hydrogen peroxide has different success in the treatment of other parasites, depending on the exposure of the parasite (i.e. if it is embedded in the gill cavity, it is more resistant) and on the stress caused to the fish by the treatment. Mansell et al. (2005) reported that hydrogen peroxide causes acute stress on *Seriola lalandi* health, but the magnitude and extent of the effects were much less than those caused by chronic infection with the monogenean *Zeuxapta seriolae*. Powell and Clark (2004) tested the effectiveness of hydrogen peroxide against amoebae, but the results from treatments were equivocal and the toxicity of hydrogen peroxide on *Salmo salar* was high. It is effective and safe against *Ambiphrya* sp. and *Gyrodactylus* sp. living on *Oncorhynchus mykiss* [Rach et al., 2000], *Ichthyophthitius* sp. on *Salmo salar* [Rintamaki-Kinnunen et al., 2005] and *Sparicotyle* sp. on *Sparus aurata* [Sitja-Bobadilla et al., 2006], but not against *Trichodina* sp. on *Oncorhynchus mykiss* [Rach et al., 2000]. Again, the parasites present differences in their resistance to hydrogen peroxide not only between species, but between developmental stages of the same species. Sitja-Bobadilla et al. (2006) observed that adults of *Sparicotyle* sp. were more sensitive to hydrogen peroxide than larvae (with only 10% of mortality) and eggs were the most resistant stage. This is in agreement with Roque (unpublished data), who exposed adult seabream infected with *Sparicotyle* sp. to 200 ppm of H_2O_2 (OxAquaculture®) for 60 min and did not observe any significant decrease in the prevalence and intensity of this parasite after the treatment.

Water Treatment

Water quality is one of the main concerns in aquaculture. Effluents from aquaculture facilities must be treated for complyiance with environmental standards, in order to reduce the organic and microbiological discharge, at the same time controlling the use of toxic compounds that could eventually affect the biota of the environment [Directive 2000/60/EC; Directive 2006/44/EC; U. S. Federal Water Pollution Control Act]. Although the use of hydrogen peroxide for this purpose is not wide spread in routine treatment of water effluents, there are patents describing products and processes based on the oxidizing power of hydrogen peroxide for treatment of organically contaminated water bodies [Patent WO2007138058; Patent TW200520683; Patent AU2005100236; Patent CN1513778]. Hydrogen peroxide decomposes organic harmful substances such as ammonia, nitrite and hydrogen sulfide, as well as negatively affects harmful microorganisms.

Water input and rearing water must also be of high quality, especially in those facilities working with closed or recirculation systems. Interactions between cultured organisms and microorganisms present naturally in water are complex, species-specific, and not fully understood [Olafsen, 2001]. Vadstein et al. (1993) and Skjermo and Vadstein (1999) suggest a bulk of strategies to control the microbial community and its interactions with cultured organisms based mainly on three principles: non-selective reduction of bacteria, selective enhancement of bacteria, and improvement of cultured organism resistance.

The suitability of hydrogen peroxide for controlling (i.e. non-selective reduction) harmful microorganisms has been tested on microalgae and bacteria with differing success. Drabkova et al (2007) have found that *Microcystis aeruginosa* (cyanobacteria) is affected by concentrations of hydrogen peroxide 10-fold lower than those toxic for *Pseudokirchneriella subcapitata* (green algae) and *Navicula seminulum* (diatom). Additionally, some authors have detected that microbial strains exposed to low levels of hydrogen peroxide are more resistant to treatments with high levels than strains not pre-exposed [Barnes et al., 1999], suggesting an activation of the enzymes involved in the degradation of hydrogen peroxide. The use of hydrogen peroxide combined with copper sulphate seems adequate for treating rearing water against *Karlodinium veneficum*, an ichthyotoxic dinoflagellate, in presence of fish [Fernandez-Tejedor et al., 2007]; these authors suggest that hydrogen peroxide inactivates karlotoxins released by *K. veneficum* cells lysed by copper sulphate.

The sensitivity of reared organisms to toxic effects of hydrogen peroxide varies among species and developmental stages, and needs to be evaluated case by case. It is difficult to determine hydrogen peroxide treatment concentrations that are harmless for cultured phytoplankton and zooplankton because they present toxic thresholds very close to those of the target harmful microorganisms. Moreno-Garrido and Cañavate (2001) tested the use of hydrogen peroxide against ciliates present in a culture of *Dunaliella salina*, but it was inappropriate since lethal doses for the ciliate also eliminated algal cells. Rotifer and *Artemia* nauplii, two zooplankton species widely used as live prey for different larval stages of cultured organisms, and one of the main inputs of microorganisms in the rearing tanks maintained in recirulation systems, present different resistances against hydrogen peroxide. Giménez et al. (2006) observed that although the lethal doses were lower in rotifer than in *Artemia* nauplii, the effective doses and exposure times that reduced the bacterial load associated to each zooplankton species were also lower in rotifer than in *Artemia* nauplii.

One of the main uses of hydrogen peroxide in aquaculture, both in freshwater and marine, is as disinfectant of fish eggs. An adequate microbial control during the egg stage and early development can improve larval survival, growth, and fish quality [Douillet and Pickering, 1999], and transmission of microbes from parents to offspring can be eliminated by disinfecting the eggs before hatching [Douillet and Holt, 1994]. Published papers dealing with the different doses and exposure times used for reducing the microbial load of eggs are summarized in Table 3. Hydrogen peroxide is currently one of the most effective therapeutants used to control fungal and bacterial infection of fish eggs [Rasowo et al., 2007], although Celada et al. (2004) and Melendre et al. (2006) observed that it was not effective for treating eggs of *Pascifastacus leniusculus*. The reduction of microorganisms attached to fish eggs will depend on the initial load, and almost all of the authors cited agree on advising the use of hydrogen peroxide even when it is less effective than other more toxic chemicals such as iodine or formaldehyde.

Aquatic organisms develop the immune system during the larval stage [Chantanachookhin et al. 1991] and some organs involved in the specific immune system such as thymus or spleen do not become active until after the metamorphosis [Manning and Tattner, 1985; Magnadottir et al. 2005].

Table 3. Scientific articles dealing with uses of hydrogen peroxide as treatment against microorganisms on aquatic eggs.

Cultured species	Target organisms	Reference
Pacifastacus leniusculus	Fungi	Celada et al., 2004
Sciaenops ocellatus	Bacteria	Douillet and Holt, 1994
Oncorhynchus mykiss	Fungi	Kitancharoen et al., 1997
Pacifastacus leniusculus	Fungi	Melendre et al., 2006
Gadus morhua *Melanogrammus aegelfinus*	Bacteria	Peck et al., 2004
Esox lucius *Stizostedion vitreum* *Perca flavescens* *Catostomus commersoni* *Acipenser fulvescens* *Polyodon spathula* *Cyprinus carpio* *Ictalurus punctatus*	Bacteria	Rach et al., 1998
Ictalurus punctatus	*Saprolegnia* sp.	Rach et al., 2004
Clarias gariepinus	Bacteria Fungi	Rasowo et al., 2007
Oncorhynchus mykiss	*Saprolegnia parasitica*	Schreier et al., 1996
Polydactylus sexfilis *Seriola rivoliana*	Bacteria	Verner-Jeffreys et al., 2007
Oncorhynchus mykiss	Bacteria	Wagner et al., 2008

Therefore, when growth of opportunistic bacteria favored by the culturing conditions coincides with a not fully developed immune system, larvae become vulnerable to diseases. Again, the use of hydrogen peroxide must be evaluated carefully for each species and developmental stage, although it is not widely used because larval stages are very sensitive not only to the toxicity of hydrogen peroxide itself but also to the protocols used for eliminating it from the rearing water (i. e. strong aeration). Giménez et al. (2009) observed that *Sparus aurata* larvae were more resistant to hydrogen peroxide than *Dentex dentex* larvae although both are Sparidae fish species. Treatments with hydrogen peroxide are mostly used on fish juveniles and adults, although with different results. It is effective against *Flavobacterium branchiophilum* in *Oncorhynchus mykiss* [Lumsden et al., 1998] and also against *Tenibacilum maritimum* in *Scophthalmus maximus* [Avendaño-Herrera et al., 2006], although the latter authors found that the stress induced in fish by hydrogen peroxide treatment made them more vulnerable to the disease. Thomas-Jinu and Goodwin (2004) observed that hydrogen peroxide was effective against *Flavobacterium columnare* in *Ictalurus punctatus* when combined with other therapeutants. Gaikowski et al. (1999) observed differences in the resistance to hydrogen peroxide when different live stages of *Oncorhynchus mykiss*, *Salmo salar*, *Salvelinus namaycush, Esox masquinongy, Esox lucius,*

Scaphirhynchus albus, Stizostedion vitreum, Catostomus commersoni, Lepomis macrochirus, Ictalurus punctatus, Pimephales promelas, Micropterus salmoides and *Perca flavescens* were exposed to levels used for the treatment against *Saprolegnia* spp. Hydrogen peroxide has also been shown to be effective against an artificial infection of *Photobacterium damselae* in Senegalese sole (*Solea senegalensis*) juveniles [Roque et al., 2011]. Repeated exposures to hydrogen peroxide increase the sensitivity to its toxicity in early developmental stages of fish [Tripi and Bowser, 2001; Giménez, unpub. res.].

Decapsulation of Artemia

Artemia sp. is a crustacean widely produced for use as live feed in the diets of aquatic animals, and it is also used as a model for toxicological studies. It is easy to store and handle because it has very resistant cysts that support extreme physiological conditions such as dessication or high salinity, due to the ability to undergo diapause, a state of profound dormancy that enhances stress tolerance [Van Stappen, 1996]. It is necessary to terminate diapause of the cysts for culturing *Artemia* sp. [Van Stappen et al., 1998; Saygi, 2003; Patent SU1472011].

Hydrogen peroxide can penetrate *Artemia* sp. cysts and influence physiological and developmental processes leading to the termination and/or modulation of diapause, in other words, it promotes development of quiescent cysts. The suggested mechanism of action is described by Robbins et al. (2010). The oxidative properties of hydrogen peroxide allow it act as a second messenger *via* enhancement of tyrosine phosphorylation and dephosphorylation. It can reversibly modify proteins posttranslationally, either by activation or inhibition. Proteins affected can be regulatory, metabolic and structural, such as mitogen-activated protein kinases (MAPKs) or those containing thiol protein residues.

Induction of Shellfish Spawning and Metamorphosis

Hydrogen peroxide activates the enzyme catalyzing oxidative synthesis of prostaglandin endoperoxide in some species of shellfish. This characteristic has been useful for obtaining broadcast spawns, because it can induce synchronous spawning in gravid male and female shellfish when it is added to seawater [Morse et al., 1977]. It can also induce the metamorphosis of shellfish larvae into the settlement stage, at similar percentages to those obtained with natural extracts of macroalgae but at low cost [Morse and Morse, 1984; Boettcher et al., 1997].

The mechanism of action is related to the generation of electronically active oxygen species in water, which are able to influence prostaglandins metabolism [Morse, 1984], and the concentration and exposure time must be optimised for the particular water properties, species specificities and individual gravity stage.

Despite the good results obtained with some species of shellfish recalcitrant to other spawning induction methods, such as *Haliotis* sp. [Morse, 1984; Peña, 1983, 1986; Taylor, 1997], or to metamorphosis induction, e.g. *Strombus gigas* [Boettcher et al., 1997], it is not widely used by the aquaculture industry because the main shellfish species produced, oysters

(*Crassostrea* sp., *Ostrea* sp.) and mussels (*Mytilus* sp., *Perna* sp.), can be successfully induced by other techniques such as thermic stress.

ONGOING RESEARCH ON USES OF HYDROGEN PEROXIDE IN AQUACULTURE AND FUTURE TRENDS

Effects on Recirculation Systems

Recirculation aquaculture systems (RAS) are the newest and most complex production systems in aquaculture. RAS are typically an indoor system provided with different filters (mechanics, organics, biological and others) and electronic and mechanic elements (pumps, heaters, data monitors) that allow the control of water conditions and quality with a certain degree of independence from the incoming water all year round. RAS do require an adequate water supply, both quantitatively and qualitatively; it is necessary to fill the system and replace the water loss due to leakages, evaporation, and discharges, Once water is in the RAS system, its quality can be kept at a desired level by means of the different filter systems. A biological filter is required for reducing the concentration of dissolved nitrites and ammonia, both of which are produced naturally by cultured organisms and are toxic to the organisms. As stated before, hydrogen peroxide can oxidize and therefore reduce the levels of these toxic compounds,. but its use can be harmful to microorganisms that colonize the biological filter. This is due to its ability not only to damage biological molecules such as DNA, but also to impair cell metabolism by inhibition/activation of proteins. In other words, it can both kill microorganisms and reduce their capacity for metabolizing nitrogen derivates.

The degradation rates of hydrogen peroxide are essential for allowing its use in the different treatments required in RAS while retaining the integrity of the microbiological filter [Madsen et al., 2000; Pedersen et al., 2006; Moller et al., 2010]. Hydrogen peroxide has usually a fast degradation rate in rearing water, because it usually contains high levels of dissolved organic matter and the aeration system produces high turbulence, but various hydrogen peroxide releasing products are also developed. These products usually contain other stabilizing substances that slow down the degradation rate of hydrogen peroxide, with the aim of increasing its effectivity. Early evaluations of Pedersen et al. (2006) evidenced adverse effects on biological filters at treatment-effective doses of hydrogen peroxide. Nevertheless, the interest in using hydrogen peroxide as a therapeutant and substitute for other traditional agents, is driving the search for management protocols that allow the use of hydrogen peroxide in a RAS system [Pedersen and Pedersen, 2012].

Welfare Considerations

The adverse effects of hydrogen peroxide have been evaluated in different commercial organisms. Traditionally, the effects on the commercial organism were linked to the evaluation of a treatment protocol against parasites, fugi and bacteria on fish and eggs. Early work of Powell and Perry (1997) concluded that hydrogen peroxide produced acute stress, and impaired oxygen transport, in rainbow trout, *Oncorhynchus mykiss* due to the increase in haematocrit, the reduction of the oxygen carrying capacity, respiratory acidosis and oxidation of haemoglobin.

Recently, studies focused only on the adverse effects of hydrogen peroxide on fish welfare have been published. Sanabria et al. (2009) listed physiological effects that can explain why larval angelfish, *Pterophyllum scalare*, exposed to hydrogen peroxide have lower success in swimbladder inflation. These effects are the relaxation and constriction of blood vessels, the induction of apoptosis, the reduction of the capacity of cells to scavenge reactive oxygen species, mitochondrial disfunction, amplification of the oxidative cell-death cycle, and DNA-strand breakage. Fish that lacked swim-bladder development required more energy in swimming for maintaining their buoyancy and thus became stressed. In terms of production, they grew slower and usually presented more deformities.

The adverse effects on adult *Salmo salar* and *Stizostediun vitreum* are usually detected on gill epithelium and in the hematology and blood chemistry, and are related to acute, short-term stress [Kiemer and Black, 1997; Bowers et al., 2002; Tort et al., 2003]. Stress indicators such as plasma cortisol, lactate, glucose, plasma calcium, phosphorous, sodium, potassium, chloride, magnesium, hematocrit, hemoglobin, total protein and albumin in adult sea bass, *Dicentrarchus labrax*, were evidenced as acute stress immediately after the exposure to hydrogen peroxide, requiring more than 24h period for full recovery [Roque et al., 2010].

Sustainability and Effects on the Environment

Although the environmental harmlessness of hydrogen peroxide has been claimed by many authors, due to its decomposition into non-toxic compounds, concerns as to its effects on the indigenous organisms surrounding aquaculture facilities are increasing [Saez and Bowser, 2001]. Hydrogen peroxide is used mainly in anti-lice treatments in fish cages embedded in the aquatic environment. No water effluent treatments are possible, so hydrogen peroxide is released directly into the water. Anti-lice treatments lack specificity and therefore may affect indigenous organisms, in particular crustaceans, in the vicinity of anti-lice treatments. Hydrogen peroxide can also negatively impact on sensitive non-target organisms by altering the population structure within the immediate surroundings [Burridge et al, 2010]. Despite these potential threats, no adverse effects on native organisms have been detected, nor on population structure, possibly due to the fast dilution and transport of hydrogen peroxide in water leading to concentrations that non-target organisms can resist [Willis et al., 2005; Schmidt et al., 2006; Burridge et al., 2010].

Conclusion

The relative amount of hydrogen peroxide used for aquaculture is insignificant compared with its industrial, commercial and municipal applications [Schmidt et al., 2006]. It is mainly used as water treatment and therapeutant against parasites and microorganisms. Its efficiency is often inconsistent in practice because its stability in water is affected by the presence of dissolved organic matter, light and turbulence; all these conditions are easily found in aquaculture environments. Every protocol of use must be tailor-made to specific situations in order to find the dose/exposure combination effective for the treatment purposes, but with regard to the welfare of the cultured organism. Some former uses, such as oxygen source for live organisms transport and induction of spawning in shellfish, have been overtaken by other systems. Environmentally, it seems that hydrogen peroxide has no negative consequences on the ecosystem. The characteristics that make hydrogen peroxide attractive for its use as a substitute of other more toxic and harmful compounds are still valid, and support future studies on its applicability in different aquaculture systems, such as RAS, or its effects on animal welfare and environment.

Acknowledgments

Gemma Giménez Papiol acknowledges the funding provided by the Agency for Administration of University and Research Grants (AGAUR) from the Generalitat de Catalunya, Spain, through the post-doctoral fellowship Beatriu de Pinós with the cofounding of the European Union by means of the Marie Curie Actions of the 7th R&R Framework Programme.

Grupo OX (Spain) has provided technical assessment regarding hydrogen peroxide.

Mrs. Xantal Romaguera (IRTA) has helped with documentation and patents search.

Dr. Lesley Rhodes (Cawthron Institute) has revised the manuscript of this chapter.

References

[1] Avendaño-Herrera, R. E.; Magariños, B.; Irgang, R. and Tozano, A. E. (2006). Use of hydrogen peroxide against the fish pathogen *Tenacibacilum maritimum* and its effects on infected turbot (*Scophthalmus maximus*). *Aquaculture*, 257, 104 – 110.

[2] Barnes, A.; Bowden, T. J.; Horne, M. T.; Ellis, A. E. (1999). Peroxide-inducible catalase in *Aeromonas salmonicida* subsp. *salmonicida* protects against exogenous hydrogen peroxide and killing by activated rainbow trous, *Oncorhynchus mykiss* L., macrophages. *Microbial Pathogenesis,* 26, 149 – 158.

[3] Boettcher, A. A.; Dyer, C.; Casey, J. and Targett, N. M. (1997). Hydrogen peroxide induced metamorphosis of queen conch, *Strombus gigas*: Tests at the commercial scale. *Aquaculture*, 148, 247 – 258.

[4] Bowers, J. M.; Speare, D. J. and Burka, J. F. (2002). The effects of hydrogen peroxide on the stress response of Atlantic Salmon (*Salmo salar*). *Journal of Veterinary Pharmacology and Theraphy*, 25, 311 – 313.

[5] Bravo, S.; Treasurer, J.; Sepulveda, M. and Lagos, C. (2010). Effectiveness of hydrogen peroxide in the control of *Caligus rogercresseyi* in Chile and implications for sea louse management. *Aquaculture,* 303, 22 – 27.

[6] Bruno, D. W. and Raynard, R. S. (1994). Studies on the use of hydrogen peroxide as a method for the control of sea lice on Atlantic salmon. *Aquaculture International*, 2, 10–18.

[7] Burridge, L.; Weis, J. S.; Cabello, F.; Pizarro, J. and Bostick, K. (2010). Chemical use in salmon aquaculture: a review of current practices and possible environmental effects. *Aquaculture,* 306, 7 – 23.

[8] Celada, J. D.; Carral, J. M.; Saez-Royuela, M.; Melendre, P. M. and Aguilera, A. (2004). Effects of different antifungal treatments on artificial incubation of the astracid crayfish (*Pacifastacus leniusculus* Dana) eggs. *Aquaculture,* 239, 249 – 259.

[9] Chantanachookhin, C.; Seikai, T. and Tanaka, M. (1991). Comparative study of the ontogeny of the lymphoid organs in three species of marine fish. *Aquaculture*, 99, 143–155.

[10] Conte, F. S. (2004). Stress and the welfare of cultured fish. *Applied Animal Behaviour Science,* 86, 205 – 223.

[11] Directive 2000/60/EC of the European Parliament and of the Council of 23 October 2000 establishing a framework for Community action in the field of water policy.

[12] Directive 2006/44/EC of the European Parliament and of the Council of 6 September 2006 on the quality of fresh waters needing protection or improvement in order to support fish life.

[13] Douillet, P. A. and Holt, G. J. (1994). Surface disinfection of red drum (*Sciaenops ocellatus* Linnaeus) eggs leading to bacteria-free larvae. *Journal of Experimental Marine Biology and Ecology*, 179, 253 – 266.

[14] Douillet, P. and Pickering, P. L. (1999). Seawater treatment for larval culture of the fish *Sciaenops ocellatus* Linnaeus (red drum). *Aquaculture*, 170, 113 – 126.

[15] Drabkova, M.; Admiraal, W.; Marsalek, B. (2007). Combined exposure to hydrogen peroxide and light-selective effects on cyanobacteria, green algae, and diatoms. *Environmental Science and Technology*, 41, 309 – 314.

[16] Duarte, C.M., Marbá, N. and Holmer, M. (2007). Rapid domestication of marine species. *Science*, 316: 382 – 383.

[17] EFSA (2007a). Scientific Opinion of the Panel on Animal Health and Welfare on a request from the European Commission on possible vector species and live stages of susceptible species not transmitting disease as regards certain fish diseases. *The EFSA Journal*, 584, 1 – 163.

[18] EFSA (2007b). Scientific Opinion of the Panel on Animal Health and Welfare on a request from the European Commission on possible vector species and live stages of susceptible species not transmitting disease as regards certain mollusc diseases. *The EFSA Journal*, 597, 1 – 116.

[19] EFSA (2007c). Scientific Opinion of the Panel on Animal Health and Welfare on a request from the European Commission on possible vector species and live stages of susceptible species not transmitting disease as regards certain crustacean species. *The EFSA Journal,* 598, 1 – 91.

[20] EFSA (2008a). Scientific Opinion of the Panel on Biological Hazards on a request from European Commission on Food Safety considerations of animal welfare aspects and husbandry systems for farmed fish. *The EFSA Journal,* 867, 1 – 24.

[21] EFSA (2009a). Statement of EFSA prepared by the AHAW Panel on: knowledge gaps and research needs for the welfare of farmed fish. *The EFSA Journal,* 1145, 1 – 7.

[22] EFSA (2009b). Scientific Opinion of the Panel on Animal Health and Welfare on a request from European Commission on General approach to fish welfare and to the concept of sentience in fish. *The EFSA Journal*, 954, 1 – 27.

[23] FAO (2006). The state of world fisheries and aquaculture 2006. FAO Fisheries and Aquaculture Department, Rome. 162 pages.

[24] FAO (2010). The state of world fisheries and aquaculture 2010. FAO Fisheries and Aquaculture Department, Rome. 218 pages.

[25] Fernandez-Tejedor, M.; Soubrier-Pedreño, M. A. and Furones, D. (2007). Mitigation of lethal effects of *Karlodinium veneficum* and *K. armiger* on *Sparus aurata*: changes in haematocrit and plasma osmolarity. *Diseases of Aquatic Organisms,* 77, 53 – 59.

[26] Gaikowski, M. P.; Rach, J. J. and Ramsay, R. T. (1999). Acute toxicity of hydrogen peroxide treatments to selected lifestages of cold-, cool-, and warmwater fish. *Aquaculture*, 178, 191 – 207.

[27] Giménez, G.; Padrós, F.; Roque, A.; Estévez, A. and Furones, D. (2006). Bacterial load reduction of live prey for fish larval feeding using Ox-Aquaculture®. *Aquaculture Research*, 37, 1130 – 1139.

[28] Giménez, G.; Padrós, F.; Roque, A.; Estévez, A. and Furones, D. (2009). Effects of a peroxide-based commercial product on bacterial load of larval rearing water and on larval survival of two species of Sparidae under intensive culture: preliminary study. *Aquaculture Research*, 40, 504 – 508.

[29] Innes Taylor, N. and Ross, L. G. (1988). The use of hydrogen peroxide as a source of oxygen for the transport of live fish. *Aquaculture*, 70, 183 – 192.

[30] Johnson, S. C.; Constible, J. M. and Richard, J. (1993). Laboratory investigations on the efficacy of hydrogen peroxide against the salmon louse *Lepeophtheirus salmonis* and its toxicological and histopathological effects on Atlantic salmon *Salmo salar* and chinook salmon *Oncorhynchus tshawytscha*. *Diseases of Aquatic Organisms*, 17, 197 – 204.

[31] Kiemer, M. C. B. and Black, K. D. (1997). The effects of hydrogen peroxide on the gill tissues of Atlantic salmon, *Salmo salar* L. *Aquaculture,* 153, 181 – 189.

[32] Kitancharoen, N.; Yamamoto, A. and Hatai, K. (1997). Fungicidal effect of hydrogen peroxide on fungal infection of rainbow trout. *Mycoscience,* 38, 375 – 378.

[33] Lumsden, J. S.; Ostland, V. E. and Ferguson, H. W. (1998). Use of hydrogen peroxide to treat experimentally induced bacterial gill disease in rainbow trout. *Journal of Aquatic Animal Health*, 10, 230 – 240.

[34] Madsen, H. C. K.; Buchmann, K. and Mellergaard, S. (2000). Treatment of trichodiniasis in eel (*Anguilla anguilla*) reared in recirculation systems in Denmark: alternatives to formaldehyde. *Aquaculture*, 186, 221 – 231.

[35] Magnadottir, B.; Lange, S.; Gudmundsdottir, S.; Bogwald, J. and Dalmo, R. A. (2005) Ontogeny of humoral immune parameters in fish. *Fish and Shellfish Immunology,* 19, 429 – 439.

[36] Manning, M. J. and Tattner, M. F. (1985). Fish Immunology. Academic Press Inc., London, 374 pages.

[37] Mansell, B; Powell, M. D.; Ernst, I. and Nowak, B. F. (2005). Effects of the gill monogenean *Zeuxapta seriolae* (Meserve, 1938) and treatment with hydrogen peroxide on pathophysiology of kingfish, *Seriola lalandi* Valenciennes, 1833. *Journal of Fish Diseases*, 28, 253 – 262.

[38] McAndrew, K. J.; Sommerville, C.; Wootten, R. and Bron, J. E. (1998). The effects of hydrogen peroxide treatment on different life-cycle stages of the salmon louse, *Lepeophtherius salmonis* (Kroyer, 1837). *Journal of Fish Diseases*, 21, 221 – 228.

[39] Melendre, P. M.; Celada, J. D.; Carral, J. M.; Saez-Royuela, M. and Aguilera, A. (2006). Effectiveness of antifungal treatments during artificial incubation of the signal crayfish eggs (*Pacifastacus leniusculus* Dana. Astracidae). *Aquaculture*, 257, 257 – 265.
[40] Mischke, C. C.; Terhune, J. S. and Wise, D. J. (2001). Acute toxicity of several chemicals to the oligochaete *Dero digitata*. *Journal of the World Aquaculture Society*, 32, 184 – 188.
[41] Moller, M. S.; Arvin, E. and Pedersen, L. F. (2010). Degradation and effect of hydrogen peroxide in small-scale recirculation aquaculture system biofilters. *Aquaculture Research,* 41, 1113 – 1122.
[42] Montgomery-Brock, D.; Sato, V. T.; Brock, J. A.and Tamaru, C. S. (2001). The application of hydrogen peroxide as a treatment for the ectoparasite *Amyloodinium ocellatum* (Brown 1931) on the Pacific threadfin *Polydactylus sexfilis*. *Journal of the World Aquaculture Society*, 32, 250 – 254.
[43] Moreno-Garrido, I. and Cañavate, J. P. (2001). Assessing chemical compounds for controlling predator ciliates in outdoor mass cultures of the green algae *Dunaliella salina*. *Aquacultural Engineering*, 24, 107 – 114.
[44] Morse, D. E.; Duncan, H.; Hooker, N. and Morse, A. (1977). Hydrogen peroxide induces spawning in mollusks, with activation of prostaglandin endoperoxide synthetase. *Science*, 196, 298 – 300.
[45] Morse, D. E. (1984). Biochemical and genetic engineering for improved production of abalones and other valuable molluscs. *Aquaculture*, 39, 263 – 282.
[46] Morse, A. N. C. and Morse, D. E. (1984). Recruitment and metamorphosis of *Haliotis* larvae induced by molecules uniquely available at the surfaces of crustose red algae. *Journal of Experimental Marine Biology and Ecology*, 75, 191 – 215.
[47] Olafsen, J. A. (2001). Interactions between fish larvae and bacteria in marine aquaculture. *Aquaculture,* 200, 223 – 247.
[48] Patent WO2007138058. Product and process for treating water bodies, sediments and soils.
[49] Patent TW200520683. Water quality controlling method of aquiculture by application of sodium percabonate.
[50] Patent AU2005100236. Water treatment process.
[51] Patent CN1513778. Method of eliminating nitrite and ammonia nitrogen in aquatic culture.
[52] Patent CN1035413. Increasing oxygen for fish and shrimpby hydrogen peroxide amide carbonate.
[53] Patent SU1472011. Method of producing nauplii from eggs of Artemia salina crustagen.
[54] Peck, M. A.; Buckley, L. J.; O'Bryan, L. M. and Lapolla, A. E. (2004). Efficacy of egg surface disinfectants in captive spawning Atlantic cod *Gadus morhua* L. and haddock *Melanogrammus aeglefinus* L. *Aquaculture Research*, 35, 992 – 996.
[55] Pedersen, L. F.; Pedersen, P. B. and Sortkjaer, O. (2006). Dose-dependent decomposition rate constants of hydrogen peroxide in small-scale bio filters. *Aquacultural Engineering*, 34, 8 – 15.
[56] Pedersen, L. F. and Pedersen, P. B. (2012). Hydrogen peroxide application to a commercial recirculating aquaculture system. *Aquacultural Engineering,* 46, 40 – 46.

[57] Peña, J. B. (1983). Algunos ensayos sobre la puesta inducida en *Haliotis discus* Reeve. Informes Tecnicos del Instituto de Investigaciones Pesqueras, 48, 441 – 454.

[58] Peña, J. B. (1986). Preliminary study on the induction of artificial spawning in *Haliotis coccinea canariensis* Nordsieck (1975). *Aquaculture,* 52, 35 – 41.

[59] Powell, M. D. and Clark, G. A. (2004). Efficacy and toxicity of oxidative disinfectants for the removal of gill amoebae from the gills of amoebic gill disease affected Atlantic salmon (*Salmo salar* L.) in freshwater. *Aquaculture Research,* 35, 112 – 123.

[60] Powell, M. D. and Perry, S. F. (1997). Respiratory and acid-base pathophysiology of hydrogen peroxide in rainbow trout (*Oncorhynchus mykiss* Walbaum). *Aquatic Toxicology*, 37, 99 – 112.

[61] Rach, J. J.; Gaikowski, M. P.; Howe, G. E. and Schreier, T. M. (1998). Evaluation of the toxicity and efficacy of hydrogen peroxide treatments on eggs of warm- and coolwater fishes. *Aquaculture*, 165, 11 – 25.

[62] Rach, J. J.; Gaikowski, M. P. and Ramsay, R. T. (2000). Efficacy of hydrogen peroxide to control infestations of hatchery-reared fish. *Journal of Aquatic Animal Health,* 12, 267 – 273.

[63] Rach, J. J.; Valentine, J. J.; Schreier, T. M.; Gaikowski, M. P. and Crawford, T. G. (2004). Efficacy of hydrogen peroxide to control saprolegniasis on channel catfish (*Ictalurus puntatus*) eggs. *Aquaculture,* 238, 135 – 142.

[64] Rae, G. H. (2002). Sea louse control in Scotland, past and present. *Pest Management Science,* 58, 515 – 520.

[65] Rasowo, J.; Okoth, O. E. and Ngugi, C. C. (2007). Effects of formaldehyde, sodium chloride, potassium permanganate and hydrogen peroxide on hatch rate of African catfish *Clarias gariepinus* eggs. *Aquaculture,* 269, 271 – 277.

[66] Rintamaki-Kinnunen, P.; Rahkonen, M.; Mannermaa-Keranen, A. L.; Suomalainen, L. R.; Mykra, H. and Valtonen, E. T. (2005). Treatment of ichthyophthiriasis after malachite green. I. Concrete tanks at salmonid farms. *Diseases of Aquatic Organisms,* 64, 69 – 76.

[67] Robbins, H. M.; Van Stappen, G.; Sorgeloos, P.; Sung, Y. Y.; MacRae, T. H.; Bossier, P. (2010). Diapause termination and development of encysted *Artemia* embryos: roles of nitric oxide and hydrogen peroxide. *The Journal of Experimental Biology,* 213, 1464 – 1470.

[68] Robertson, P. K. J.; Black, K. D.; Adams, M.; Willis, K.; Buchan, F.; Orr, H.; Lawton, L. and McCullagh, C. (2009). A new generation of biocides for control of crustacea in fish farms. *Journal of Photochemistry and Photobiology B: Biology*, 95, 58 – 63.

[69] Roque, A.; Lacuesta, B.; Caballero, A.; Duncan, N. and Estevez, A. (2011). Uso del Ox-Aquaculture© en el control de una infección experimental con *Photobacterium damselae* en el lenguado senegalensis *Solea senegalensis.* XIII Congreso Nacional de Acuicultura, 21-24 November 2011. Barcelona, Spain. Abstract in English.

[70] Roque, A.; Yildiz, H. Y.; Carazo, I. and Duncan, N. (2010). Physiological stress responses of sea bass (*Dicentrarchus labrax*) to hydrogen peroxide (H_2O_2) exposure. *Aquaculture,* 304, 104 – 107.

[71] Saez, J. A. and Bowser, P. R. (2001). Hydrogen peroxide concentrations in hatchery culture units and effluent during and after treatment. *North American Journal of Aquaculture,* 63, 74 – 78.

[72] Sanabria, C.; Diamant, A. and Zilberg, D. (2009). Effects of commonly used disinfectants and temperature on swim bladder non-inflation in freshwater angelfish, *Pterophyllum scalare* (Lichtestein). *Aquaculture,* 292, 158 – 165.

[73] Saygi, Y. B. (2003). Effects of hydrogen peroxide, cold storage and decapsulation on the hatching success of *Artemia* cysts. The Israeli *Journal of Aquaculture-Bamidgeh,* 55, 107 – 113.

[74] Schreier, T. M.; Rach, J. J. and Howe, G. E. (1996). Efficacy of formalin, hydrogen peroxide, and sodium chloride on fungal-infected rainbow trout eggs. *Aquaculture,* 140, 323 – 331.

[75] Schmidt, L. J.; Gaikowski, M. P. and Gingerich, W. H. (2006). Environmental assessment for the use of hydrogen peroxide in aquaculture for treating external fungal and bacterial diseases of cultured fish and fish eggs. USGS report, 180 pp.

[76] Sitja-Bobadilla, A.; Conde de Felipe, M. and Alvarez-Pellitero, P. (2006). *In vivo* and *in vitro* treatments against *Sparicotyle chrysophrii* (Monogenea: Microcotylidae) parasitizing the gills of gilthead sea bream (*Sparus aurata* L.). *Aquaculture*, 261, 856 – 864.

[77] Skjermo, J. and Vadstein, O. (1993). Characterization of the bacterial flora of mass cultivated *Brachionus plicatilis*. *Hydrobiologia*, 255/256, 185 – 191.

[78] Taylor, B. E. (1997). Protein utilization, hormone treatment and nutrient metabolism as they apply to culture of abalone *Haliotis kamtschatkana*. PhD Thesis, Department of Zoology, University of British Columbia, Vancouver, Canada. 132 pp.

[79] Thomas-Jinu, S. and Goodwin, A. E. (2004). Acute columnaris infection in channel catfish, *Ictalurus punctatus* (Rafinesque): efficacy of practical treatments for warmwater aquaculture ponds. *Journal of Fish Diseases*, 27, 23 – 28.

[80] Tort, M. J.; Wooster, G. A. and Bowser, P. R. (2003). Effects of hydrogen peroxide on hematology and blood chemistry parameters of walleye *Stizostedion vitreum*. *Journal of the World Aquaculture Society,* 34, 236 – 242.

[81] Treasurer, J. W. and Grant, A. (1997). The efficacy of hydrogen peroxide for the treatment of farmed Atlantic salmon, *Salmo salar* L. infected with sea lice (Copepoda: Caligidae). *Aquaculture*, 148, 265 – 275.

[82] Treasurer, J. W.; Wadsworth, S. and Grant, A. (2000). Resistance of sea lice, *Lepeophtherius salmonis* (Kroyer), to hydrogen peroxide on farmed Atlantic salmon, *Salmo salar* L. *Aquaculture Research*, 31, 855 – 860.

[83] Tripi, C. and Bowser, P. R. (2001). Toxicity of hydrogen peroxide to pre-exposed young-of-the-year walleye: effects of water hardness and age of fish. *Journal of the World Aquaculture Society,* 32, 416 – 421.

[84] U. S. Federal Water Pollution Control Act. (2002). 33 U. S. C. 1251 et seq.

[85] Van Stappen, G. (1996). Artemia: Use of cysts. In: Manual on the Production and Use of Live Food for Aquaculture. FAO Fisheries Technical Paper No. 361, Food and Agriculture Organization (ed. P. Lavens and P. Sorgeloos), pp. 107 – 136, Rome, Italy.

[86] Van Stappen, G.; Lavens, P. and Sorgeloos, P. (1998). Effects of hydrogen peroxide treatment in Artemia cysts of different geographical origin. *Achives of Hydrobiology Special Issues Advanced Limnology*, 52, 281 – 296.

[87] Verner-Jeffreys, D. W.; Nakamura, I. and Shields, R. J. (2007). Surface disinfection of Pacific threadfin, *Polydactylus sexfilis*, and amberjack, *Seriola rivoliana*, eggs. *Aquaculture Research,* 38, 605 – 612.

[88] Wagner, E. J.; Arndt, R. E.; billman, E. J.; Forest, A. and Cavender, W. (2008). Comparison of the efficacy of iodine, formalin, salt, and hydrogen peroxide for control of external bacteria on rainbow trout eggs. *North American Journal of Aquaculture*, 70, 118 – 127.

[89] Willis, K. J.; Gillibrand, P. A.; Cromey, C. J. and Black, K. D. (2005). Sea lice treatments on salmon farms have no adverse effects on zooplankton communities: A case study. *Marine Pollution Bulletin*, 50, 806 – 816.

In: Hydrogen Peroxide
Editors: Gilberto Aguilar and Raphael A. Guzman
ISBN: 978-1-62257-414-8

Chapter 6

FUNCTIONS OF HYDROGEN PEROXIDE IN REGULATING FUNGAL VIABILITY AND PATHOGENICITY

Shiping Tian, Guozheng Qin, Boqiang Li and Zhang Zhanquan
Key Laboratory of Plant Resources, Institute of Botany,
Chinese Academy of Sciences, Beijing, China

ABSTRACT

As known as well, the active oxygen species hydrogen peroxide (H_2O_2) has been proven to play critical roles in various cellular activities, including the direct cause of membrane damage, the component of structural defense, the signal molecule operating within the challenged cell and the antimicrobial compound because of H_2O_2 acting as a signal molecule in the induction of systemic acquired resistance.

In addition, H_2O_2 is relatively stable, can diffuse freely throughout the cell, and has a similarly low reactivity toward biological molecules. In this review, we mainly summarize the functions of hydrogen peroxide in regulating the viability of antagonistic yeasts and the pathogenicity of fungal pathogens, as well as the molecular targets of hydrogen peroxide in fungal cells.

1. INTRODUCTION

Hydrogen peroxide (H_2O_2), as a stable form of reactive oxygen species (ROS), is indispensable for eukaryotic and prokaryotic cell growth, development and resistance responses (Foreman et al. 2003), and plays an important role during many life progresses, particularly in the interaction of microbe-plant (Møller et al., 2007). H_2O_2 produced in plant tissue can directly restraint pathogen infection, and reinforce the plant cell walls, such as lignification and cross-linking of cell wall structural proteins (Apel and Hirt, 2004; Quan et al., 2008), as well as induce the programmed cell death (PCD) of plant to limit the spread of disease from the infection point (Gechev and Hille, 2005). Generation of ROS is a characteristic feature of defense reactions of eukaryotic cells against pathogen attack (Mellersh et al., 2002). When fungal pathogens encroach plant, they often encounter the attack of ROS, primarily superoxide (O_2^-) and H_2O_2, generated by different host enzyme systems of a plant at the site of infection (Tanaka et al., 2004). Oxidative burst is one of the earliest host responses after pathogen attack (Mellersh et al., 2002). Moreover, pathogens often develop several defense responses, including the inducement of the antioxidant enzymes and the nonenzymatic protective molecules to protect their cells from ROS damage (Macarisin et al., 2007). Therefore, ROS have been considered to be cytotoxic and mutagenic and for their proper function in signaling their production must be carefully controlled.

H_2O_2 are continuously produced as byproducts of various metabolic pathways mainly localized in mitochondria, peroxisomes, chloroplasts or the membrane-bound Nicotinamide adenine dinucleotide (NADPH) oxidases complex (Figure 1). In mammalian cells, mitochondria are the major source of ROS. The relative contribution of mitochondria to ROS production in plant is low (Apel and Hirt, 2004). In plant mitochondria superoxide anion radical production occurs mainly by the electron transport chain. This process results in the formation of H_2O_2, primarily through the action of a mitochondrion-specific manganese SOD (Ślesak *et al.* 2007). Chloroplasts are a major source for H_2O_2 production in plant (Quan *et al.* 2008). The photosynthetic electron transport (PET) chain in the chloroplast is responsible for H_2O_2 production. The electron transport components are auto-oxidizable and under conditions limiting the availability of NADP, superoxide anion radical can be formed. The superoxide anion is catalyzed by SOD to produce H_2O_2 (Ślesak *et al.* 2007).

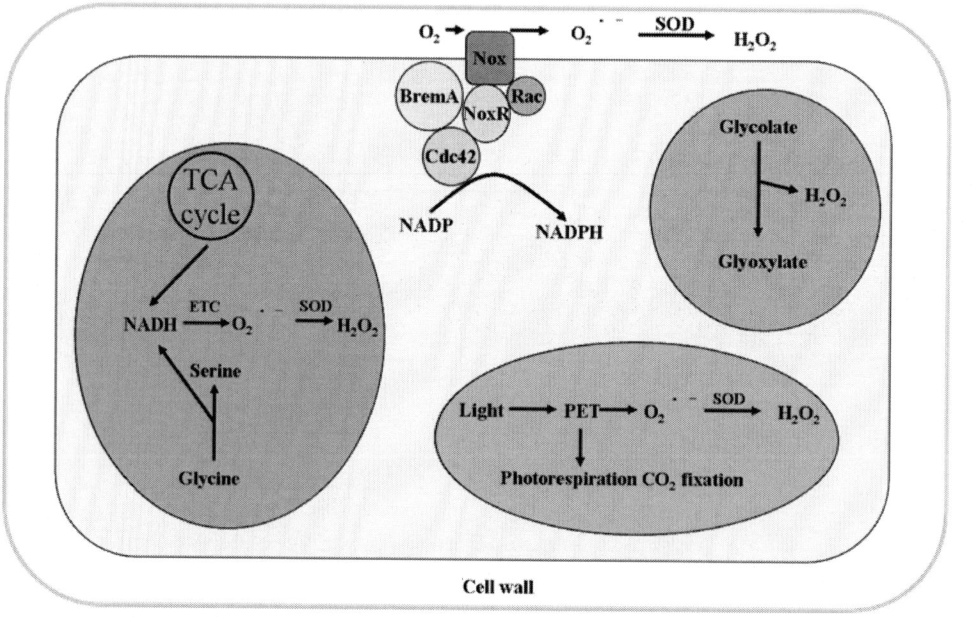

Figure 1. Source of H_2O_2 in eukaryotic cell. ETC, electron transport chain; PET, photosyntheticelectron transport chain; TCA, tricarboxylic acid cycle.

The main function of peroxisomes in the plant cell is photorespiration, which is light-dependent uptake of O_2 and the associated release of CO_2 connected with the generation of H_2O_2. H_2O_2 is generated in the peroxisome mainly by glycolate oxidation (Dat *et al.* 2000).

In eukaryotic cell, organelles-mitochondria, peroxisomes, and chloroplasts- contain multiple enzymes or enzyme systems for removing H_2O_2. The balance between H_2O_2 production and the activities of these H_2O_2-removing systems determines concentration of H_2O_2 present and thus to what extent signaling and/or damage will occur (Møller *et al.* 2007).

NADPH oxidases are enzymes inserted in the plasma membrane, that produce ROS using electrons derived from intracellular NADPH and present in a large array of mainly multicellular eukaryotes (Lalucque and Silar, 2003; Sumimoto, 2008). According to current knowledge, the membrane-bound NADPH oxidases complex is the major source in microbial eukaryotes (Figure 1). Regulated synthesis of ROS by membrane bound fungal NADPH oxidases also plays a key role in fungal morphogenesis, growth, and development (Mellersh et al., 2002). The key enzyme in defense reactions against pathogens that triggers the oxidative burst (Foreman et al., 2003). Synthesis of ROS by specific NADPH oxidases has been widely proved to be involved in both defence and cellular differentiation processes in fungi (Aguirre et al., 2005, Takemoto et al., 2011). Filamentous fungi generate ROS during conidial germination, appressorium development, and hyphal tip growth (Shinogi et al., 2003). H_2O_2 can modulate the activities of many components in signaling, such as protein phosphatases, protein kinases and transcription factors (Cheng and Song, 2006).

2. Effect of Hydrogen Peroxide on Viability and Pathogenicity of Fungal Pathogens and the Mechanisms Involved

How the host cells respond to the invasion of fungi in a self-protective process is an important biological problem attracting a wide range of interests. The induced production of toxic molecules within the host upon fungal attack, such as H_2O_2, represents a challenge to an invading fungus (Soundararajan et al., 2004). The localized generation of H_2O_2 and its antimicrobial effect has been reported in a number of host-fungal interactions, but the details about the regulatory mechanisms of H_2O_2 production, the targets of H_2O_2, and the signal transduction pathway of H_2O_2 in fungal pathogens remains largely unknown.

2.1. Influence of Hydrogen Peroxide on Viability of Fungal Pathogens

Throughout the life cycle, plants and animals have to react to various threats coming from the outside environment. Pathogen infection represents a major problem. Upon recognition of a microbial pathogen such as fungal pathogen, a plant or an animal host starts to release high levels of reactive oxygen species (ROS), mostly superoxide anion and H_2O_2 (Bolwell et al., 1995; Iles and Forman, 2002). This rapid production of ROS, called the oxidative burst (also known as the respiratory burst in phagocytes of mammals), would provide an extremely hostile environment for pathogens (Hamann et al., 2008). These ROS are produced by different host enzyme systems, among which NADPH oxidases represent the most important component. NADPH oxidases, localized in the host plasma membrane, catalyze the conversion of O_2 to superoxide anions, which are converted to H_2O_2 either by spontaneous dismutation or by the catalytic activity of a cell wall superoxide dismutase

(Babior, 2004). The resulting H_2O_2 can not only destroy the invading pathogens directly (Mellersh et al., 2002; Phillips et al., 2003) but also function as signaling molecules in the regulation of host defense response, eventually restrict the spread of pathogens from the infection sites (Torres et al., 2005).

It has been shown that H_2O_2, produced by plant cell interaction with a fungal pathogen, inhibited fungal spore germination and germ tube growth (Peng and Kuc, 1992; Lu and Higgins, 1999). Mellersh et al. (2002) reported that H_2O_2 generation in plant in response to attack of fungal pathogens was necessary and sufficient to account for fungal penetration failure. Phillips et al. (2003) showed that low dose of exogenous H_2O_2 could induce programmed cell death (PCD) in the pathogenic fungi, *Candida albicans*. Many key markers of PCD including chromatin margination and condensation, nuclear envelope separation, and nuclear fragmentation were observed in *C. albicans*, accompanied by the elevated levels of intracellular ROS.

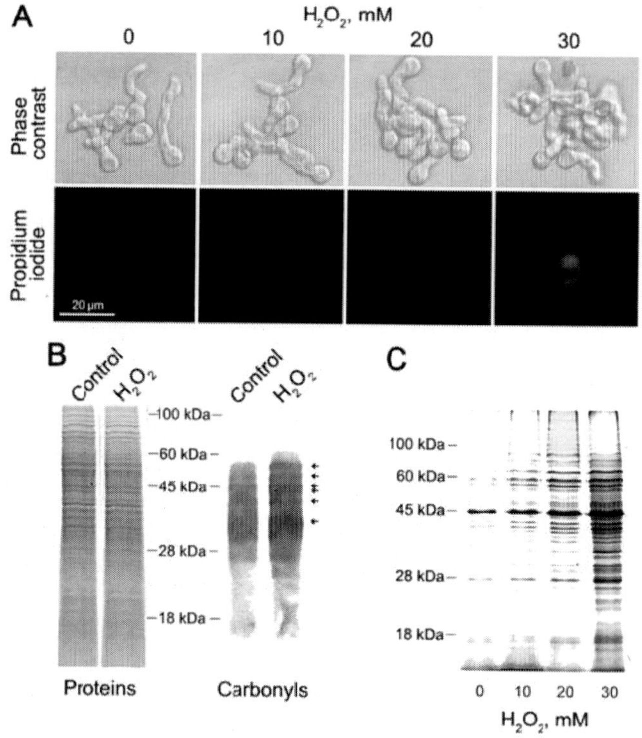

Figure 2. *Penicillium expansum* exhibited a limited loss of plasma membrane integrity, but displayed protein oxidative damage under H_2O_2 stress. (A) Propidium iodide (PI) staining indicates that loss of plasma membrane integrity is not the main reason for H_2O_2-induced fungal death. Before fluorescent staining, the fungal spores were cultured in potato dextrose broth medium until germination, and then treated with indicated concentrations of H_2O_2 for 60 min at 25°C. Scale bar, 20 µm. (B) H_2O_2 at 30 mM increases oxidative damage to total cellular proteins as indicated by immunodetection of protein carbonyl groups. The oxidatively modified proteins were separated by SDS-PAGE, blotted and detected using anti-dinitrophenyl-group antibodies (right panel). The arrowheads indicate protein bands that are highly carbonylated in spores under H_2O_2 stress. Coomassie Brilliant Blue staining was used to monitor equal loading of samples (left panel). (C) Detection of protein aggregation in *P. expansum* after H_2O_2 exposure.

In our previous study, we have shown that intracellular accumulation of ROS in *Penicillium expansum*, a widespread fungal pathogen that causes decay in a variety of fruits, was involved in cell damage and destruction (Qin et al., 2007). In recent, we investigated the effects of nitric oxide (NO) on spore germination of *P. expansum* and the possible mechanism, and found an increased level of intracellualr ROS and enhanced carbonylation damage in spores of *P. expansum* under NO stress with the use of an oxidant sensitive probe and western blot analysis (Lai et al., 2011). Exogenous superoxide dismutase (SOD) and ascorbic acid (Vc) could increase the resistance of the spore to the inhibitory effect of NO. These results indicate that NO can induce the generation of ROS which subsequently causes severe oxidative damage to proteins crucial to the process of spore germination of *P. expansum*. Additionally, we have found that the mechanisms, by which potassium tetraborate inhibits germination and germ tube elongation of *Colletotrichum gloeosporioides* conidia, are greatly associated with delaying nuclear division, impairing endocytosis and mitochondria of the pathogen (Shi et al., 2011). Further study was conducted to investigate the molecular changes that lead to death of *P. expansum* in response to H_2O_2 stress (Qin et al., 2011). We found that the loss of plasma membrane integrity was not evident in *P. expansum* upon exposure to half lethal dose of H_2O_2, indicating that membrane damage was not the main reason for H_2O_2-induced fungal death. By analyzing the level of protein oxidation and aggregation in whole-cell lysates, we demonstrated that protein oxidative modification might be associated with H_2O_2-induced fungal death (Figure 2) (Qin et al., 2011).

2.2. Target of Hydrogen Peroxide

The mechanisms whereby H_2O_2 cause death in pathogenic fungi are poorly understood. Previous studies suggested that H_2O_2 has the ability to translocate across the cytoplasmic membrane of the fungi and react with the cellular components. However, the specific intracellular component(s) targeted by H_2O_2 remain to be determined. ROS may cause oxidative damage to biological macromolecules including proteins, lipids, and nucleic acids (Yan et al., 2005). Analysis of the antifungal peptide histatin 5 against *C. albicans* indicated that ROS induced by antifungal peptide might lead to respiration inhibition and loss of mitochondrial membrane potential, thus caused killing of *C. albicans* (Helmerhorst et al., 1999). These findings have pointed to a direct involvement of mitochondria in the process of cell death in *C. albicans* (Helmerhorst et al., 2001). However, although mitochondria have been shown to be sensitive to ROS, the specific site of inhibition of ROS in mitochondria is still unclear.

Mitochondria are vital organelles within the eukaryotic cell that perform a variety of biochemical functions, such as energy supply, signaling, cellular differentiation, cell death, as well as the control of the cell cycle and cell growth (Qin et al., 2009; McBride et al., 2006). In recent years, mitochondria have been shown to be important for the infection process of fungal pathogens. Mitochondria were necessary for *Cryptococcus neoformans* to survive in low oxygen conditions in the host during infection (Ingavale et al., 2008). Mitochondrial Mrb1 protein and Fow1 protein were demonstrated to be specifically required for pathogenicity of *Ustilago maydis* and *Fusarium oxysporum*, respectively (Bortfeld et al., 2004; Inoue et al., 2002).

In our previous study, we found that H_2O_2 induced oxidative stress in the fungal pathogen *P. expansum* and mitochondria might be sensitive to H_2O_2 (Qin et al., 2011). Mitochondrial components such as proteins are particularly vulnerable to oxidative damage (Das et al., 2001). In order to identify the targets of H_2O_2 proteins in mitochondria, we developed a procedure for high-purity mitochondrial separations from *P. expansum* and examined the variations of the mitochondrial sub-proteome under oxidative stress of H_2O_2 (Figure 3). A set of mitochondrial proteins including respiratory enzyme complexes, tricarboxylic acid cycle enzymes, and membrane carriers were identified which showed differential expression in response to H_2O_2 (Qin et al., 2011). Particular attention was paid to the respiratory complex III, which was shown to contribute to ROS production in animals and yeasts (Guzy et al., 2005; Guzy et al., 2007). By using the inhibitor of respiratory chain complex III, we proved that mitochondrial complex III is a major source of ROS production upon exposure to H_2O_2. These results suggest that complex III serves as the target of H_2O_2 in *P. expansum* (Qin et al., 2011). Besides mitochondrial complex III, we observed reduced levels and degradation of several subunits of mitochondrial ATP synthase complex, the last enzyme complex of the respiratory enzyme complexes located in the mitochondrial inner membrane (Yu et al., 2003).

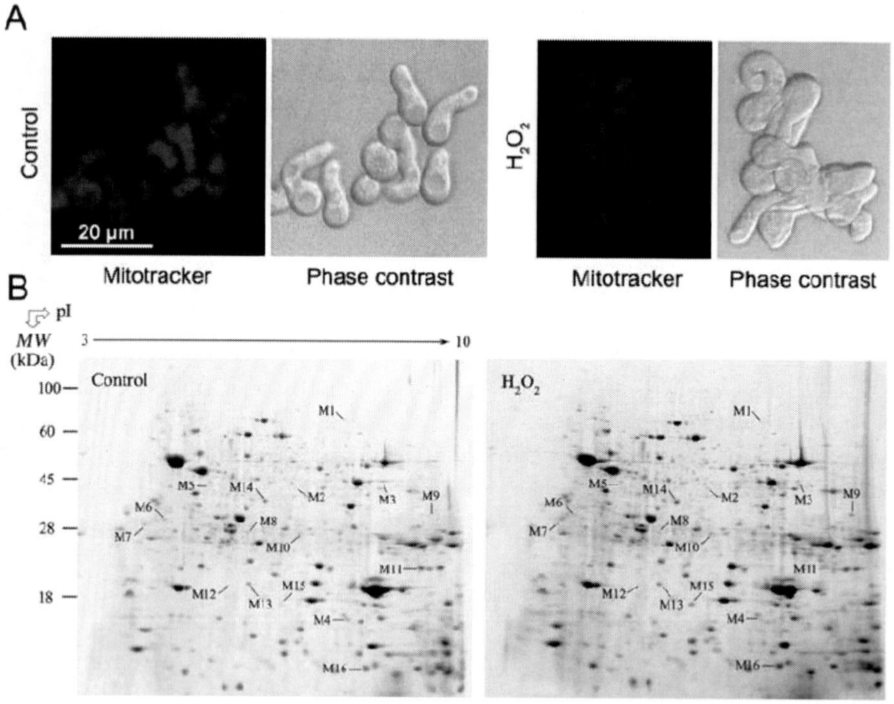

Figure 3. Two-dimensional protein pattern revealed the changes of mitochondrial proteins in response to H_2O_2. (A) H_2O_2 induced collapse of mitochondrial membrane potential in the fungal pathogen as indicated by Mitotracker orange staining. The scale bar represents 20 μm. (B) The fungal spores were treated with 0 mM (control) or 30 mM H_2O_2 for 60 min at 25°C after the spores were germinated in potato dextrose broth medium. Mitochondria were purified from the fungi through a two-step Percoll density gradient centrifugation. Mitochondrial proteins (500 μg) were separated by two-dimensional gel electrophoresis using 13 cm Immobiline Drystrip with a pH 3–10 nonlinear gradient. After electrophoresis, proteins were visualized by Coomassie Blue staining. Numbers indicate proteins that were differentially expressed under H_2O_2 stress and subsequently identified by mass spectrometry.

The alterations in ATP synthase are likely to affect oxidative phosphorylation and limit the synthesis of ATP. Our data showed that ATP levels declined quickly in the response of *P. expansum* to H_2O_2 stress. Reduction of intracellular ATP levels by the inhibitor of ATP synthase decreased viability of the fungus under oxidative stress, suggesting that ATP synthase may serve as the target of H_2O_2 and contributes to death of *P. expansum* (Qin et al., 2011).

2.3. Affect of Hydrogen Peroxide on Pathogenicity of Fungal Pathogens

The damaging effect of ROS on cell components has been well established (Soundararajan et al., 2004). However, increasing evidences suggest that ROS produced in fungal pathogen by specific enzymes, such as NADPH oxidases (NOX), regulate different cellular functions, including cell proliferation, sexual reproduction, and formation of penetration structures (Aguirre et al., 2005). NADPH oxidase is a conjugated enzyme complex involving membrane-bound and cytosolic compounds that catalyzes the formation of O_2^-. The produced superoxide anions could be converted to H_2O_2 either by O_2^- or by the catalytic activity of a cell wall superoxide dismutase (Babior, 2004). Filamentous fungi possess three distinct subfamilies of Nox enzymes, among which NoxA and NoxB are homologs of the mammalian catalytic subunit gp91phox. The third one (NoxC) is the homolog of human Nox5 and the plant Nox homologs, containing putative calcium-binding EF-hand motifs. In addition, a homolog of the regulatory subunit p67phox (NoxR) and the small GTPase Rac are found in fungi (Takemoto et al., 2006). Other components of the fungal Nox complex remain to be determined.

Analyses of NADPH oxidases function in *Aspergillus nidulans* (Lara-Ortiz et al., 2003) and *Podospora anserina* (Malagnac et al., 2004) showed that these enzymes are involved in sexual differentiation. In the phytopathogenic fungus *Claviceps purpurea*, a biotrophic pathogen, the NADPH oxidase Cpnox1 is required for normal pathogenic development. However, deletion of another NADPH oxidase Cpnox2 had no influence on pathogenicity (Giesbert et al., 2008). By comparison, in the rice blast fungus *Magnaporthe grisea*, both Nox1 and Nox2 are required independently for pathogenicity (Egan et al., 2007). Mutants of either *Nox1* or *Nox2* in *M. grisea* are impaired in appressorium-mediated cuticle penetration, and lead to the loss of pathogenicity. Assay of ROS production during differentiation of appressoria indicated that O_2^- generation was not reduced in either *nox1* or *nox2* mutants, whereas less O_2^- was produced in the appressorium of the double knock-out mutant compared with the wild-type strain.

Similarly, in the phytopathogenic fungus *Botrytis cinerea*, NADPH oxidase play an important part in virulence of by generating O_2^-, which can be converted further by superoxide dismutases to H_2O_2. The two NADPH oxidases BcnoxA and BcnoxB have significant impact on differentiation and pathogenesis. Analyses of single and double knock-out mutants of *BcnoxA* and *BcnoxB* showed that both genes are involved in formation of sclerotia in *B. cinerea*. The *bcnoxB* mutants were retarded in the formation of primary lesions, probably due to an impaired formation of penetration structures, whereas the *bcnoxA* mutant showed much slower in colonizing the host tissue, although it were able to penetrate host tissue in the same way as the wild type. Furthermore, *B. cinerea* contains a homolog of a regulatory subunit of the mammalian Nox complex, bcnoxR, which is required for activation of both BcnoxA and BcnoxB. Kim et al. (2009) reported that ROS generation and homoeostasis was associated with fungal pathogenicity. The authors demonstrated that mutation of *TmpL* encoding a novel transmembrane protein had major impact on redox homeostasis and impaired the virulence in both a plant pathogen (*Alternaria brassicicola*) and an animal pathogen (*A. fumigatus*). Overexpression of the Yap1p homolog in the mutant, which is involved in cellular defense against ROS, suppressed the accumulation of intracellular ROS and partly restored virulence. Collectively, these data suggest that ROS, including H_2O_2 and O_2^-, play a complex effect on fungal virulence.

2.4. Signaling of Hydrogen Peroxide

ROS have long been recognized as toxic molecules because ROS cause damage to macromolecules and membranes. However, ROS also represent important signaling molecules acting as secondary messengers and are involved in various cellular processes. ROS are ideally suited to be signaling molecules, because ROS are small, and can diffuse short distances. Additionally, ROS are produced by several mechanisms, some of which are rapid and controllable and there are numerous mechanisms for their rapid removal. Hydrogen peroxide is now recognized as important signaling molecule, being able to cross cellular membranes (Van Breusegem et al., 2008). In the context of fungal reactions to hydrogen peroxide, the question arises: how does a fungal cell accept the hydrogen peroxide signal, and how is the response realized? Signal transduction from hydrogen peroxide to the nucleolus in fungi might follow different pathways such as mitogen-activated protein kinase (MAPK) cascade, protein tyrosine phosphatases, G-proteins and transcription factors. It has been shown that ROS can regulate MAPK signal transduction and gene expression. In *B. cinerea*, the MAPK Bmp3 protein is the major transcriptional regulator of the nox genes (Segmüller et al., 2008). Bmp3 has opposite effect on the two nox gene expression: nox A is downregulated in bmp3 deletion mutants, whereas nox B is strongly upregulated. This data suggest that there is a link between MAPK signaling and the Nox complexes, although the exact nature of this link remains largely unknown.

Increasing evidence indicates that thioredoxins, peroxiredoxins, protein tyrosine phosphatases and other proteins contain critical cysteines in the thiolate form can react specifically with H_2O_2. It is shown that protein tyrosine phosphatases activity was regulated through reversible oxidation of the catalytic cysteine to sulfenic acid by H_2O_2 (Denu and Tanner, 1998). Fungal G protein pathways are used to sense and respond to environmental stimuli. The interaction of heterotrimeric G proteins with a class of transmembrane fungal receptors (G protein-coupled receptors, GPCRs) regulates the activity of second messengers and protein kinases, leading to the transduction of pheromone and nutrient signals (Bahn et al., 2007). Jones et al. (2007) reported that the G protein mutants of *Neurospora crassa* are sensitive to H_2O_2 treatment, suggesting that G protein may be involved in signal transduction of H_2O_2. The transmission of information from H_2O_2 to the nucleus could be mediated by direct oxidation of proteins involved in signaling cascades (Aguirre et al., 2005). H_2O_2 shows drastic effect on protein structure via modification of thiol (SH) groups, which is important for the function of many crucial proteins (Heller et al., 2011). Although H_2O_2 are now recognized as important signaling molecules in fungi, the details regarding signal transduction from H_2O_2 to the nucleolus remain largely unknown.

3. Effects of Hydrogen Peroxide on Viability and Biocontrol Efficacy of Antagonistic Yeasts and Involved Regulating Mechanisms

Postharvest decay in fruits results in great economic losses every year. So far, synthetic fungicides are still the primary means to control postharvest diseases. However, development of resistant strains of pathogens, public concern and regulatory restrictions about the fungicide residues on plants have emphasized the need to find alternative methods for disease control (Tian, 2006). In recent years, biological control was considered as a promising alternative to synthetic fungicides. Many biocontrol microbial agents, including bacteria and yeasts, have been exploited and widely investigated against different postharvest fungal pathogens (Spadaro and Gullino, 2004). Compared with antagonistic bacteria, yeasts have a number of attributes which make them suitable as biocontrol agents of postharvest diseases, such as simple nutritional requirements, survival in adverse environmental conditions, lack of pathogenicity for the host plant (Tian et al., 2002; Qin et al., 2004). Biological control is only effective when antagonistic yeasts can successfully colonize and form a high concentration of population in wounds of fruits (Fan and Tian, 2001). Wounding of fruit tissue will cause the accumulation of reactive oxygen species (ROS), such as hydrogen peroxide, which has an important effect on biocontrol efficacy of antagonistic yeasts (Castoria et al., 2003; Tolaini et al., 2010). In addition, environmental stresses during formulation and distribution may also cause accumulation of ROS, and decrease the viability and biocontrol efficacy of antagonistic yeasts. Here, we will introduce the effects of ROS, mainly hydrogen peroxide, on viability and biocontrol efficacy of various antagonistic yeasts, the main acting targets of ROS, and methods enhancing the tolerance against oxidative stress.

3.1. Effects of Hydrogen Peroxide on Viability and Biocontrol Efficacy in Antagonistic Yeasts

Generally, hydrogen peroxide has adverse effect on viability and biocontrol performance of antagonistic yeasts. Liu et al. (2011b) found that addition of hydrogen peroxide in cell suspension distinctly decreased the viability of antagonistic yeast *Cystofilobasidium infirmominiatum*, which has demonstrated biocontrol efficacy against postharvest diseases on several fruits, including pear, apple and sweet cherry. Moreover, a dose- and time-dependent effect was clearly observed. Survival rate of the yeast was 93% after 20 min of treatment with 5 mM of hydrogen peroxide, while, the survival rate rapidly declined to 5.8% after 60 min of treatment with 20 mM of hydrogen peroxide. Castoria et al. (2003) observed that growth rate of two antagonistic yeasts, *Cryptococcus laurentii* LS-28 and *Rhodotorula glutinis* LS-11, were markedly reduced after 1 h of treatment with 2 and 20 mM of hydrogen peroxide *in vitro*. Meanwhile, the *C. laurentii* showed a higher stress resistance against hydrogen peroxide than *R. glutinis*.

In vivo, wounding of plant hosts will trigger a peak of accumulation of hydrogen peroxide. On apple fruit, the peak of hydrogen peroxide occurs at 30-60 min after wounding, companying with the accumulation of superoxide anion. The peak value of concentration of hydrogen peroxide per wound reach about 1.8 nmoles for apple fruit stored for 7 months, compared with only about 0.6 nmoles for fruit stored for 1 month (Castoria et al. 2003). Further, Macarisin et al. (2010) found that inoculating antagonistic yeast *Metschnikowia fructicola* into wound of lemon fruit continued to stimulate accumulation of hydrogen peroxide up to 66 h after inoculation. The accumulation of hydrogen peroxide in fruit wounds will make adverse effects on viability of antagonistic yeasts. It was found that population of *R. glutinis* LS-11 and *C. laurentii* LS-28 showed a marked drop at the first 1 or 2 hours after inoculation, and addition of catalase, an enzyme that catalyze the decomposition of hydrogen peroxide to water and oxygen, completely eliminated the drop in population. Accordingly, the biocontrol activities of the two yeasts against *Botrytis cinerea* and *Penicilium expansum* on apple fruit were improved by adding mixture of SOD (superoxidase dismutase) and CAT.

3.2. Acting Targets of Hydrogen Peroxide in Antagonistic Yeasts

Hydrogen peroxide-induced damage has been extensively and systematically studied in a few model species of yeasts, such as *Saccharomyces cerevisiae*, *Schizosaccharomyces pombe*, and *Candida* species (Madeo et al., 2004; Chauhan et al., 2006; Pekmez et al., 2008). Exogenous hydrogen peroxide acts as an oxidative stress to induce accumulation of intracellular ROS, and a series of physiological and biochemistry changes in yeast cells. ROS in yeast cells can directly attack macromolecules such as proteins, lipids and nucleic acids. Reactions of ROS with proteins result in the oxidation of amino acid side chains, such as the formation of protein carbonyl groups by oxidation of the side chains of lysine, proline, arginine, and threonine residues (Stadman and Berkett, 1997). ROS can cause the peroxidation of unsaturated fatty acid chains of membrane lipids, yield lipid peroxides, lipid alcohols, and aldehydic by-products such as 4-hydroxynonenal and malondialdehyde (MDA) (Antunes et al., 1996). Moreover, both nuclear and mitochondrial DNA are the key targets of ROS, which is probably mediated by hydroxyl radicals generated by the Fenton reaction, and lead to an increase in the frequency of intrachromosomal recombination (Moradas-Ferreira and Costa, 2000). The damage of these macromolecules results in the malfunction of biochemical activities, and thereby compromises cell viability. On the other hand, a low dose of exogenous hydrogen peroxide can trigger apoptosis, a highly regulated cellular suicide program (Madeo et al., 1999). During the process of apoptosis, yeast cells show some characteristic markers, including DNA cleavage and apoptosis-typical chromatin condensation, externalization of phosphatidylserine to the outer leaflet of the plasma membrane, and cytochrome c release from mitochondria (Madeo et al., 2004). In yeast apoptosis, intracellular ROS signal act as a central regulators.

Compared with these model species of yeasts, there are few studies focused on the oxidative damage caused by hydrogen peroxide in antagonistic yeasts. Liu et al. (2011b) found that antagonistic yeast *C. infirmominiatum* exposed to 10 mM exogenous hydrogen peroxide showed a distinct accumulation of endogenous hydrogen peroxide. The ROS level in the cells was gradually increased along with the incubation time. Meanwhile, the level of carbonylated proteins was also enhanced. Accumulation of hydrogen peroxide has also been detected during the formulation of antagonistic yeasts. Freeze drying is one of the successful methods for preserving yeast cells. However, we found that accumulation of endogenous hydrogen peroxide in antagonistic yeast *C. laurentii* gradually increased after freeze drying and subsequent storage at 4 °C (Figure 4A-H). Accordingly, the viability of the yeast decreased from more than 95% to less than 15% (Figure 4I). Meanwhile, distinct drop in membrane integrity of *C. laurentii* was detected using propidium iodide (PI), a membrane impermeable fluorescent stain (Figure 4J).

These results suggested that accumulation of intracellular ROS, including hydrogen peroxide, might relate to the damage of plasma membrane. So far, studied on the target of hydrogen peroxide in antagonistic yeasts during the interaction of them with plant hosts and pathogens, are still lack.

3.3. Improvement of Oxidative Stress Tolerance of Antagonistic Yeasts against Hydrogen Peroxide

As hydrogen peroxide causes decrease in viability of antagonistic yeasts during application and formulation, improvement of oxidative stress tolerance is very important to enhance or maintain biocontrol performance of antagonistic yeasts. Liu et al. (2011b) reported that oxidative stress tolerance of antagonistic yeast *C. infirmominiatum* was improved by incubation with glycine betaine (GB), which is one of the most common compatible osmoprotectant.

They found that incubation of *C. infirmominiatum* with 1 mM GB for 2 h significantly reduced the accumulation of hydrogen peroxide and oxidative damage of proteins in yeast cells subsequently exposed to 10 mM hydrogen peroxide.

Compared with untreated control, GB-treated yeast cells exhibited a faster growth in wounds of apple fruit and better biocontrol performance against *P. expansum*. In the previous study, we found that four antagonistic yeasts (*Pichia membranaefaciens, C. laurentii, Candida guilliermondii* and *R. glutinis*) could significantly reduce oxidant damage caused by *Monilinia fructicola* producing ROS in peach fruits, as these antagonistic yeasts could stimulate the activities of chitinase, β-1,3-glucanase, catalase, peroxidase and the expressions of corresponding genes during all storage periods (Xu et al., 2008). Moreover, *P. membranaefaciens* has been proved to alleviate oxidative stress caused by *Penicillium expansum* in sweet cherry fruit, because of inducing higher activities of catalase, glutathione peroxidase and β-1,3-glucanase, as well as the transcript expressions of *CAT, GPX* and *Gns1*genes in yeast-treated fruit (Xu and Tian, 2008).

Treatment with biocontrol agents is related to alleviating proteins carbonylation and mitigating pathogen-induced oxidative damage. Induction of antioxidant defense response may be an important mechanism of antagonistic yeast in mitigating pathogen-induced oxidative stress to postharvest fruit and controlling postharvest disease.

Exogenous chemical compounds can improve resistance biocontrol agents to environment condistion. Trehalose, a nonreducing disaccharide and a major reserve carbohydrate in yeast cells, is considered as a protection metabolite against various stresses, such as heat, dehydration, freezing and hyperosmotic stress (Kwon et al. 2003). In their study, a two fold increase in content of intracellular trehalose was determined after HS treatment at 40 °C for 30 min, compared with that in un-treated yeast cells.

They suggested that increase of intracellular trehalose may play an important role in enhancing hydrogen peroxide stress tolerance in yeast cells. Induced accumulation of intracellular trehalose can also lower the production of endogenous hydrogen peroxide during formulation of antagonistic yeasts.

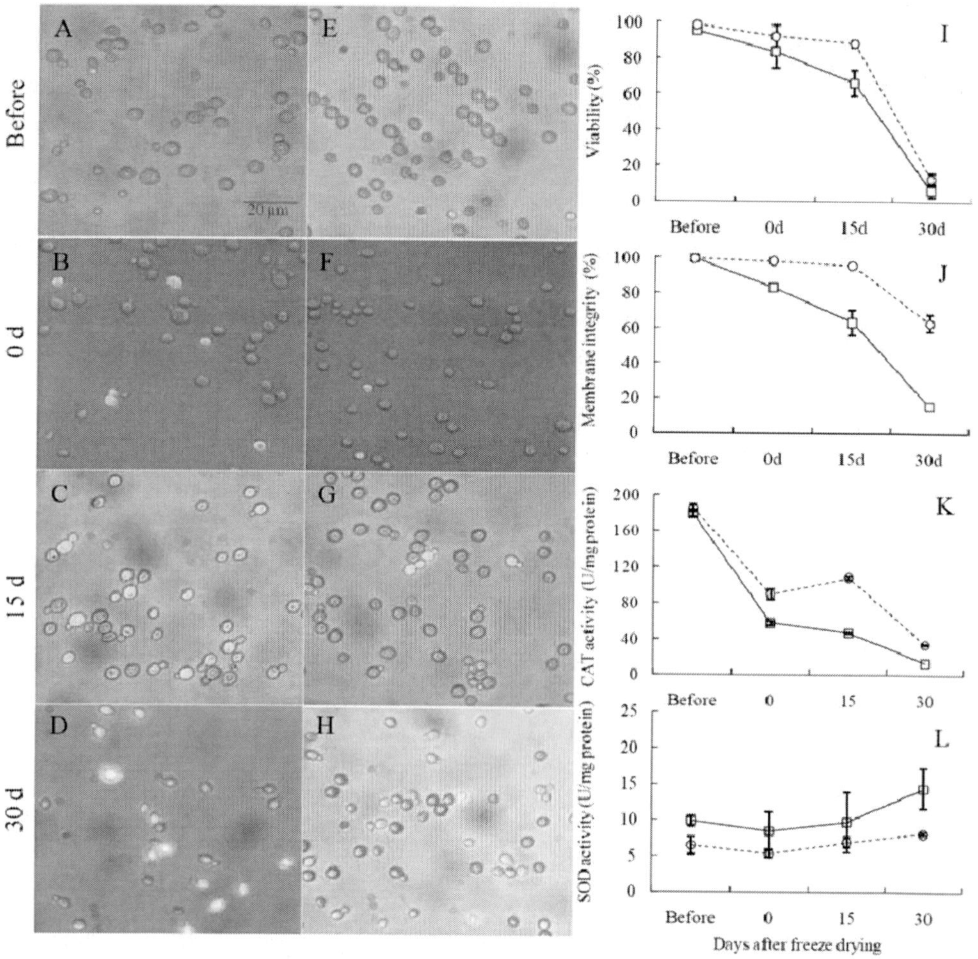

Figure 4. Effects of endogenous trehalose on accumulation of intracellular ROS, cell viability, membrane integrity, and activities of antioxidant enzyme CAT and SOD in *C. laurentii*. Yeast cells of antagonistic yeast *C. laurentii* with lower (real line) or higher (dotted line) level of endogenous trehalose were freeze dried and subsequently stored at 4 °C using 5% skimmed milk + 10% glucose as protectant. (A)-(D), accumulation of intracellular ROS in yeast cells with lower level of trehalose; (E)-(H), accumulation of intracellular ROS in yeast cells with higher level of trehalose; (I)-(L), changes in viability, membrane integrity, and activities of CAT and SOD. Error bars indicate standard deviations of the means.

Li et al. (2006) found incubation with a trehalose-containing medium could increase the content of intracellular trehalose in antagonistic yeast *C. laurentii*, which resulted in increase in survival rate after freeze-drying and biocontrol efficacy against *P. expansum* on apple fruit. Further, it was observed that yeast cells of *C. laurentii* with a higher content of internal trehalose showed a lower level of hydrogen peroxide accumulation and higher activities of antioxidant enzymes (CAT) after freeze drying and subsequent storage (Figure 4). Besides trehalose, we also found that using citric acid as carbon source could also induce accumulation of intracellular trehalose in *C. laurentii* (Li et al., 2007). Citric acid as a cheaper carbon source has a potential value to improve stress tolerance of antagonistic yeasts in practical application.

In recent, Liu et al. (2011a) reported that a mild heat shock (HS) pretreatment (30 min at 40 ºC) improved the tolerance of *M. fructicola* to subsequent oxidative (0.4 M hydrogen peroxide, 20–60 min) and high temperature stress (45 ºC, 20–30 min). HS-pretreatment significantly decreased the accumulation of hydrogen peroxide in *M. fructicola*, and increased viability of the yeast under both stresses and populations in wounds of apple fruit. As a result, HS-treated yeast cells exhibited significantly greater biocontrol performance against *P. expansum* on apple fruit stored at 25ºC compared with that with untreated yeast cells. Additionally, they found that a gene *TPS*1, which encodes trehalose-6-phosphate synthase (TPS1), was up-regulated during the mild heat shock at 40 ºC. TPS1 is a key enzyme in the synthesis of endogenous trehalose in organisms (De Silva-Udawatta and Cannon, 2001).

Conclusion

Hydrogen peroxide, as a signal molecule in the induction of systemic acquired resistance, plays critical roles in the interaction of microbe-plant. Many reports have been indicated that some fungi and bacteria are very sensitive to oxidative stress of H_2O_2. The oxidative stress reduces the viability of bacterial and fungal cells by acting on special transporters. Undesirable accumulation of ROS can cause oxidative injury to biological macromolecules in fungal cells, including the direct cause of redox signaling, the oxidative damage of respiratory chain, the collapse of mitochondrial membrane potential, mtDNA deletion and lipid peroxidation, resulting in mitochondrial dysfunction and apoptosis. The complex III of the mitochondrial respiratory chain, which serves as the major site for ROS production in fungal pathogen under H_2O_2 stress, has been proven to be molecular targets of H_2O_2 in fungal cells. Moreover, the special bacterial membrane transporters, such as PSPTO_1720, play an important role in response to H_2O_2 stress. However, exploring the detail mechanisms how ROS signals are transferred, and identifying the ROS generated by this complex, control fungal polarized growth, and cellular differentiations will be important areas for further investigation.

Acknowledgments

We thank the National Natural Science Foundation of China (31030051), the National High Technology Research (863) Program of China (2012AA101607) and by the Chinese Academy of Sciences (KSCX2-EW-G-6) for supporting the funds on research works related to this review.

References

Aguirre J, Ríos-Momberg M, Hewitt D, Hansberg W. 2005. Reactive oxygen species and development in microbial eukaryotes. *Trends Microbiol.* 13:111-118.

Aguirre J, Ríos-Momberg M, Hewitt D, Hansberg W. 2005. Reactive oxygen species and development in microbial eukaryotes. *Trends Microbiol.* 13:111-118.

Antunes F, Salvador A, Marinho HS, Alves R, Pinto RE. 1996. Lipid peroxidation in mitochondrial inner membranes I. An integrative kinetic model. *Free Radic Biol Med.* 21:917-943.

Apel K, Hirt H. 2004. Reactive oxygen species: Metabolism, oxidative stress, and signal transduction. Annual Review of Plant Biol., 55: 373-399.

Babior BM. 2004. NADPH oxidase. *Curr. Opin. Immunol.* 16:42-47.

Bahn YS, Xue C, Idnurm A, Rutherford J, Heitman J, Cardenas M. 2007. Sensing the environment: lessons from fungi. *Nature Rev. Microbiol.* 5:57-69.

Bolwell GP, Butt VS, Davies DR, Zimmerlin A. 1995. The origin of the oxidative burst in plants. *Free Radic. Res.* 23:517-532.

Bortfeld M, Auffarth K, Kahmann R, Basse CW. 2004. The *Ustilago maydis* a2 mating-type locus genes *lga2* and *rga2* compromise pathogenicity in the absence of the mitochondrial p32 family protein Mrb1. *Plant Cell* 16:2233-2248.

Castoria R, Caputo L, De Curtis F, De Cicco V. 2003. Resistance of postharvest biocontrol yeasts to oxidative stress: a possible new mechanism of action. *Phytopathol.* 93:564-572.

Chauhan N, Latge JP, Calderone R. 2006. Signalling and oxidant adaptation in *Candida albicans* and *Aspergillus fumigatus*. *Nat. Rev. Microbiol.* 4:435-444.

Cheng Y, Song C. 2006. Hydrogen peroxide homeostasis and signaling in plant cells. *Sci. in China C- Life Sci.,* 49:1-11.

Das N, Levine RL, Orr WC, Sohal RS. 2001. Selectivity of protein oxidative damage during aging in *Drosophila melanogaster*. *Biochem J.* 360:209-216.

Dat J, Vandenabeele S, Vranová E, Van Montagu M, Inzé D, Breusegem F. 2000. Dual action of the active oxygen species during plant stress responses. *Cell Mole Life Sci.* 57: 779-795.

De Silva-Udawatta MN, Cannon JF. 2001. Roles of trehalose phosphate synthase in yeast glycogen metabolism and sporulation. *Mol. Microbiol.* 40:1345-1356.

Denu JM, Tanner KG. 1998. Specific and reversible inactivation of protein tyrosine phosphatases by hydrogen peroxide: evidence for a sulfenic acid intermediate and implications for redox regulation. *Biochem.* 37:5633-5642.

Egan MJ, Wang ZY, Jones MA, Smirnoff N, Talbot NJ. 2007. Generation of reactive oxygen species by fungal NADPH oxidases is required for rice blast disease. *Proc. Natl. Acad. Sci. U S A.* 104:11772-11777.

Fan Q, Tian SP. 2001. Postharvest biological control of grey mold and blue mold on apple by *Cryptoccus albidus* (Saito) Skinner. Postharvest *Biol Technol.* 21:341-350.

Foreman J, Demidchik V, Bothwell JHF, Mylona P, Miedema H et al. 2003. Reactive oxygen species produced by NADPH oxidase regulate plant cell growth. *Nature*, 422:442-446.

Gechev TS, Hille J. 2005. Hydrogen peroxide as a signal controlling plant programmed cell death. *J. Cell Biol.*, 168: 17-20.

Giesbert S, Schürg T, Scheele S, Tudzynski P. 2008. The NADPH oxidase Cpnox1 is required for full pathogenicity of the ergot fungus *Claviceps purpurea*. *Mol. Plant Pathol.* 9:317-327.

Guzy RD, Hoyos B, Robin E, Chen H, Liu L, Mansfield KD, Simon MC, Hammerling U, Schumacker PT. 2005. Mitochondrial complex III is required for hypoxia-induced ROS production and cellular oxygen sensing. *Cell Metab.* 1:401-408.

Guzy RD, Mack MM, Schumacker PT. 2007. Mitochondrial complex III is required for hypoxia-induced ROS production and gene transcription in yeast. *Antioxid Redox Signal.* 9:1317-1328.

Hamann A, Brust D, Osiewacz HD. 2008. Apoptosis pathways in fungal growth, development and ageing. *Trends Microbiol.* 16:276-283.

Heller J, Tudzynski P. 2011. Reactive oxygen species in phytopathogenic fungi: signaling, development, and disease. *Annu. Rev. Phytopathol.* 49:369-390.

Helmerhorst EJ, Breeuwer P, van't Hof W, Walgreen-Weterings E, Oomen LC, Veerman EC, Amerongen AV, Abee T. 1999. The cellular target of histatin 5 on *Candida albicans* is the energized mitochondrion. *J. Biol. Chem.* 274:7286-7291.

Helmerhorst EJ, Troxler RF, Oppenheim FG. 2001. The human salivary peptide histatin 5 exerts its antifungal activity through the formation of reactive oxygen species. *Proc. Natl. Acad. Sci. U S A.* 98:14637-14642.

Iles KE, Forman HJ. 2002. Macrophage signaling and respiratory burst. *Immunol. Res.* 26: 95-105.

Ingavale SS, Chang YC, Lee H, McClelland CM, Leong ML, Kwon-Chung KJ. 2008. Importance of mitochondria in survival of *Cryptococcus neoformans* under low oxygen conditions and tolerance to cobalt chloride. *PLoS Pathogen.* 4:e1000155.

Inoue I, Namiki F, Tsuge T. 2002. Plant colonization by the vascular wilt fungus *Fusarium oxysporum* requires *FOW1*, a gene encoding a mitochondrial protein. *Plant Cell* 14:1869-1883.

Jones CA, Greer-Phillips SE, Borkovich KA. 2007. The response regulator RRG-1 functions upstream of a mitogen-activated protein kinase pathway impacting asexual development, female fertility, osmotic stress, and fungicide resistance in *Neurospora crassa. Mol. Biol. Cell.* 18:2123-2136.

Kim KH, Willger SD, Park SW, Puttikamonkul S, Grahl N et al. 2009. TmpL, a transmembrane protein required for intracellular redox homeostasis and virulence in a plant and an animal fungal pathogen. *PLoS Pathog.* 5: e1000653.

Kwon HB, Yeo ET, Hahn SE, Bae SC, Kim DY., Byun MO. 2003. Cloning and characterization of genes encoding trehalose-6-phosphate synthase (TPS1) and trehalose-6-phosphate phosphatase (TPS2) from *Zygosaccharomyces rouxii. FEMS Yeast Res.* 3:433-440.

Lai TT, Li BQ, Qin GZ, Tian SP. 2011. Oxidative damage involves in the inhibitory effect of nitric oxide on spore germination of *Penicillium expansum. Current Microbiol.* 62: 229-234.

Lalucque H, Silar P. 2003. NADPH oxidase: an enzyme for multicellularity? *Trends Microbiol.,* 11:9-12.

Lara-Ortiz T, Riveros-Rosas H, Aguirre J. 2003. Reactive oxygen species generated by microbial NADPH oxidase NoxA regulated sexual development in *Aspergillus nidulans. Mol. Microbiol.* 50:1241-1255.

Li BQ, Tian SP. 2006. Effects of trehalose on stress tolerance and biocontrol efficacy of *Cryptococcus laurentii. J. Appl. Microbiol.* 100:854-861.

Li BQ, Tian SP. 2007. Effect of intracellular trehalose in *Cryptococcus laurentii* and exogenous lyoprotectants on its viability and biocontrol efficacy on *Penicillium expansum* in apple fruit. *Lett. Appl. Microbiol.* 44:437-442.

Liu J, Wisniewski M, Droby S, Tian S, Hershkovitz V, Tworkoski T. 2011a. Effect of heat shock treatment on stress tolerance and biocontrol efficacy of *Metschnikowia fructicola. FEMS Microbiol. Ecol.* 76:145-155.

Liu J, Wisniewski M, Droby S, Vero S, Tian S, Hershkovitz V. 2011b. Glycine betaine improves oxidative stress tolerance and biocontrol efficacy of the antagonistic yeast *Cystofilobasidium infirmominiatum. Int. J. Food Microbiol.* 146:76-83.

Lu H, Higgins VJ. 1999. The effect of hydrogen peroxide on the viability of tomato cells and of the fungal pathogen *Cladosporium fulvum. Physiol. Mol. Plant Pathol.* 54:131-143.

Macarisin D, Cohen L, Eick A, Rafael G, Belausov E et al. 2007. *Penicillium digitatum* suppresses production of hydrogen peroxide in host tissue during infection of citrus fruit. *Phytopathol.* 97:1491–1500.

Macarisina D, Droby S, Bauchanc G, Wisniewski M. 2010. Superoxide anion and hydrogen peroxide in the yeast antagonist–fruit interaction: A new role for reactive oxygen species in postharvest biocontrol? *Postharvest Biol Technol.* 58:194-202.

Madeo F, Fröhlich E, Ligr M, Grey M, Sigrist SJ, Wolf DH, Fröhlich KU. 1999. Oxygen stress: a regulator of apoptosis in yeast. *J. Cell Biol.* 145:757-767.

Madeo F, Herker E, Wissing S, Jungwirth H, Eisenberg T, Fröhlich KU. 2004. Apoptosis in yeast. *Curr. Opin. Microbiol.* 7:655–660.

Malagnac F, Laluque H, Lepere G, Silar P. 2004. Two NADPH oxidase isoforms are required for sexual reproduction and ascospore germination in the filamentous fungus *Podospora anserina*. *Fungal. Gen. Biol.* 41:982-997.

McBride HM, Neuspiel M, Wasiak S. 2006. Mitochondria: more than just a powerhouse. *Curr. Biol.* 16:R551-R560.

Mellersh DG, Foulds IV, Higgins VJ, Heath MC. 2002. H_2O_2 plays different roles in determining penetration failure in three diverse plant-fungal interactions. *Plant J.* 29: 257-268.

Mellersh DG, Foulds IV, Higgins VJ, Heath MC. 2002. H_2O_2 plays different roles in determining penetration failure in three diverse plant-fungal interactions. *Plant J.* 29: 257-268.

Molina L, Kahmann R. 2007. An *Ustilago maydis* gene involved in H_2O_2 detoxification is required for virulence. *Plant Cell.* 19:2293-2309.

Møller IM, Jensen PE, Hansson A. 2007. Oxidative modifications to cellular components in plants. *Ann. Rev. Plant Biol.* 58:459-481.

Moradas-Ferreira P, Costa V. 2000. Adaptive response of the yeast *Saccharomyces cerevisiae* to reactive oxygen species: defences, damage and death. *Redox Rep.* 5:277-285.

Pekmez M, Arda N, Hamad İ, Kiğ C, Temizkan G. 2008. Hydrogen peroxide-induced oxidative damages in *Schizosaccharomyces pombe*. *Biologia.* 63:151-155.

Peng M, Kuc J. 1992. Peroxidase-generated hydrogen peroxide as a source of antifungal activity *in vitro* and on tobacco leaf discs. *Phytopathol.* 82:696-699.

Phillips AJ, Sudbery I, Ramsdale M. 2003 Apoptosis induced by environmental stresses and amphotericin B in *Candida albicans*. *Proc. Natl. Acad. Sci. U S A.* 100:14327-14332.

Qin G, Liu J, Cao B, Li B, Tian S. 2011. Hydrogen peroxide acts on sensitive mitochondrial proteins to induce death of a fungal pathogen revealed by proteomic analysis. *PLoS One* 6:e21945.

Qin G, Meng X, Wang Q, Tian S. 2009. Oxidative damage of mitochondrial proteins contributes to fruit senescence: a redox proteomics analysis. *J. Proteome Res.* 8:2449-2462.

Qin G, Tian S, Chan Z, Li B. 2007. Crucial role of antioxidant proteins and hydrolytic enzymes in pathogenicity of *Penicillium expansum*: analysis based on proteomics approach. *Mol. Cell Proteomics* 6:425-438.

Qin QZ, Tian SP, Xu Y. 2004. Biocontrol of postharvest diseases on sweet cherries by four antagonistic yeasts in different storage conditions. *Postharvest Biol. Technol.* 31:51-58.

Quan LJ, Zhang B, Shi WW, Li HY. 2008. Hydrogen peroxide in plants: a versatile molecule of the reactive oxygen species network. *J. Integrative Plant Biol.* 50:2-18.

Segmüller N, Kokkelink L, Giesbert S, Odinius D, Van Kan JAL, Tudzynski P. 2008. NADPH oxidases are involved in differentiation and pathogenicity in *Botrytis cinerea*. *Mol. Plant-Microbe Interact.* 21:808-819.

Shi XQ, Li BQ, Qin GZ, Tian SP. 2011. Antifungal activity of borate against *Colletotrichum gloeosporioides* and its possible mechanism. *Plant Dis.* 95:63-69.

Shinogi T, Suzuki T, Kurihara T, Narusaka Y, Park P. 2003. Microscopic detection of reactive oxygen species generation in the compatible and incompatible interactions of *Alternaria alternata* Japanese pear pathotype and host plants. *J. Gen. Plant Pathol.* 69: 7-16.

Ślesak I, Libik M, Karpinska B, Karpinski S, Miszalski Z. 2007. The role of hydrogen peroxide in regulation of plant metabolism and cellular signalling in response to environmental stresses. *Acta Biochimica Polonica*, 54:39-50.

Soundararajan S, Jedd G, Li X, Ramos-Pamploña M, Chua NH, Naqvi NI. 2004. Woronin body function in *Magnaporthe grisea* is essential for efficient pathogenesis and for survival during nitrogen starvation stress. *Plant Cell.* 16:1564-1574.

Spadaro D, Gullino ML. 2004. State of the art and future prospects of the biological control of postharvest fruit diseases. *Int. J. Food Microbiol.* 91:185-194.

Stadman ER, Berkett BS. 1997. Reactive oxygen-mediated protein oxidation in aging and disease. *Chem Res Toxicol.* 10:485-494.

Sumimoto H. 2008. Structure, regulation and evolution of Nox-family NADPH oxidases that produce reactive oxygen species. *FEBS J.,* 275:3249-3277.

Takemoto D, Kamakura S, Saikia S, Becker Y, Wrenn R et al. 2011. Polarity proteins Bem1 and Cdc24 are components of the filamentous fungal NADPH oxidase complex. *Proc. Natl. Acad. Sci. US.* 108:2861-2866.

Takemoto D, Tanaka A, Scott B. 2006. A p67phox-like regulator is recruited to control hyphal branching in a fungal-grass mutualistic symbiosis. *Plant Cell.* 18:2807-2821

Tanaka A, Christensen MJ, Takemoto D, Park P, Scott B. 2006. Reactive oxygen species play a role in regulating a fungus-perennial ryegrass mutualistic interaction. *Plant Cell.* 18:1052-1066.

Tian SP, Fan Q, Xu Y, Liu HB. 2002. Biocontrol efficacy of antagonist yeasts to grey mold and blue mold on apples and pears in controlled atmospheres. *Plant Dis.* 86:848-853.

Tian SP. 2006. Microbial control of postharvest diseases of fruits and vegetables: current concepts and future outlook. In *Microbial Biotechnology in Horticulture*, Volume 1, Edited by Dr Ranesh C. Ray and Owen P. Ward, Published by Science Publishers Inc. Enfield NH USA. pp.163-202.

Tolaini V, Zjalic S, Reverberi M, Fanelli C, Fabbri AA, Del Fiore A, De Rossi P, Ricelli A. 2010. Lentinula edodes enhances the biocontrol activity of *Cryptococcus laurentii* against *Penicillium expansum* contamination and patulin production in apple fruits. *Int. J. Food Microbiol.* 138:243-249.

Torres MA, Jones JD, Dangl JL. 2005. Pathogen-induced, NADPH oxidase-derived reactive oxygen intermediates suppress spread of cell death in *Arabidopsis thaliana. Nat. Genet.* 37:1130-1134.

Van Breusegem F, Bailey-Serres J, Mittler R. 2008. Unraveling the tapestry of networks involving reactive oxygen species in plants. *Plant Physiol.* 147:978-984.

Xu XB, Qin GZ, Tian SP. 2008. Effect of microbial biocontrol agents on alleviating oxidative damage of peach fruit subjected to fungal pathogen. *Int. J. Food Microbiol.* 126:153-158.

Xu XB, Tian SP. 2008. Reducing oxidative stress in sweet cherry fruit by *Pichia membranaefaciens*: a possible mode of action against *Penicillium expansum. J. Appl. Microbiol.* 105:1170-1177.

Yan S, Tang Z, Su W, Sun W. 2005. Proteomic analysis of salt stress-responsive proteins in rice root. *Proteomics.* 5:235-244.

Yu JH, Yun SY, Lim JW, Kim H, Kim KH. 2003 Mass spectrometry and tandem mass spectrometry analysis of rat mitochondrial ATP synthase: up-regulation in pancreatic acinar cells treated with cerulein. *Proteomics*. 3:2437-2445.

In: Hydrogen Peroxide
Editors: Gilberto Aguilar and Raphael A. Guzman
ISBN: 978-1-62257-414-8

Chapter 7

Electrochemical Characterization of O_2 Metabolism in Isolated Mitochondria Revealed Massive Production of H_2O_2 upon ATP Synthase Activation

Raluca Marcu[1,*], Stefania Rapino[1], Francesco Paolucci[2] and Marco Giorgio[1]

[1]European Institute of Oncology, Department of Experimental Oncology, Milan, Italy

[2]University of Bologna, Department of Chemistry "G. Ciamician," Bologna, Italy

Abstract

The mitochondrial reactive oxygen species (ROS) production is a crucial issue that has been involved in a variety of physio-pathological processes and aging. However, at the present measurements of mitochondrial ROS production is controversial. Indeed, the available tools are poorly specific and not direct.

Now taking advantage of the progress in microelectrode production it has been possible to approach the specific and direct measurement of O_2 species in a suspension of isolated mitochondria.

Such electrochemical determinations confirm active H_2O_2 release by mitochondria above the scavenging activity and reveal a significant role of ATP synthase activation in the control of mitochondrial ROS production.

These findings offer a new explanation for the effects of mitochondrial respiration on oxidative stress. Regardless the damaging effect of ROS that has been already associated to intense mitochondrial work during exercise in muscle for example, H_2O_2 spikes may represent specific cellular signal. The crosstalk between the mitochondrion and the cell is indeed an emergent issue. About this argument our observation discloses a novel basic mechanism of mitochondrial ROS production and highlights the involvement of intracellular redox balance on health.

---

* Authors contact info: marco.giorgio@ifom-ieo-campus.it ; Tel: +39-02-94375040; Fax +39-02-94375990.

BIOCHEMISTRY OF REACTIVE OXYGEN SPECIES

Reactive oxygen species (ROS) comprise a variety of molecules and free radicals (chemical species with unpaired electrons) derived from molecular oxygen. In the ground state, molecular oxygen has two unpaired electrons with the same spin in the outer orbitals (triplet state). Spin restriction rules limit the reactivity of molecular oxygen and therefore the reactions involving O_2 are one electron transfer reactions. The reduction of O_2 by one electron at a time produces less stable species and with increased reactivity towards other biological molecules (Halliwell 2007)

Addition of one electron to the ground-state O_2 molecule generates the superoxide anion, $O_2^{\cdot -}$, a non diffusible, short lived specie with reduced reactivity towards biological substrates. $O_2^{\cdot -}$ is the precursor for other ROS through direct or enzyme/metal-catalysed reactions with other molecules. $O_2^{\cdot -}$ undergoes a dismutation reaction producing hydrogen peroxide, H_2O_2, reaction that can be catalysed by superoxide dismutases enzymes (SOD):

$$O_2^{\cdot -} + 2\,H^+ \rightarrow H_2O_2 + O_2.$$

Among ROS, H_2O_2 is the most stable and longest lived specie, is membrane permeable ad diffusible (Giorgio 2007). H_2O_2 can be reduced to water or partially reduced to hydroxyl radical OH^- through the interaction with transition metals such as free iron or copper (Fenton reaction):

$$Fe(II) + H_2O_2 \rightarrow Fe(III) + {}^{\cdot}OH + OH^-.$$

Although hydroxyl radical is very short lived (half life less than 1 nsec), if produced in the close proximity of DNA can react with the DNA bases or backbone producing damaged bases or strand breaks. However the significance of this reaction is conditioned by the availability of free iron, generally sequestered by different metal-binding proteins (Halliwell 2007)

In vivo $O_2^{\cdot -}$ and subsequently H_2O_2 are produced by both enzymatic and non-enzymatic systems. There are several enzymes that generate deliberately and in a regulated manner $O_2^{\cdot -}$ in different cellular compartments: the phagocytic oxidases and NADP/H oxidases (NOX family) in the plasma membrane, amino oxidases in mitochondria, peroxisomal oxidases in peroxisomes, sulphydryl oxidases in endoplasmic reticulum, cyclooxygenase, lipid oxygenase and xanthine oxidase in cytosol (Giorgio 2007) In the case of NOX family of enzymes, the electron needed to reduce O_2 to $O_2^{\cdot -}$ travels from NADPH and FAD bound to the enzyme on one side of the lipid bilayer to a heme prosthetic group localised in the membrane and from this site to molecular oxygen on the other side of the membrane.

The non-enzymatic production of $O_2^{\cdot -}$ occurs when a single electron is transferred to O_2 directly by reduced coenzymes or prosthetic groups (flavins, iron sulphur clusters) (Halliwell 2007) The electron transfer chain localised in the inner mitochondrial membrane contains several redox centres that can leak electrons to O_2 and represents the most important source of $O_2^{\cdot -}$ in cells of aerobic organisms. It has been estimated that about 1 to 3% from the O2 reduced in mitochondria may form $O_2^{\cdot -}$ (Halliwell 2007) Mitochondrial ETC is the site of ATP production through the process of oxidative phosphorylation. Electrons from different catabolic pathways are funnelled into universal electron acceptors (nicotinamid and flavins nucleotides) and then pass through a series of membrane bound carriers up to molecular oxygen that is reduced to water. The energy of the electron transfer is conserved in a proton gradient that drives the synthesis of ATP from ADP and inorganic phosphate. The electron carrying molecules that function in the respiratory chain (ubiquinone, cytochromes and iron-sulphur proteins) are organised into supramolecular complexes: complex I (NADH dehydrogenase) catalyses electron transfer from NADH, complex II (succinate dehydrogenase) collects electrons from succinate, complex III (ubiquinone cytochrome c oxidoreductase) carries electrons from ubiquinone to cytochrome c and complex IV (cytochrome c oxidase) transfers the electrons to O_2 (Murphy 2009).

Complex I produces $O_2^{\cdot-}$ from the reaction between O_2 and the fully reduced flavin mononucleotide centre (FMN) while the fraction of reduced FMN is set by the $NADH/NAD^+$ ratio. Another mechanism through which complex I can produce $O_2^{\cdot-}$ is by reverse electron transport from the reduced ubiquinone: in the presence of a significant proton motive force electrons are forced back through complex I to the FMN site. Complex III produces $O_2^{\cdot-}$ which is formed as a result of autooxidation of ubisemiquinone, an intermediate compound formed during the ubiquinone cycle (Q-cycle) (Zorov 2006, Murphy 2009).

There are several factors that control the rate of superoxide formation and that are modulated by intracellular metabolic reactions: the concentration of available O_2 at the level of mitochondria, the concentration of the electron carrier able to interact with O_2 to generate $O_2^{\cdot-}$, the fraction of this carrier that exists in a redox form that can interact with O_2 and the rate of the electron transfer form the donor to O_2 (Bell 2007, Hoffman 2009). Accordingly, three modes of operation of mitochondria have been identified for the efflux of H_2O_2 from these organelles. The first two modes result in a high efflux of H_2O_2 and involve a high $NADH/NAD^+$ ratio in the matrix (first mode) and a highly reduced ubiquinone pool without synthesis of ATP (second mode). The third operational mode occurs when mitochondria are synthesizing ATP or use the proton motive force for other functions such as thermogenesis (state 3) and is associated with a low efflux of H_2O_2 (Hoffman 2007).

Mitochondria produce ROS on both sides of inner mitochondrial membrane. While generally $O_2^{\cdot-}$ is not diffusible, H_2O_2 can pass mitochondrial membranes and reach the cytosol and nucleus with putative deleterious effects. Therefore ROS produced by mitochondria are quickly inactivated by a set of enzymes at both intra- and extra-mitochondrial level. Two forms of superoxide dismutases catalyse the dismutation of $O_2^{\cdot-}$ to H_2O_2: manganese SOD (MnSOD) eliminates superoxide in the mitochondrial matrix while copper and zinc SOD (CuZnSOD) control the levels of $O_2^{\cdot-}$ in the intermembrane space and cytosol. The product of superoxide dismutation, H_2O_2, is then decomposed by the combined activity of glutathione peroxidases, peroxiredoxins and catalases in both mitochondria and cytosol (Halliwell 2007).

H_2O_2, THE SIGNALLING FACE OF ROS

ROS, including $O_2^{\cdot-}$, H_2O_2 and OH^- are all potentially oxidizing agents but not all of them are equally implicated in the different cellular functions. H_2O_2, for example, is membrane-permeable and diffusible, less-reactive and longer-lived than OH· or $O_2^{\cdot -}$, and, as such, it is best suited for intra- and even intercellular signaling (Stone 2006).

Accumulating evidence indicate that living systems generate H_2O_2 intentionally for different purposes such as host defense, apoptosis, oxidative biosynthesis and signal transduction (Wallace 2005, Orrenius 2007). All aerobic organisms maintain an intracellular steady state level of H_2O_2 in a range that is conserved among different species and that varies from a low 0.001 μM to 0.5-0.7 μM. The functions of H_2O_2 are concentration dependent: high levels of H_2O_2 generate oxidative stress resulting in growth arrest and apoptosis while low physiologic levels of H_2O_2 can function as signaling agents and stimulate cell proliferation and differentiation (stone 2006).

A well characterized system of controlled ROS generation in response to stimuli (cytokines, growth factors, hormones, neurotransmitters) is the NOX family of NADPH oxidases located in the plasma membrane of most mammalian cells that produce $O_2^{\cdot -}$ and H_2O_2 in the extracellular space and have a pro-growth effect. It has been hypothesized that also mitochondrial H_2O_2 is generated in a regulated manner such as to reflect the overall metabolic activity of the cell. Another system of controlled ROS production is the $p66^{Shc}$ protein which in response to specific signals functions as a redox enzyme that generates mitochondrial H_2O_2 and triggers apoptosis supporting the hypothesis that mitochondrial ROS are not just by-products of respiration but instead have a definite role in cellular physiology (Giorgio 2007).

At low concentrations, H_2O_2 can act as a signaling messenger by producing reversible oxidative modifications of proteins. Proteins poses redox groups that can switch between oxidation and reduction in a ROS-dependent manner participating thus in the H_2O_2 mediated signaling. The best candidates in this sense are the cysteine residues that can form a disulfide bond in the presence of H_2O_2 resulting in a conformational change of the protein and alteration of its function. Alternatively, the thiol group of a single cysteine (-SH) can be oxidised to intermediate sulphenic (-SOH), sulphinic ($-SO_2H$) or sulphonic ($-SO_3H$) acids depending on the concentration of the oxidant. Hence the oxidation state of the cysteine residues is an indication of the overall intracellular redox state. The disulfide bonds can be reduced to their original state by thioredoxins and glutaredoxins while sulphinic acid groups can be reduced back by sulfiredoxins. These rapid and reversible modifications of the proteins can impact on phosphorylation cascades stimulating signalling. Examples of proteins that can be modulated by H_2O_2 are the protein tyrosine phosphatases (PTP) that contain an essential cysteine residue sensitive to oxidation and that are functionally inactivated as a result of endogenous H_2O_2 production. Through oxidation of specific cysteine residues H_2O_2 can activate or inactivate various transcription factors such as HIF1α and NF-kB (Halliwell 2007).

The dual character of H_2O_2 as mediator of growth and apoptosis suggests specificity in its biological activity (Giorgio 2007). This could be determined by both intensity of the pro-oxidant challenge and the intracellular site of production, increasing the complexity of redox signaling. Although H_2O_2 can diffuse across membranes, the presence of abundant and very efficient antioxidant system limits the effects of H_2O_2 close to its site of generation.

Electrochemical Revealing of Mitochondrial H_2O_2

Traditional assays for H_2O_2 measurement relay on fluorescent probes that lack specificity and often result in artefactual signals (Votyakova 2004, Gomes 2005).

Below (Figure 1) is reported an example of results obtained from the usage of such approaches.

Recently, reactive species produced by living cells in response to mechanical challenges have been measured directly and selectively through electrochemistry at platinised carbon microelectrode (Arbault 2004, Amatore 2008). To overcome limitations of current methodologies, we implemented a novel electrochemical assay for the simultaneous evaluation of O_2 reduction metabolism and H_2O_2 production in suspensions of isolated mitochondria. Here we show the results obtained with this approach on mouse liver mitochondria in the presence of different energetic substrates, respiratory inducers and inhibitors.

The study was performed using the "O_2 metabolism electrochemical analyzer (OMEA)". The OMEA consists of an assay chamber bearing the Clark type electrode for O_2 detection and the platinised carbon fibre electrode used for H_2O_2 measurement. The potentials for the detection of both species (-0.7 V for O_2 and 0.6 V for H_2O_2) were supplied by two independent potentiostats. In this system, the reduction and oxidation of O_2 and respectively H_2O_2 present in the assay chamber generate electrical currents at the detection electrodes (named shortly O_2 signal and H_2O_2 signal) that are amplified and analysed (see next Figure 2).

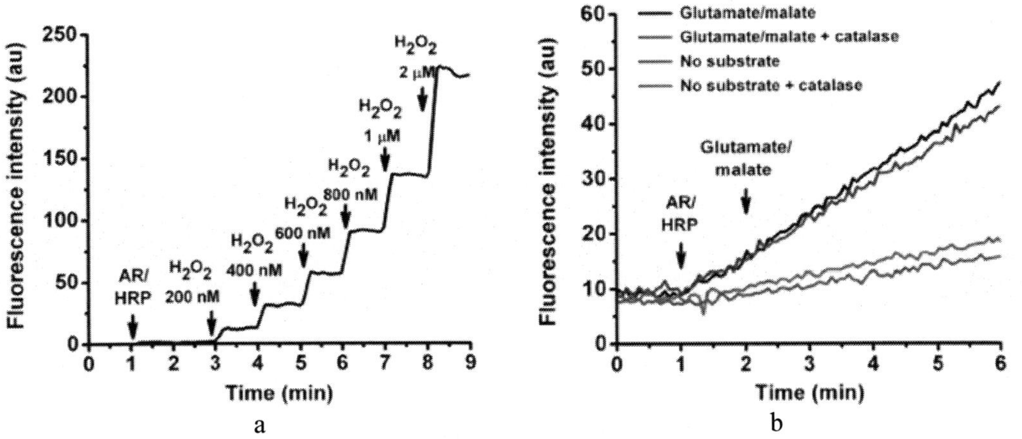

Figure 1. Amplex Red/HRP measurements of H_2O_2 production by isolated liver mitochondria. a) Calibration curve for Amplex Red/HRP measurements of H_2O_2 production by isolated liver mitochondria. The calibration curve was obtained in the presence of non-respiring isolated liver mitochondria in mitochondrial assay buffer containing 1 μM Amplex Red/ 0.5 U/ml HRP by adding known concentrations of $H2O_2$. The curve was corrected for the non-specific oxidation of the dye in the absence of H_2O_2. b) Amplex Red/HRP measurements of H_2O_2 production by isolated liver mitochondria in the absence and presence of excess of catalase (1 μM bovine liver catalase)

To test the sensitivity and specificity of the system, we measured H_2O_2 and O_2 signals obtained following the addition of known amounts of H_2O_2 into the assay chamber, in the absence or presence of catalase. Results, shown upon normalization in the Figure 3, revealed that platinised carbon electrodes were able to detect sub micromolar concentration of H_2O_2 and to provide a linear response for the entire range of concentrations tested (0.5-5 μM). The H_2O_2 signal increased after each single addition of H_2O_2, and, in the absence of catalase, remained stable and summed upon sequential additions, indicating the accumulation of H_2O_2 in the assay chamber (Figure 3a, black line). On the contrary, when catalase was present, while still increasing after each addition, the H_2O_2 signal was not stable in time and rapidly disappeared, reflecting the specific scavenging of H_2O_2 by catalase (Figure 3a, blue line). Simultaneously, the Clark type electrode revealed an O_2 formation rate according to a stoichiometry of 1 mol O_2 generated per 2 mol H_2O_2 degraded, in agreement with the reaction mechanism of catalases (Oshino 1973). As expected, H_2O_2 scavenging rate by catalase increased when higher concentrations of this enzyme were used.

These results validated OMEA as a highly sensitive monitoring system that can be used for the simultaneous detection of H_2O_2 and O_2 in biological samples. Therefore, we analyzed with OMEA the respiration and H_2O_2 production of freshly isolated mouse liver mitochondria, in the presence of different respiratory substrates and inhibitors of mitochondrial respiration. When the mitochondrial electron transfer chain operates within the assay chamber, OMEA reveals reduction in O_2 concentration, due to active mitochondrial respiration, and formation of H_2O_2, due to partial reduction of O_2, which diffuses outside the mitochondria (See reaction scheme in Figure 2c).

It is known that both inside and outside mitochondria H_2O_2 is efficiently buffered by scavenging enzymes, mainly catalases and glutathione peroxidases. Since at equilibrium OMEA electrodes showed net H_2O_2 concentration after scavenging, we verified the presence of these enzymes in our mitochondrial preparations and, as expected, we confirmed their

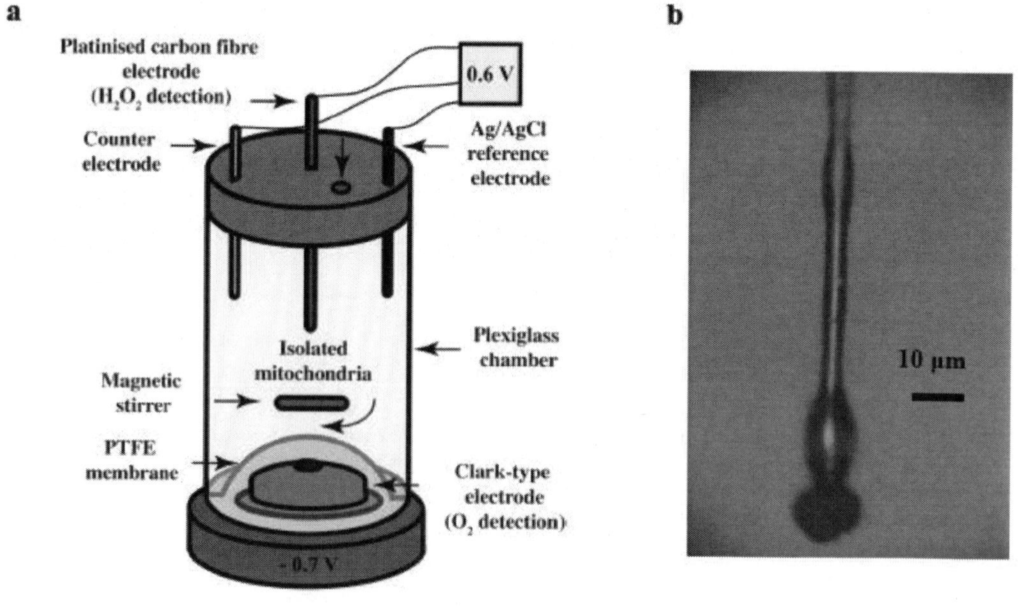

Figure 2. (Continued on next page).

c

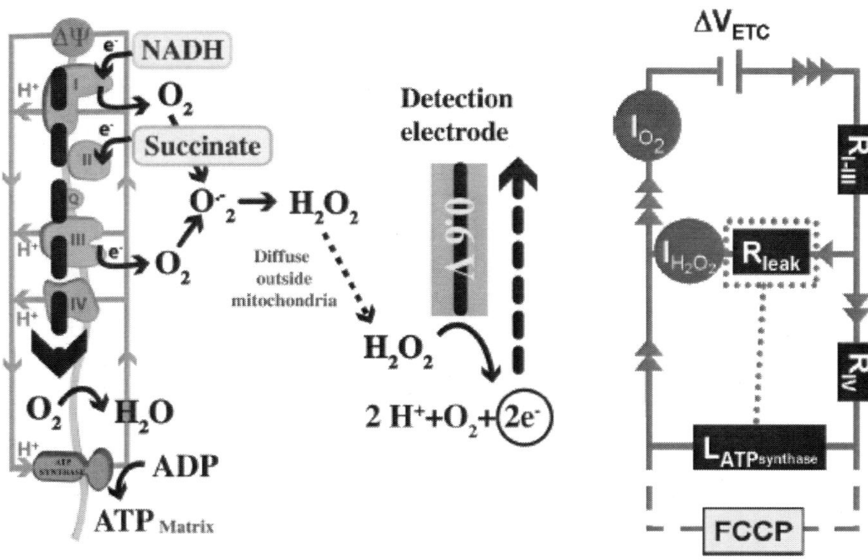

Figure 2. Electrochemical instrumentation for simultaneous measurement of H_2O_2 and O_2. a) Simultaneous detection of H_2O_2 and O_2 was achieved with a two electrodes setup located in the same assay chamber. Electrochemical detection of H_2O_2 was performed with a platinised carbon fibre electrode at an oxidation potential of 0.6 V against an Ag/AgCl reference electrode (or 0.4 V for selectivity control experiments). O_2 was detected electrochemically using a Clark-type electrode at a reduction potential of -0.7 V against an Ag/AgCl reference electrode. The electrochemical cells for the detection of these species were separated by the polytetrafluoroethylene (PTFE) membrane covering the Clark-type electrode. b) Platinised carbon fibre electrode employed for the electrochemical detection of H_2O_2: detail of the polymer and platinum depositions on the carbon fibre. c) Electrochemical detection of mitochondrial H_2O_2. Left: incomplete reduction of O_2 during electron transfer reactions at complex I and III of mitochondrial ETC leads to the formation of $O_2{\cdot}-$ on both sides of inner mitochondrial membrane. $O_2{\cdot}-$ is quickly dismutated by SODs into H_2O_2 that is either degraded inside mitochondria by glutathione peroxidases, peroxiredoxins and catalases or can diffuse outside mitochondria being removed by extramitochondrial scavenging enzymes. The efflux of mitochondrial H_2O_2 that escapes buffering is measured through amperometric assays. Right: electrical equivalent circuit, "orthogonal" to the proton circuit shown on the left, schematically representing the circulation of electrons in the ETC. The model includes (i) the source of electron motive force (emf = ΔVETC) associated to the respiratory chain; (ii) sinks of emf (RI-III and RIV) associated to the internal electron transfer resistances of respiratory chain complexes and mediators; (iii) a load (LATPase) that corresponds to ATP synthase controlled intermembrane H+ pumping; (iv) an electron leak pathway represented by a (variable) resistance Rleak. This resistance, and hence the amount of H_2O_2-generating current, receives a feedback control by the ATP synthase activity (dotted line). Two measurement gauges (amperometers) are fitted in the circuit and correspond to the two amperometric devices used to measure simultaneously the total current IO2 (coupled to O2 consumption) and the leak current IH H_2O_2 (coupled to electron leaking from the respiratory chain). Also shown in the circuit is the alternative current pathway which, upon the addition of protonophore FCCP, bypasses the ATP synthase associated load.

presence by Western blotting (not shown). Moreover, standard enzymatic activity assay measurements of total H_2O_2 scavenging together with catalase and Gpx activities validated an elevated H_2O_2 scavenging activity in isolated mouse liver mitochondrial preparations. In particular, we found a catalase activity of 80 ± 0.3 U/mg proteins and respectively a GPx activity of 0.58 ± 0.05 mU/mg proteins, suggesting a significant contribution of catalases in the scavenging process. Although these enzymes are present in mitochondrial matrix, it is likely that peroxisomal contaminations represent an important source of active catalases in preparations of isolated mouse liver mitochondria. Indeed, testing for OMEA response to pulses of H_2O_2 in a mitochondrial suspension revealed transient H_2O_2 signals and accumulating O_2 signals (Figure 3b), indicating that catalases were actively involved in the removal of H_2O_2.

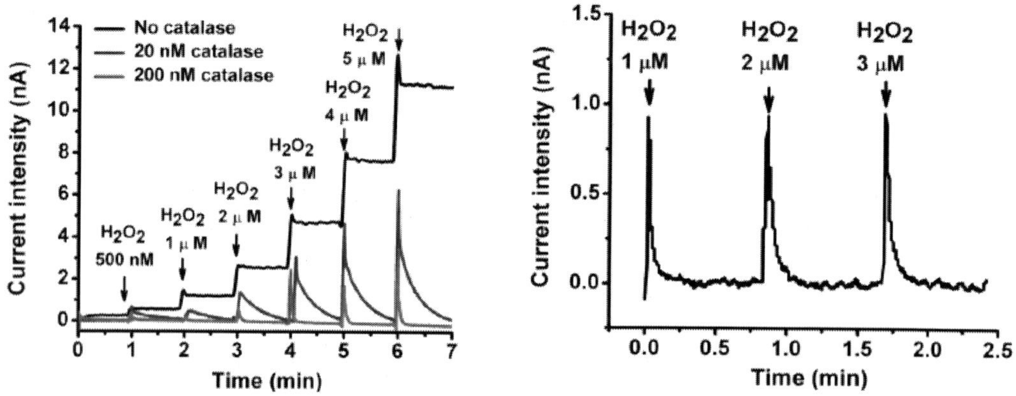

a b

Figure 3. Electrochemical detection of H_2O_2 and the contribution of scavenging enzymes present in mitochondrial preparation to the levels of H_2O_2 measured at platinised carbon fibre electrodes. a) Platinised carbon fibre electrodes detect sub-micromolar concentrations of H_2O_2 and the signal measured at the electrode is directly proportional to the concentration of H_2O_2 in the assay chamber. In the presence of different concentrations of catalase (20 nM catalase blue line and 200 nM catalase red line), single additions of H_2O_2 are rapidly removed and the kinetic of these events is measured by the electrode. b) Calibration of platinised carbon electrodes with known concentrations of H_2O_2 (1 μM, 2 μM and 3 μM) in the presence of non-energized mitochondria is accompanied by generation of O_2 with a 1:2 stoichiometry, indicating the presence of catalases in the mitochondrial preparation.

Addition of mitochondria into the assay chamber resulted in an H_2O_2 signal that was set to zero with respect to the electrode equilibration baseline. In fact, once a potential has been applied, platinised carbon fibre microelectrodes undergo an equilibration phase due to diffusion effects, slow partial oxidation and modification of the active sites of the platinised surface (Bard 2001). The outcome of this equilibration processes is a progressive and continuous reduction of the oxidation current registered at the electrode.

Then, the following addition of energetic substrates, such as glutamate/malate or succinate that provide electrons to complex I and respectively complex II of ETC, stimulated stable respiration but without inducing any sensible change in the H_2O_2 signal measured by OMEA (Figure 4a). Nevertheless, induction of state 3 respiration by addition of 200 μM ADP to mitochondria energized with either glutamate/malate or succinate, revealed by the massive and transient increase of O_2 consumption (1-1.5 minutes long), resulted in a synchronized increase of H_2O_2 signals. Despite the presence of scavenging enzymes, mainly catalase, as previously indicated, H_2O_2 accumulated in the assay chamber as long as state 3 lasted. Yet, once all ADP was phosphorylated by ATP synthase and O_2 consumption resumed to state 4 levels, the H_2O_2 signals diminished simultaneously and eventually reached levels attained prior to state 3 induction (Figure 4b).

Notably, inhibition of ATP synthase activity with oligomycin suppressed the H_2O_2 rise induced by ADP, (Figure 4b, red lines) indicating that ATP synthase activation was required for triggering H_2O_2 production observed previously. Moreover, H_2O_2 production increased as long as state 3 lasted since higher amounts of ADP prolonged both state 3 and H_2O_2 boost duration. Then, addition of exogenous catalase reduced the H_2O_2 peak level but not its duration. These findings indicate that ATP synthase activity triggers an autocatalytic/self-propagating mechanism that accelerates H_2O_2 formation that progressively prevails over scavenging.

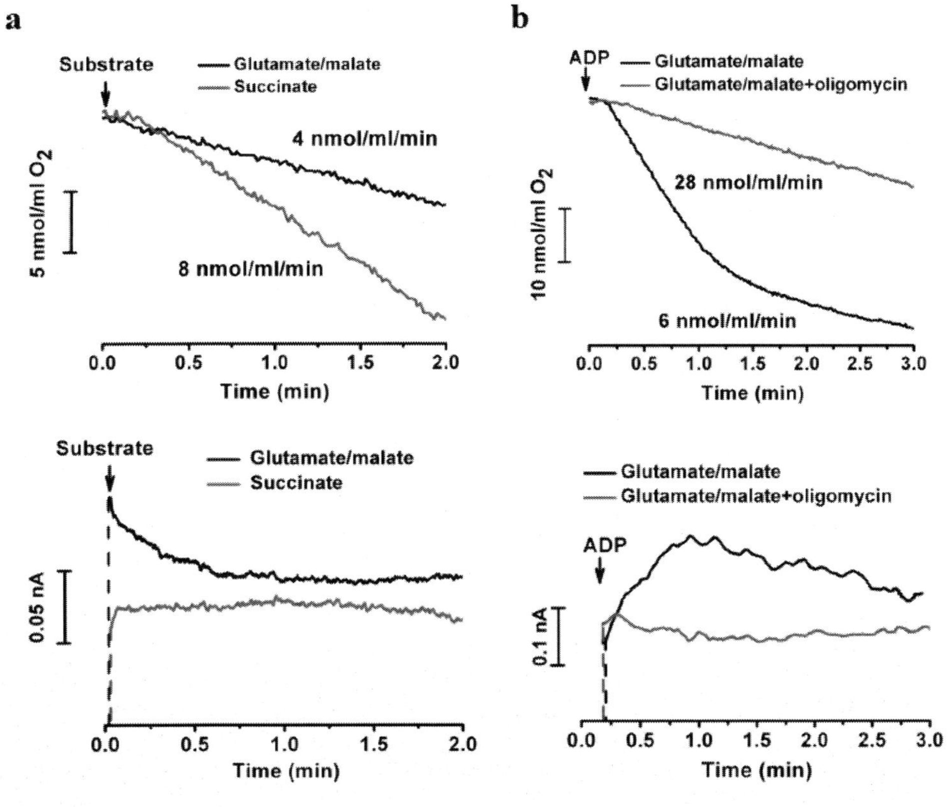

Figure 4. Simultaneous measurement of H_2O_2 production and O_2 consumption by isolated mouse liver mitochondria. Mitochondrial H_2O_2 production and O_2 consumption were measured simultaneously at the platinised carbon fibre electrode and respectively Clark-type electrode. Isolated mouse liver mitochondria, 1 mg/ml, were assayed in mitochondrial assay buffer. a) H_2O_2 production and respiration of isolated mitochondria after the addition of either glutamate/malate (5/2,5 mM) or succinate (5 mM). b) H_2O_2 production and respiration of state 4 and state 3 mitochondria ± oligomycin 200 μg/ml. Where indicated by arrow 200 μM ADP was added.

Interestingly, quantification of H_2O_2 production rate upon state 3 induction revealed a higher production of H_2O_2 when mitochondria were supplied with glutamate/malate than with succinate, despite the fact that succinate resulted in increased O_2 consumption with respect to glutamate/malate. In these conditions, the ratio H_2O_2 production/O_2 consumption (i.e. the percentage of O_2 that is reduced partially to H_2O_2) was higher in state 3 mitochondria that were utilizing glutamate and malate (2.46% ± 0.62) than succinate (0.38% ± 0.14). These data suggest that state 3 H_2O_2 formation does not strictly correlate with the overall amount of O_2 consumed.

Regardless the effect on respiration, ATP synthase activation affects transmembrane potential by inducing a moderate depolarization of mitochondrial membrane. Thus, to investigate the correlation between mitochondrial membrane depolarization and H_2O_2 production, we measured H_2O_2 signals after stimulation of mitochondria with p-trifluoromethoxy carbonyl cyanide phenyl hydrazone (FCCP), which dissipates the proton gradient across the mitochondrial inner membrane and maximizes mitochondrial respiration (uncoupled state). Addition of low concentrations of FCCP (10-30 nM) to energised mitochondria, that increased moderately the respiration rate with respect to state 4 respiration, resulted in a transient increase of H_2O_2 production (2-3 min long) followed by the removal of this specie (not shown). High concentration of FCCP (1 μM) maximised respiration rate (15.72 ± 2.3 nmol/ml/min for glutamate/malate or 52.75 ± 7.3 nmol/ml/min for succinate) and generated an instantaneous burst of H_2O_2 that was removed gradually (H_2O_2 removal rate: -0.291 ± 0.04 nmol/ml/min for glutamate/malate and -0.293 ± 0.07 nmol/mol/min for succinate). However, inducing the collapse of mitochondrial membrane potential by uniport-mediated monovalent cation influx (Nicolli 1991) (additions of A23 to mitochondria in KCl-based medium plus EDTA or NaCl to mitochondria in sucrose-based medium plus EDTA) doubled respiration rate without generating H_2O_2.Therefore it appears that a moderate depolarization rather than the collapse of mitochondrial membrane potential are effective in sustaining continuous H_2O_2 production from ETC.

How ATP synthase activation affects the rate of partial reduction of O_2 by ETC redox centers is unclear. It has already been shown that several processes can modulate the rate of electron leakage from ETC, including mitochondrial membrane potential (Starkov 2003), membrane fluidity (Schoenfeld 2008) and conformational changes in proteins of the ETC (Batandier 2004). Recent data indicate that supramolecular organization of respiratory complexes into supercomplexes (respirasomes) favors the optimal functioning of mitochondria with possible effects also on mitochondrial ROS production (Wittig 2009, Frenzel 2010). An interesting hypothesis could be that membrane depolarization associated with ATP synthesis or conformational changes occurring to FoF1-ATPase during ADP phosphorylation modify the architecture of complex I/III supercomplex altering its reactivity toward O_2 and favouring ROS production through a mechanism that competes efficiently with scavenging. Regardless of its molecular basis, ATP synthase-synchronized ROS production above the scavenging fence may be relevant if occurring *in vivo* since this event can be linked to both physiological oscillating (Zorov 2006) and pathological acute (Duchen 2000) ROS generating processes. Thus ATP synthase activation can signal through H_2O_2 to the surrounding cellular components the energetic state of mitochondria. This phenomenon can contribute to oxidative stress as well, explaining why intense mitochondrial phosphorylation, as it occurs during exercise, increases levels of intracellular ROS (Powers 2008).

CONCLUSION

Up to date, H_2O_2 production in living samples has been investigated only indirectly, mostly by virtue of its oxidative effects on endogenous molecules or exogenously added synthetic markers. Experiments using such approaches described reduced ROS formation from state 3 and uncoupled mitochondria (Korshunov 1997, Starkov 2003) while supporting the idea that hyperpolarization favours electron leakage from ETC. Nevertheless, other evidence indicates that ROS production is associated with transient mitochondrial depolarization (Huser 1998, Wang 2008).

In our conditions, measurements of H_2O_2 production by isolated mouse liver mitochondria with the Amplex Red/HRP method revealed a moderate increase in the fluorescence rate of the dye upon ADP addition with respect to state 4 respiring mitochondria (Figure 5). Although this phase appeared to be synchronous with state 3 respiration, its precise duration could not be appreciated in the absence of a system for the simultaneous measurement of O_2 consumption.

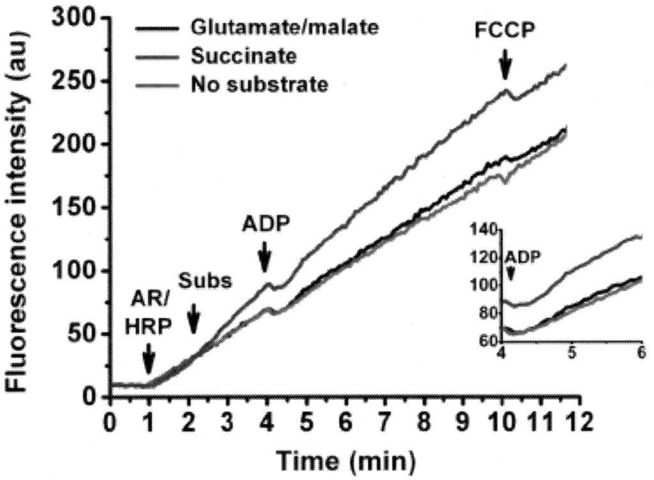

Figure 5. Amplex Red/HRP measurements of H_2O_2 production by state 3 mitochondria.
H_2O_2 production of liver mitochondria, 1 mg/ml in mitochondrial assay buffer, was assayed with a fluorimeter (excitation 560 nm, emission 590 nm) in the presence of 1 μM Amplex Red/ 0.5 U/ml HRP. Where indicated by arrows 5/2.5 mM glutamate/malate (black line) or 5 mM succinate (blue line), 200 μM ADP and 1 μM FCCP were added. As control, the additions of ADP and FCCP were also performed in the absence of respiratory substrate (black line).

Moreover we observed a significant contribution to the fluorescent signal from mitochondria in the absence of respiratory substrates most probably due to unspecific side reactions of HRP (Votyakova 2004). Since fluorescent probes such as Amplex Red are irreversibly oxidised by ROS processes related for instance to H_2O_2 scavenging cannot be appreciated and therefore the dynamic behaviour of H_2O_2 cannot be investigated. For this reason clear kinetic quantification of ROS production by respiring mitochondria was not possible.

On the contrary, OMEA allowed for the specific, sensitive, fast and reversible measurements of aerobic metabolism. The possibility to control the electrode potential ensures the selectivity and specificity of our system for the measured species. Then, since the efficiency of O_2 reduction during aerobic metabolism is usually described as the ratio between the ROS generated by mitochondria and the total amount of O_2 reduced, it is important to obtain synchronised measurements of both H_2O_2 production and O_2 consumption at the same time. Finally, electrochemical measurements are better suited for a comprehensive description of the real behaviour of the system since they allow for quantifications of both generation and scavenging processes.

Our results obtained with specific H_2O_2 and O_2 electrodes establish that isolated mouse liver mitochondria actively generate H_2O_2 above the scavenging fence upon ATP synthase activation. Under basal respiration conditions and in the presence of energetic substrates the accumulation of H_2O_2 was prevented by scavenging enzymes that at the equilibrium buffered to the minimum the concentration of H_2O_2 in the mitochondrial suspension. In particular, we found that catalases present in mitochondrial preparations are mainly responsible for the removal of H_2O_2. On the opposite, induction of state 3 respiration resulted in fast H_2O_2 generation leading to the accumulation of this specie despite the scavenging.

The molecular mechanism triggered by ATP synthase activation that favours persistent H_2O_2 production by isolated mitochondria remains puzzling. The depolarization of mitochondrial membrane could play a role in H_2O_2 generation since same phenomenon was seen upon moderate dosage of uncoupler. Besides changes in mitochondrial membrane potential, it has been shown that other processes can modulate the rate of electron leakage from ETC, including, membrane fluidity and conformational changes in proteins of the ETC. Recent data indicate that supramolecular organization of respiratory complexes into supercomplexes (respirasomes) favors the optimal functioning of mitochondria with possible effects also on mitochondrial ROS production. An interesting hypothesis could be that membrane depolarization associated with ATP synthesis or conformational changes occurring to F_0F_1-ATPase during ADP phosphorylation modify the architecture of I/III supercomplex altering its reactivity toward O_2 and enhancing $O_2^{\cdot-}$ production.

Regardless of its molecular basis, ATP synthesis-synchronized ROS production above the scavenging fence may be relevant if occurring *in vivo*. This event, in fact, can be linked to both physiological and pathological ROS generating processes. Thus ATP synthase activation can in principle signal through H_2O_2 to the surrounding cellular components the energetic state of mitochondria. This phenomenon can contribute to oxidative stress as well explaining why intense oxydative phosphorylation, as it occurs during exercise, increases levels of intracellular ROS

REFERENCES

Amatore, C. et al. Real-time amperometric analysis of reactive oxygen and nitrogen species released by single immunostimulated macrophages. *Chembiochem* 9, 1472-1480 (2008).

Arbault, S. et al. Oxidative stress in cancer prone xeroderma pigmentosum fibroblasts. Real-time and single cell monitoring of superoxide and nitric oxide production with microelectrodes. *Carcinogenesis* 25, 509-515 (2004).

Bard, A.J. & Faulkner, L.R., *Electrochemical Methods: fundamentals and applications* (Edt John Wiley & Sons Inc., New York, 2001).

Batandier, C., Leverve, X. & Fontaine, E. Opening of the mitochondrial permeability transition pore induces reactive oxygen species production at the level of the respiratory chain complex I. *J Biol Chem* 279, 17197-17204 (2004).

Bell, E.L. et al. The Qo site of the mitochondrial complex III is required for the transduction of hypoxic signaling via reactive oxygen species production. *J Cell Biol* 177, 1029-1036 (2007).

Duchen, M.R. Mitochondria and Ca(2+)in cell physiology and pathophysiology. *Cell Calcium* 28, 339-348 (2000).

Frenzel, M., Rommelspacher, H., Sugawa, M.D. & Dencher, N.A. Ageing alters the supramolecular architecture of OxPhos complexes in rat brain cortex. *Exp Gerontol* (Epub ahead of print) (2010).

Giorgio M, Trinei M, Migliaccio E, Pelicci PG. Hydrogen peroxide: a metabolic by-product or a common mediator of ageing signals? *Nat Rev Mol Cell Biol.* 8,722-8 (2007).

Gomes, A., Fernandes, E. & Lima, J.L. Fluorescence probes used for detection of reactive oxygen species. *J Biochem Biophys Methods* 65, 45-80 (2005).

Halliwell, B., Gutteridge, John M. C. (ed.) *Free radicals in biology and medicine*, Edn. Fourth (Oxford University Press Inc., New York; 2007).

Han, D., Williams, E. & Cadenas, E. Mitochondrial respiratory chain-dependent generation of superoxide anion and its release into the intermembrane space. *Biochem J* 353, 411-416 (2001).

Hoffman, D.L., Salter, J.D. & Brookes, P.S. Response of mitochondrial reactive oxygen species generation to steady-state oxygen tension: implications for hypoxic cell signaling. *Am J Physiol Heart Circ Physiol* 292, H101-108 (2007).

Hoffman, D.L. & Brookes, P.S. Oxygen sensitivity of mitochondrial reactive oxygen species generation depends on metabolic conditions. *J Biol Chem* 284, 16236-16245 (2009).

Huser, J., Rechenmacher, C.E. & Blatter, L.A. Imaging the permeability pore transition in single mitochondria. *Biophys J* 74, 2129-2137 (1998).

King, M.S., Sharpley, M.S. & Hirst, J. Reduction of hydrophilic ubiquinones by the flavin in mitochondrial NADH:ubiquinone oxidoreductase (Complex I) and production of reactive oxygen species. *Biochemistry* 48, 2053-2062 (2009).

Korshunov, S.S., Skulachev, V.P. & Starkov, A.A. High protonic potential actuates a mechanism of production of reactive oxygen species in mitochondria. *FEBS Lett* 416, 15-18 (1997).

Lambert, A.J., Buckingham, J.A., Boysen, H.M. & Brand, M.D. Diphenyleneiodonium acutely inhibits reactive oxygen species production by mitochondrial complex I during reverse, but not forward electron transport. *Biochim Biophys Acta* 1777, 397-403 (2008).

Murphy, M.P. How mitochondria produce reactive oxygen species. Biochem J 417, 1-13 (2009).

Nicolli, A., Redetti, A. & Bernardi, P. The K+ conductance of the inner mitochondrial membrane. A study of the inducible uniport for monovalent cations. J *Biol Chem* 266, 9465-9470 (1991).

Orrenius, S., Gogvadze, V. & Zhivotovsky, B. Mitochondrial oxidative stress: implications for cell death. *Annu Rev Pharmacol Toxicol* 47, 143-183 (2007).

Oshino, N., Oshino, R. & Chance, B. The characteristics of the "peroxidatic" reaction of catalase in ethanol oxidation. *Biochem J* 131, 555-563 (1973).

Powers, S.K. & Jackson, M.J., Exercise-induced oxidative stress: cellular mechanisms and impact on muscle force production. *Physiol Rev* 88, 1243-1276 (2008).

Schonfeld, P. & Wojtczak, L. Fatty acids as modulators of the cellular production of reactive oxygen species *Free Radic Biol Med* 45, 231-241 (2008).

Starkov, A.A. & Fiskum, G. Regulation of brain mitochondrial H2O2 production by membrane potential and NAD(P)H redox state. *J Neurochem* 86, 1101-1107 (2003).

Stone, J.R. & Yang, S. Hydrogen peroxide: a signaling messenger. *Antioxid Redox Signal* 8, 243-270 (2006).

Turrens, J.F. Mitochondrial formation of reactive oxygen species. *J Physiol* 552, 335-344 (2003).

Votyakova, T.V. & Reynolds, I.J. Detection of hydrogen peroxide with Amplex Red: interference by NADH and reduced glutathione auto-oxidation. *Arch Biochem Biophys* 431, 138-144 (2004).

Wallace, D.C. A mitochondrial paradigm of metabolic and degenerative diseases, aging, and cancer: a dawn for evolutionary medicine. *Annu Rev Genet* 39, 359-407 (2005).

Wang, W. et al. Superoxide flashes in single mitochondria. *Cell* 134, 279-290 (2008).

Wittig, I. & Schagger, H. Supramolecular organization of ATP synthase and respiratory chain in mitochondrial membranes. *Biochim Biophys Acta* 1787, 672-680 (2009).

Zorov, D.B., Juhaszova, M. & Sollott, S.J. Mitochondrial ROS-induced ROS release: an update and review. *Biochim Biophys Acta* 1757, 509-517 (2006).

In: Hydrogen Peroxide
Editors: Gilberto Aguilar and Raphael A. Guzman

ISBN: 978-1-62257-414-8

Chapter 8

DETERMINATION OF SCAVENGING CAPACITY AGAINST HYDROGEN PEROXIDE: RECENT TRENDS ON CHEMICAL METHODS

Marcela A. Segundo*, Carla Castro, Luís M. Magalhães and Salette Reis

REQUIMTE, Department of Chemical Sciences,
Faculty of Pharmacy, University of Porto, Porto, Portugal

ABSTRACT

Hydrogen peroxide is one of the most important reactive oxygen species and it has gained an enormous attention due to its high permeability within cell and across cellular membranes, as an in vivo precursor of strong oxidants (namely hydroxyl radical and hypochlorous acid) and due to its implication in the generation of superoxide radical anion. Therefore, the development of reliable analytical methods to determine the scavenging capacity against hydrogen peroxide has been a topic of increased interest within pharmaceutical and biomedical area. Currently, several methods are available for performing the determination of scavenging capacity against hydrogen peroxide using UV–vis spectrophotometry, fluorescence and chemiluminescence as detection systems. In this chapter, a critical review of these methods will be undertaken, by comparing the figures of merit and limitations provided by some of the most applied methods (measurement of intrinsic absorption of H_2O_2 in UV region, fluorimetric assays based in scopoletin and homovanillic acid, chemiluminescence quenching of luminol and lucigenin, among others). Moreover, special emphasis will be given to methods developed in the last three years, based on automatic systems and nanotechnology breakthroughs.

* Corresponding author. Phone +351 220428676; fax +351 226093483, E-mail: msegundo@ff.up.pt.

1. Introduction

Reactive oxygen species (ROS) are able to display a dual role in biological systems as they show positive and negative effects under physiological and pathological levels, respectively. In fact, at low concentrations ROS act as signaling and regulatory molecules while at higher concentrations they are known to behave as cytotoxic agents [1]. Indeed, when an imbalance between the ROS formation and the ability of antioxidant protection against them occurs, oxidative stress takes place and, consequently, a large number of clinical conditions including cancer, cardiovascular diseases and inflammatory states are seriously promoted [2]. Among the ROS produced in biological systems, including here superoxide anion, hydroxyl radical, peroxyl radical, and hydrogen peroxide, this last species has gained attention due to its high permeability within cell and across cellular membranes. Hydrogen peroxide is generated in vivo, under physiological conditions by peroxisomes, by several oxidative enzymes including glucose oxidase and D-amino acid oxidase, and by dismutation of superoxide radical, catalyzed by superoxide dismutase. Despite its low reactivity concerning direct oxidation of DNA and lipids, H_2O_2 is capable of inactivating important cellular enzymes by oxidation of catalytic –SH groups [3].

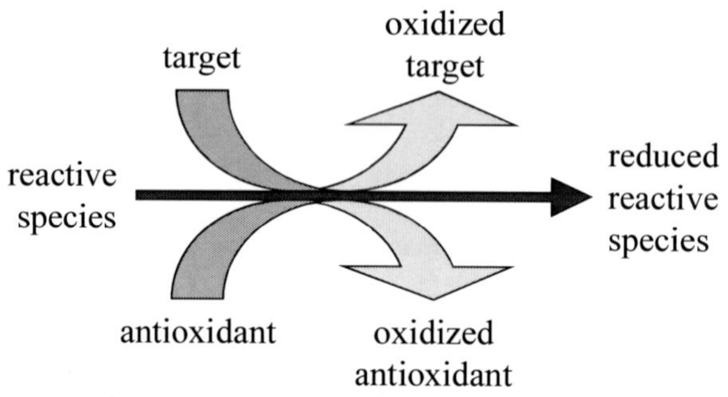

a) competitive scheme

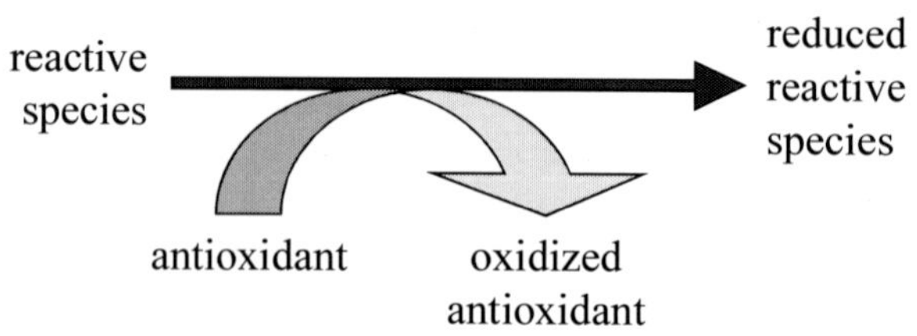

b) non-competitive scheme

Reprinted from Analytica Chimica Acta, Vol. 613, Magalhães, L. M.; Segundo, M. A.; Reis, S.; Lima, J. L. F. C. Methodological Aspects About in Vitro Evaluation of Antioxidant Properties, 1-19, Copyright (2008), with permission from Elsevier.

Figure 1. Schematic representation of competitive (a) and non-competitive (b) approaches for in vitro determination of antioxidant capacity.

Furthermore, H_2O_2 is an in vivo precursor of hydroxyl radical and it can also activate the production of superoxide anion radical. As both radicals can cause even more damage in cellular macromolecules, scavenging activity against H_2O_2 may be an important feature on treatment of pathologies where ROS generation takes place, such as inflammatory processes [1, 4]. Therefore, the development of reliable analytical methods to determine the scavenging capacity against H_2O_2 has been a topic of increased interest within pharmaceutical and biomedical area, especially concerning characterization of drug effects and screening of pharmaceutical properties in plant extracts [5].

The assessment of antioxidant capacity may rely on simple (non-competitive) or more complex (competitive) reaction schemes [5] as depicted schematically in Figure 1. In the competitive scheme, the target species, defined as any compound that represents a biomolecule (or its residue) which may be attacked in vivo, and the antioxidant compounds compete for the reactive species (H_2O_2 in our case). The assessment of antioxidant capacity is based on the quantification of a compound that facilitates the analytical measurement, defined as "probe". In most of the competitive assays, the probe is the target species or its oxidized form. Nevertheless, the probe can also be a compound added after the competitive reaction, allowing here the quantification of the remaining H_2O_2 or of the remaining target molecules.

In the non-competitive assays, antioxidant compounds will interact with H_2O_2 without the presence of any other competing target molecule. In this way, these assays involve two components in the initial reaction mixture: the antioxidant compound(s) and H_2O_2, which may also be the probe for reaction monitoring. Otherwise, the remaining H_2O_2 will be measured after addition of some derivative reagent.

In this chapter, we intend to review the methods applied to determination of scavenging activity against H_2O_2, with special emphasis on automation by flow injection based techniques.

2. Assessment of Scavenging Capacity by Batch Methods

The determination of H_2O_2 can be performed using volumetric methods or instrumental procedures based on spectrophotometric, fluorimetric, electrochemical or chemiluminescence detection [6]. The application of these techniques to assessing the scavenging capacity against H_2O_2 is summarized in Table 1. Undeniably, one of the most common applied method is based on the intrinsic absorption of H_2O_2 in the UV region [7-8]. Because the H_2O_2 concentration is decreased in the presence of scavenger compounds, the absorbance value of the reaction media at 230 nm is also decreased. Though this is a very simple non-competitive method, it presents important limitations because samples frequently absorb at the detection wavelength, requiring the performance of a "blank" measurement. The existence of sample "blanks" can compromise both the precision and the accuracy of methodologies. In this particular case, it may be difficult to distinguish between small absorbance changes (caused by weak or diluted scavengers) when there is a large background absorption. Moreover, the "blank" contribution may change after the reaction between H_2O_2 and sample components, as their structure may be modified.

Table 1. Summary of analytical features of some in vitro chemical methods for determination of scavenging capacity against H_2O_2

Target compound	Detection probe	Principle of measurement	Quantification	Recent applications
n.a.	H_2O_2	Absorbance decrease due to consumption of H_2O_2 is inhibited by antioxidants	inhibition percentage	grape extracts, lignans
HVA	biphenyl dimmer of HVA	Fluorescence increase due to oxidation of HVA is inhibited by antioxidants and measured after a fixed time period	inhibition percentage	herbal extract
n.a.	Ti-H_2O_2 complex	Absorbance increase due to formation of Ti-H_2O_2 complex is inhibited by antioxidants and measured after a fixed time period	inhibition percentage expressed as ascorbate equivalents	deerberry fruit, strawberries and blackberries
aryl oxalate ester	9,10-diphenylanthracene	CL emission due to fluorophore excitation is inhibited by antioxidants	reciprocal of inhibition ratio	oxidation products of quercetin
lucigenin	lucigenin (oxidized)	CL emission due to oxidation of lucigenin is inhibited by antioxidants	inhibition percentage	NSAIDs, pharmaceutical beta-blockers compounds
luminol	luminol (oxidized)	CL emission due to oxidation of luminol is inhibited by antioxidants	inhibition percentage	pomegranate extracts, carotenoids, retinal carotenoids

n.a.: not applicable, HVA: homovanillic acid, CL: chemiluminescence, NSAIDs: non-steroidal anti-inflammatory drugs.

Table 2. Flow-based methods for determination of scavenging capacity against H_2O_2

Flow method	Detection probe	pH value	Detection system	Application to samples	Det. rate (h^{-1})	RSD (%)	Ref.
FIA	Luminol	10.0	CL	Indolamines	n. g.	n. g.	[29]
FIA	Luminol	7.4	CL	Plant extracts	n. g.	n. g.	[30]
FIA	PSA + 4-AAP + HRP	7.4	Spect.	Pharmaceutical compounds	26	<1.0	[31]
FIA-microfluidic	CCPO + BPEA	n. g.	CL	–	14	<2	[34]
SIA	Luminol	9.5	CL	Human serum	20	<4.9	[32]
MPFS	Lucigenin	7.4	CL	Pharmaceutical formulations and tea	70	<2.0	[33]
MPFS	Luminol	Alkaline	CL	Pharmaceutical formulations and tea	160	<2.0	[33]

4-AAP, 4-aminoantipyrine; BPEA, 9,10-bis-(phenylethynyl)anthracene; CCPO, 2-carbopentyloxy-3,5,6-trichlorophenyl oxalate; CL, chemiluminescence; FIA, flow injection analysis; HRP, horseradish peroxidase; MPFS, multipumping flow system; n. g., not given; PSA, phenol-4-sulfonic acid; SIA, sequential injection analysis; Spect., Spectrophotometry.

Other frequently employed assays are based on detection of H_2O_2 upon the oxidation of colorimetric or fluorimetric substrates, catalyzed by horseradish peroxidase. Considering both types of detection, fluorimetry provides lower detection limits and the determination at lower concentration ranges, which are more suitable for assessment of biological effects. Among the fluorimetric substrates applied so far, scopoletin [9] and homovanillic acid [10] are those most frequently employed. Utilization of homovanillic acid has been reported as advantageous because its fluorescent biphenyl dimer is more stable than scopoletin [10]. For both substrates, the presence of substances with H_2O_2 scavenging capacity prevents their oxidation, which is monitored by fluorimetry.

However, peroxidase-based methods do not allow determining whether the antioxidants are reacting directly with H_2O_2 (translating their scavenging capacity), or if antioxidants are reacting with intermediates formed from the enzyme and H_2O_2 (it is possible that the superoxide radical is produced during enzyme activity). Moreover, antioxidant compounds such as ascorbic acid, quercetin dihydrate, and thiols can be a substrate for the peroxidase enzyme, introducing errors in the determination of scavenging capacity. Therefore, if the compound(s)/sample components do interfere with peroxidase-based system, other methods for determination of H_2O_2 should be used. For instance, the direct reaction of H_2O_2 and titanium(IV) has been successfully applied as it originates a complex (Ti-H_2O_2) [11] in acidic medium, assessed spectrophotometrically at 410 nm [12].

Another enzyme-free methodology was implemented through application of peroxyoxalate chemiluminescence using 9,10-diphenylanthracene as a fluorophore (probe) and imidazole as catalyst [13]. In this case, peroxyoxalate chemiluminescence involves hydrogen peroxide imidazole-catalyzed oxidation of an aryl oxalate ester yielding a high-energy intermediate (dioxetanedione) that transfers its energy to the fluorophore. The transition of the excited state of the fluorophore to its ground state causes the emission of light. Therefore, any compound with capacity to scavenge H_2O_2 promotes chemiluminescence inhibition. Due to the non-polar environment employed (ethyl acetate/acetonitrile (9:1)) this method is tailored to evaluate mainly the H_2O_2 scavenging capacity of lipophilic antioxidants, present in hydrophobic extracts of biological samples, for instance. The chemiluminometric detection was also the basis for determining H_2O_2 scavenging capacity using either luminol [14] or lucigenin [15]. Although the application of chemiluminescence-based methods for assessment of scavenging capacity against ROS is frequently, because it implies simple procedures, fast reaction and short detection times, it should be emphasized that they are not selective towards H_2O_2 as other ROS (superoxide anion, hydroxyl and peroxyl radicals) also elicits luminol and lucigenin chemiluminescence [16]. Furthermore, luminol derived radicals may also deplete the antioxidant pool in the samples, originating erroneous lower results [17].

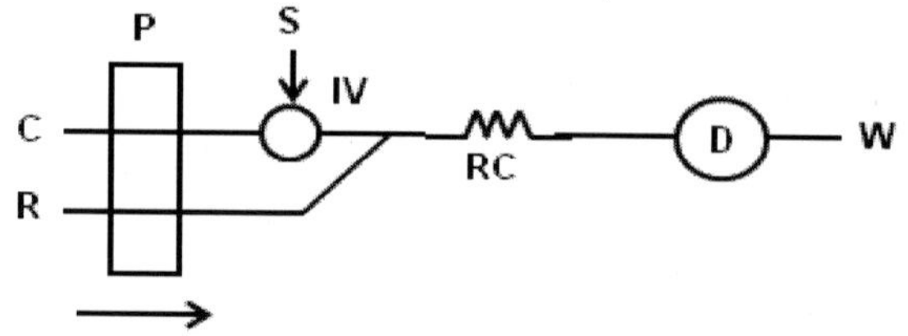

Figure 2. Schematic representation of a flow injection analysis (FIA) system. C: carrier; IV: injection valve; P: pump; D: detector; RC: reaction coil; S: sample; R: reagent; W: waste.

3. Assessment of Scavenging Capacity by Automatic Flow Injection Methods

Flow Injection Analysis (FIA) was proposed in 1975 by Ruzicka and Hansen [18] as an automation tool for chemical analysis. After 37 years, this technique is well-known and accepted by the scientific community, and it also allows the performance of assays that are not feasible when carried out manually, by taking advantage of the reproducible timing attained in these systems. The most simple FIA assembly comprehends a pumping element (peristaltic or piston driven), an injection valve, a reaction coil and a detection system, as schematically depicted in Figure 2. Flow injection methods rely on three principles: (1) reproducible sample injection or insertion in a flowing carrier stream; (2) controlled dispersion of the sample zone; and (3) reproducible timing of its movement from the injector point to the detection system [19]. The increased availability of computers in the chemical lab during the 1990's fostered the development of sequential injection analysis (SIA) [20] and also the development of flow systems relying on the flow network concept, including here multi-commutation [21] and multisyringe flow injection analysis (MSFIA) [22]. More recently, multipumping flow systems have been proposed [23]. Although based on the same principles of FIA, this new generation of computer-controlled flow systems offers more flexibility concerning method development and operation because the manifold channels are connected to computer-controlled devices (solenoid valves or micro-pumps) that enable the flexible access to reagent(s), sample and carrier in any software-defined combination [24].

In general, the reported flow based automatic methods presented several features, such as simplicity, versatility, low cost, and high sample throughput that are advantageous for rapid and reliable determination of reducing/antioxidant capacity in both pure compounds and complex matrices. Besides this, the reaction/determination takes place in a contained environment, minimizing the operator exposition to organic solvents applied essentially as matrices of sample extracts. Furthermore, the reaction conditions (time, mixing, pH) are strictly controlled, improving the repeatability of results [25-27]. A summary about the characteristics of flow injection based methods proposed for assessment of scavenging capacity against H_2O_2 is given in Table 2.

A versatile and simple FIA system based on chemiluminescence detection was developed for the determination of scavenging capacity against several ROS, including H_2O_2 [28]. A three-channel manifold (Figure 3a) was designed to accommodate the chemiluminogenic reagent (luminol), the reactive species (H_2O_2 in our case, provided from a diluted commercial solution), and the carrier stream, which transported the injected sample with antioxidant properties. After a residence time of about 5 s in the reaction coil, this mixture was subsequently merged with luminol, just before entering in the chemiluminescence-flow cell (reaction time estimated to be 1.8 s). Thus, the FIA-design exploits the consumption of H_2O_2 by antioxidant(s), which results in the appearance of negative chemiluminescence signal proportional to the scavenging ability of the compound(s). This flow system was applied for assessing the scavenging capacity of indolamine compounds [29]. However, the luminol solution applied contained added Co(II). The presence of Co(II) causes the concomitant generation of other oxygen reactive species (hydroxyl radical and superoxide anion, for instance), derived from a Fenton-like reaction [14, 16]. Although luminol was added to the mixture sample + H_2O_2 just before detection, this system may not be considered selective to this ROS.

A similar flow injection system was applied for determination of scavenging capacity against H_2O_2 of plant extracts from the Lamiaceae family [30]. In this case, the sample was injected directly in the H_2O_2 stream (Figure 3b). The mixture between the reactive species and antioxidants in the plant extracts were attained through dispersion of the sample in the carrier stream, which may constitute a drawback as a concentration gradient of H_2O_2 is formed along the sample plug. Compared to the systems described before, a pH value closer to that found in vivo (7.4 instead of 10.0) is applied, which is more adequate to study in vivo putative effects. As mentioned before, the presence of Co(II) leads to the generation of other ROS, which makes this methodology not selective towards scavenging of H_2O_2.

In order to circumvent the lack of selectivity of chemiluminesce-based methods, several colorimetric reactions were implemented in flow injection systems (Figure 4) for the determination of H_2O_2, including the oxidation of iodide to iodine, the formation of $[TiO_2]^{2+}$ complex, the formation of [Ti-xylenol orange-$O_2]^{2+}$ complex, the oxidation of 3,3',5,5'-tetramethylbenzidine and the co-oxidation of phenol-4-sulfonic acid and 4-aminoantipyrine, where these two last reactions were catalyzed by horseradish peroxidase [31].

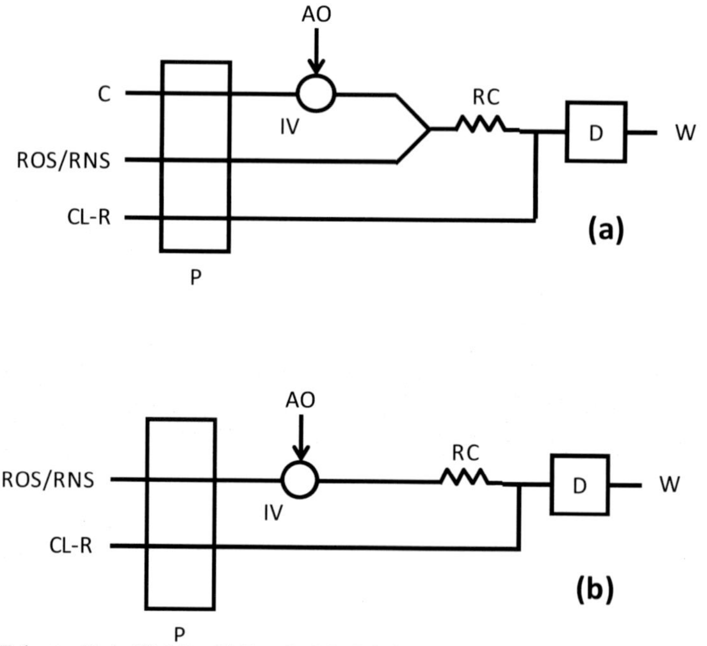

Reprinted from Talanta, Vol. 78, Magalhães, L. M.; Lúcio, M.; Segundo, M. A.; Reis, S.; Lima, J. L. F. C. Automatic Flow Injection Based Methodologies for Determination of Scavenging Capacity against Biologically Relevant Reactive Species of Oxygen and Nitrogen, 1219-1226, Copyright (2009), with permission from Elsevier.

Figure 3. Schematic representation of flow systems where the reaction between H_2O_2 and antioxidants from sample takes place before addition of a chemiluminescence detection probe. (a) FIA system adapted from ref. [28, 29]; (b) FIA system adapted from ref. [30, 34]. AO, antioxidant sample; C, carrier; CL-R, chemiluminescence reagent; D, detector; IV, injection valve; P, pump; RC, reaction coil; ROS/RNS; reactive species of oxygen or nitrogen (H_2O_2 in our case); W, waste.

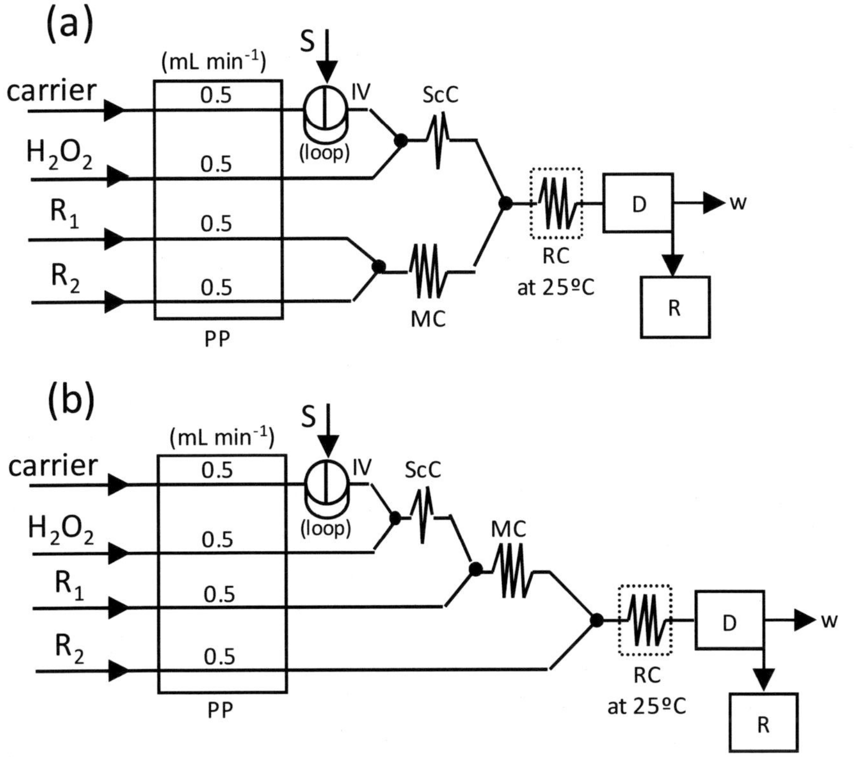

Figure 4. FIA manifold configurations applied in scavenging studies. R1 and R2, reagents; S, antioxidant solution; PP, peristaltic pump; IV, injection valve; ScC, scavenging coil; MC, mixing coil; RC, reaction coil; D, detector (spectrophotometer); R, recorder; w, waste.

The FIA methods based on the formation of titanium complexes were not suitable because no scavenging effect was observed for glutathione, a known H_2O_2 scavenger, up to 100 mM. Furthermore, the FIA method based on the oxidation of 3,3',5,5'-tetramethylbenzidine originated a high IC_{50} value (2511 μM) compared to the other two methods. The IC_{50} value represents the concentration of scavenging compound necessary to attain the decrease to half (50%) of the analytical signal provided by the initial concentration of H_2O_2, prior to the scavenging reaction. An analytical method has improved sensitivity when a lower IC_{50} value is attained for the same scavenger. When applying the same method to different samples or compounds, the lowest IC_{50} value belongs to the best scavenger.

For all reactions tested, the lowest IC_{50} value for glutathione (344 μM) was obtained for the iodide method, indicating a higher sensitivity. However, further application of this method was discarded due to the lower wavelength of detection (350 nm), which originated high blank values for other tested pharmaceutical compounds. Hence, the co-oxidation of phenol-4-sulfonic acid and 4-aminoantipyrine was further applied to samples.

The control of reaction media pH was also emphasized, where IC_{50} values from unbuffered and buffered media at pH 7.4 were different, showing the importance of this parameter when extrapolation to possible in vivo effects are expected. Finally, the enzyme superoxide dismutase was assayed as antioxidant compound and no effect was found, showing that the FIA system was selective towards H_2O_2, despite the presence of horseradish peroxidase.

The importance of assessing the scavenging capacity against different reactive species in clinical studies was highlighted by Kishikawa and co-workers [32]. Using a sequential injection system, carrier (600 µL), H_2O_2 (20 µL) and sample (5 µL) aliquots were sequentially aspirated into the holding coil and, after flow reversal, sent into a reaction coil, where they further merged with a luminol stream, just before the chemiluminescence detector. Using this methodology, serum from healthy, diabetic and rheumatic patients were analyzed and it was found a significantly higher antioxidative activity against H_2O_2 in the serum from patients with pathologies. For rheumatic patients, this was justified by the increase of catalase production upon inflammation. For diabetic patients, a low H_2O_2 production by neutrophils has been reported in previous studies, justifying the enhanced (or, more correctly, less depleted) capacity against H_2O_2 of serum samples from these patients.

Automation through multipumping flow system (MPFS) has also been pursued to determine "total antioxidant capacity", applying H_2O_2 as reactive species [33]. In this flow system, the active components are solenoid micropumps that enabled the insertion and efficient mixing of sample and reagents (H_2O_2 and chemiluminogenic reagent) as well as the transportation of the reaction zone towards the chemiluminescence-flow cell. Two chemiluminogenic reagents were tested, revealing luminol as more sensitive than lucigenin because IC_{50} values for ascorbic acid and trolox (a soluble analogue of vitamin E) were ten times lower for luminol than those obtained using lucigenin. As the detection takes place before end-point conditions, this difference may be due to the highest reaction rate observed for luminol-H_2O_2 elicited chemiluminescence than for lucigenin-H_2O_2 elicited chemiluminescence. Both strategies were applied for the determination in pharmaceutical formulations and tea extracts.

Other flow injection based systems have been developed for assessment of scavenging activity against H_2O_2. Worth of mention is the microfluidic system based on a modified peroxyoxalate chemiluminescence assay [34]. Chemiluminescence was generated from the reaction of 2-carbopentyloxy-3,5,6-trichlorophenyl oxalate (CCPO) with H_2O_2 in the presence of the fluorophore 9,10-bis-(phenylethynyl)anthracene (BPEA) and sodium salicylate used as a catalyst. Thus, when an antioxidant (β-carotene, quercetin or α-tocopherol) was present in the assay mixture, it scavenged the H_2O_2 and quenched the production of light. In the 2-inlet microfluidic system devised, the antioxidant plugs were injected into the hydrogen peroxide stream, which afterwards was merged with the flow stream containing CCPO + BPEA + sodium salicylate. It should be highlighted that this system is suitable for compounds with poor solubility in aqueous media as the reaction takes place in a 3:7 mixture of ethyl acetate:acetonitrile that was used for preparation of all solutions.

4. Recent Trends

As mentioned in the introduction section, competitive evaluation of scavenging activity may be more relevant than measuring the simple interaction between antioxidants and H_2O_2.

Therefore, new trends on this field were brought recently by an automatic multisyringe flow injection analysis system, where the scavenging capacity against H_2O_2 was assessed through a non-competitive (antioxidant + H_2O_2) or a competitive (antioxidant + H_2O_2 + biological target) scheme without physical reconfiguration of the manifold [35]. Moreover, the developed method was applied to in vitro evaluation of scavenging capacity against H_2O_2 using reaction conditions similar to those found in vivo, concerning the oxidant concentration (100 μM), temperature (37°C) and pH 6.9 and also a low scavenging reaction time (20 s). The results obtained demonstrated that, despite the fact of endogenous antioxidant compounds having higher ability to scavenge H_2O_2 when compared to biological targets (showed through application of the non-competitive scheme), their antioxidant profile in the presence of an extra competitive molecule depends on the concentration of the antioxidant as well as on the target molecule present.

In fact, when a non-competitive scheme was applied, IC_{50} values for the antioxidant compounds glutathione (1191 μM) and pyruvate (446 μM) were lower than those obtained for biological targets such as cysteine (2616 μM), taurine (359 mM) and adenine (2224 μM), indicating that reactivity towards H_2O_2 was higher for antioxidant compounds than for biological targets. However, when a competitive scheme was applied, the scavenging effectiveness against H_2O_2 depended on the biological molecule present (Figure 5), showing clearly that antioxidant assessment should also take into consideration the concomitant reactivity of biological molecules or structures that are prone to oxidative damage. For instance, glutathione did not exert any scavenging effect against H_2O_2 in the presence of taurine or adenine, but a different profile was observed as depicted in Figure 5A and B. Some interaction was observed between glutathione and cysteine (Figure 5C), but it was not possible to assess which species reacted in a lower extension with H_2O_2.

Nanotechnology is undeniably changing several aspects in Chemistry, including in Analytical Chemistry. The assessment of scavenging activity against H_2O_2 is not an exception and a methodology based on gold nanoshells as an enzyme-free optical nanoprobe for assessing H_2O_2 has been recently proposed [36]. For this, gold nanoparticles were preadsorbed in silica nanoparticles, forming nanocomposites where the gold nanoparticles acted as nucleation sites for Au deposition through reduction of Au(III) by H_2O_2. As the concentration of H_2O_2 increases, the gold nanoparticles on the nanocomposites are enlarged until continuous nanoshells are formed. During the growth procedure, the spectra changes (increase of absorption intensity and shift of absorption maximum towards larger wavelengths) are correlated to the concentration of H_2O_2 present in the reaction media. Therefore, the formation of Au nanoshells is a suitable procedure to measure low concentrations of H_2O_2 (at μM range) using spectrophotometry. Furthermore, as the concentration of H_2O_2 was directly related to the wavelength shift, a direct visualization of the scavenging effect was possible by simply accompanying the color change in the reaction media from light pink (high amount of antioxidants) towards blue (low amount of antioxidants), depending upon the quantity of antioxidant (or remaining H_2O_2).

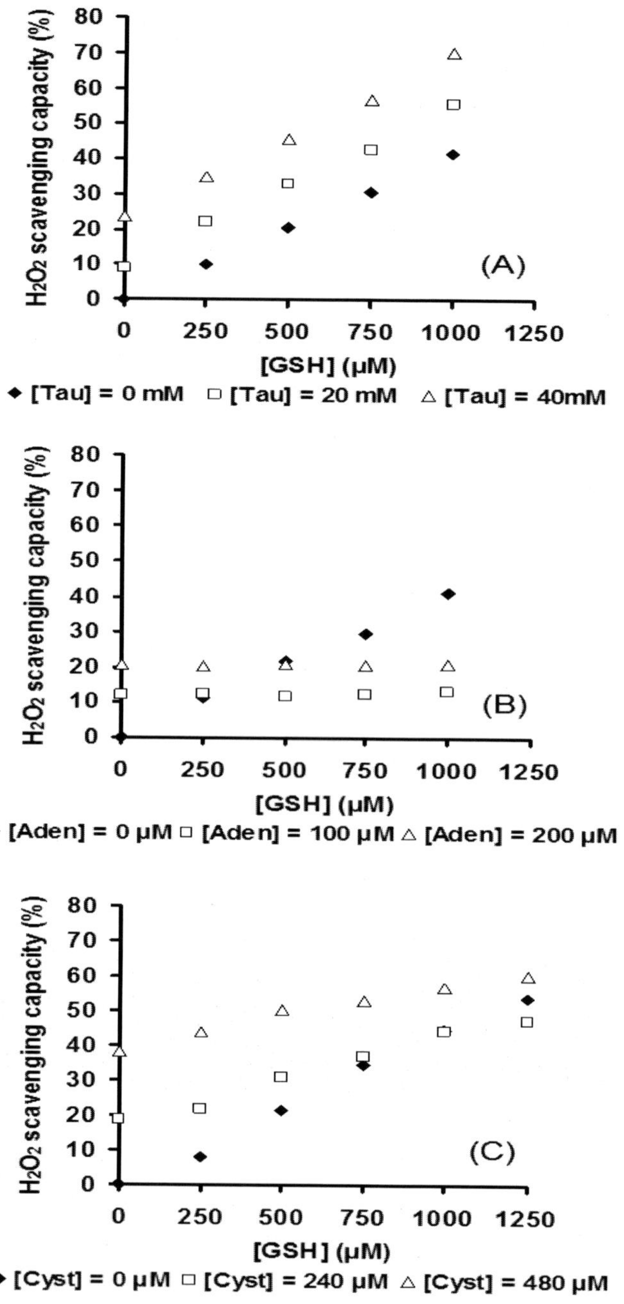

Reprinted from Talanta, Vol. 81, Ribeiro, J. P. N.; Magalhães, L. M.; Segundo, M. A.; Reis, S.; Lima, J. L. F. C. Hydrogen Peroxide, Antioxidant Compounds and Biological Targets: An in Vitro Approach for Determination of Scavenging Capacity Using Fluorimetric Multisyringe Flow Injection Analysis, 1840-1846, Copyright (2010), with permission from Elsevier.

Figure 5. Scavenging capacity against H_2O_2 by glutathione (GSH) using taurine (A, Tau), adenine (B, Aden) and cysteine (C, Cyst) as competitive molecule.

The same nanoprobes were employed on a surface-enhanced Raman scattering (SERS) method for estimation of antioxidant activity using self-assembled three-dimensionally (3D) ordered gold nanoparticles arrays [37]. In this case, the 3D ordered nanostructure arrays were a technically important breakthrough as they can lead to improved and more practical applications of optical, catalytic, and biosensing performances of nano-structures. The construction of 3D arrays started by modifying the silica spheres, covering their surfaces with positively charged amino groups. Subsequently the modified spheres formed closely packed arrays on the glass substrates by the vertical deposition method and the silica templates were dipped into a beaker of aqueous gold nanoparticles with vigorous stirring. During this process, the negatively charged gold nanoparticles became attached onto the amino groups of the modified silica sphere surfaces by electrostatic interactions. Then, these attached gold nanoparticles worked as nucleation sites for the growth of a gold shell layer, which was mediated by H_2O_2 and whose extension depended on H_2O_2 concentration. During this process, concomitant with the increase of gold coverage on silica cores, the SERS spectrum of Raman report molecule Nile blue A, which used gold nanoshell arrays as substrates, presented a clear change in intensity, which was quantitatively correlated with the concentration of H_2O_2. Both methodologies represent a step forward the detection of lower amounts of H_2O_2 and they were successfully applied to pure antioxidant compounds, but application to real samples is still lacking.

CONCLUSION

Current methods for assessment of scavenging activity against H_2O_2 rely essentially in chemiluminescence methods, because it provides low detection limits required for studying low concentrations of H_2O_2 and antioxidants found in vivo. However, in most of the methodologies, results may be biased because of concomitant formation of other ROS, namely superoxide anion. For excluding this problem, testing with superoxide dismutase is mandatory to evaluate if the chemiluminescence signal is elicited or quenched by this last species.

Although some technological developments through nanosciences have brought new perspectives for application of spectrophotometric methods or more sophisticated detection techniques, such as surface-enhanced Raman scattering, the improvement of reaction conditions, capable to cope with real samples, are still needed.

If in vivo meaningful results are expected, reaction conditions closer to those found in vivo should be implemented, concerning pH, temperature, concentration of H_2O_2, time frame of its generation and, most importantly, the presence of biological targets. Therefore, though more complex, methods based on competitive schemes, where a biological target molecule is also present, should be pursued in the future.

ACKNOWLEDGMENTS

Fundação para a Ciência e a Tecnologia (FCT) is acknowledged for the financial support through Strategic Project PEst-C/EQB/LA0006/2011. L. M. Magalhães thanks FSE (Fundo

Social Europeu) and MCTES (Ministério da Ciência, Tecnologia e Ensino Superior) for the financial support through the POPH-QREN program. C. Castro also thanks FCT for PhD grant SFRH/BD/41627/2008.

References

[1] Halliwell, B.; Gutteridge, J. M. C., *Free Radicals in Biology and Medicine*. 4th ed.; Oxford University Press: Oxford, 2007.

[2] Rees, M. D.; Kennett, E. C.; Whitelock, J. M.; Davies, M. J. Oxidative Damage to Extracellular Matrix and Its Role in Human Pathologies. *Free Radic. Biol. Med.* 2008, *44*, 1973-2001.

[3] Valko, M.; Leibfritz, D.; Moncol, J.; Cronin, M. T. D.; Mazur, M.; Telser, J. Free Radicals and Antioxidants in Normal Physiological Functions and Human Disease. *Int. J. Biochem. Cell Biol.* 2007, *39*, 44-84.

[4] Halliwell, B. Phagocyte-Derived Reactive Species: Salvation or Suicide? *Trends Biochem. Sci.* 2006, *31*, 509-515.

[5] Magalhães, L. M.; Segundo, M. A.; Reis, S.; Lima, J. L. F. C. Methodological Aspects About in Vitro Evaluation of Antioxidant Properties. *Anal. Chim. Acta* 2008, *613*, 1-19.

[6] Kohen, R.; Nyska, A. Oxidation of Biological Systems: Oxidative Stress Phenomena, Antioxidants, Redox Reactions, and Methods for Their Quantification. *Toxicol. Pathol.* 2002, *30*, 620-650.

[7] Beers, R. F.; Sizer, I. W. A Spectrophotometric Method for Measuring the Breakdown of Hydrogen Peroxide by Catalase. *J. Biol. Chem.* 1952, *195*, 133-140.

[8] Ruch, R. J.; Cheng, S. J.; Klaunig, J. E. Prevention of Cyto-Toxicity and Inhibition of Intercellular Communication by Antioxidant Catechins Isolated from Chinese Green Tea. *Carcinogenesis* 1989, *10*, 1003-1008.

[9] Corbett, J. T. The Scopoletin Assay for Hydrogen-Peroxide - a Review and a Better Method. *J. Biochem. Biophys. Methods* 1989, *18*, 297-307.

[10] Pazdzioch-Czochra, M.; Widenska, A. Spectrofluorimetric Determination of Hydrogen Peroxide Scavenging Activity. *Anal. Chim. Acta* 2002, *452*, 177-184.

[11] Sellers, R. M. Spectrophotometric Determination of Hydrogen-Peroxide Using Potassium Titanium(IV) Oxalate. *Analyst* 1980, *105*, 950-954.

[12] Wang, S. Y.; Jiao, H. J. Scavenging Capacity of Berry Crops on Superoxide Radicals, Hydrogen Peroxide, Hydroxyl Radicals, and Singlet Oxygen. *J. Agric. Food Chem.* 2000, *48*, 5677-5684.

[13] Arnous, A.; Petrakis, C.; Makris, D. P.; Kefalas, P. A Peroxyoxalate Chemiluminescence-Based Assay for the Evaluation of Hydrogen Peroxide Scavenging Activity Employing 9,10-Diphenylanthracene as the Fluorophore. *J. Pharmacol. Toxicol. Meth.* 2002, *48*, 171-177.

[14] Yildiz, G.; Demiryurek, A. T.; Sahin-Erdemli, I.; Kanzik, I. Comparison of Antioxidant Activities of Aminoguanidine, Methylguanidine and Guanidine by Luminol-Enhanced Chemiluminescence. *Br. J. Pharmacol.* 1998, *124*, 905-910.

[15] Chang, M. C.; Uang, B. J.; Wu, H. L.; Lee, J. J.; Hahn, L. J.; Jeng, J. H. Inducing the Cell Cycle Arrest and Apoptosis of Oral Kb Carcinoma Cells by Hydroxychavicol:

Roles of Glutathione and Reactive Oxygen Species. *Br. J. Pharmacol.* 2002, *135*, 619-630.

[16] Parejo, L.; Codina, C.; Petrakis, C.; Kefalas, P. Evaluation of Scavenging Activity Assessed by Co(II)/EDTA-Induced Luminol Chemiluminescence and DPPH$^{\cdot +}$ (2,2-Diphenyl-1-Picrylhydrazyl) Free Radical Assay. *J. Pharmacol. Toxicol. Meth.* 2000, *44*, 507-512.

[17] Lu, C.; Song, G.; Lin, J. M. Reactive Oxygen Species and Their Chemiluminescence-Detection Methods. *Trac-Trends Anal. Chem.* 2006, *25*, 985-995.

[18] Ruzicka, J.; Hansen, E. H. Flow Injection Analyses. 1. New Concept of Fast Continuous-Flow Analysis. *Anal. Chim. Acta* 1975, *78*, 145-157.

[19] Ruzicka, J.; Hansen, E. H., *Flow Injection Analysis*. 2nd ed.; John Wiley and Sons: New York, 1988.

[20] Ruzicka, J.; Marshall, G. D. Sequential Injection - a New Concept for Chemical Sensors, Process Analysis and Laboratory Assays. *Anal. Chim. Acta* 1990, *237*, 329-343.

[21] Reis, B. F.; Gine, M. F.; Zagatto, E. A. G.; Lima, J. L. F. C.; Lapa, R. A. Multicommutation in Flow-Analysis. 1. Binary Sampling - Concepts, Instrumentation and Spectrophotometric Determination of Iron in Plant Digests. *Anal. Chim. Acta* 1994, *293*, 129-138.

[22] Cerdà, V.; Estela, J. M.; Forteza, R.; Cladera, A.; Becerra, E.; Altimira, P.; Sitjar, P. Flow Techniques in Water Analysis. *Talanta* 1999, *50*, 695-705.

[23] Lapa, R. A. S.; Lima, J. L. F. C.; Reis, B. F.; Santos, J. L. M.; Zagatto, E. A. G. Multi-Pumping in Flow Analysis: Concepts, Instrumentation, Potentialities. *Anal. Chim. Acta* 2002, *466*, 125-132.

[24] Segundo, M. A.; Magalhães, L. M. Multisyringe Flow Injection Analysis: State-of-the-Art and Perspectives. *Anal. Sci.* 2006, *22*, 3-8.

[25] Magalhães, L. M.; Lúcio, M.; Segundo, M. A.; Reis, S.; Lima, J. L. F. C. Automatic Flow Injection Based Methodologies for Determination of Scavenging Capacity against Biologically Relevant Reactive Species of Oxygen and Nitrogen. *Talanta* 2009, *78*, 1219-1226.

[26] Magalhães, L. M.; Santos, M.; Segundo, M. A.; Reis, S.; Lima, J. L. F. C. Flow Injection Based Methods for Fast Screening of Antioxidant Capacity. *Talanta* 2009, *77*, 1559-1566.

[27] Magalhães, L. M.; Ribeiro, J. P. N.; Segundo, M. A.; Reis, S.; Lima, J. L. F. C. Multi-Syringe Flow-Injection Systems Improve Antioxidant Assessment. *Trac-Trends Anal. Chem.* 2009, *28*, 952-960.

[28] Sariahmetoglu, M.; Wheatley, R. A.; Cakici, I.; Kanzik, I.; Townshend, A. Flow Injection Analysis for Monitoring Antioxidant Effects on Luminol Chemiluminescence of Reactive Oxygen Species. *Anal. Lett.* 2003, *36*, 749-765.

[29] Sariahmetoglu, M.; Wheatley, R. A.; Cakycy, Y.; Kanzyk, Y.; Townshend, A. Evaluation of the Antioxidant Effect of Melatonin by Flow Injection Analysis-Luminol Chemiluminescence. *Pharmacol. Res.* 2003, *48*, 361-367.

[30] Erdemoglu, N.; Turan, N. N.; Cakici, I.; Sener, B.; Aydin, A. Antioxidant Activities of Some Lamiaceae Plant Extracts. *Phytother. Res.* 2006, *20*, 9-13.

[31] Ribeiro, J. P. N.; Segundo, M. A.; Reis, S.; Lima, J. L. F. C. Spectrophotometric FIA Methods for Determination of Hydrogen Peroxide: Application to Evaluation of Scavenging Capacity. *Talanta* 2009, *79*, 1169-1176.

[32] Kishikawa, N.; Ohyama, K.; Yao, J.; Miyamoto, A.; Imazato, T.; Ueki, Y.; Nakashima, K.; Maehata, E.; Kuroda, N. Automated Analysis of the Serum Antioxidative Activities against Five Different Reactive Oxygen Species by Sequential Injection System with a Chemiluminescence Detector. *Clin. Chim. Acta* 2010, *411*, 1111-1115.

[33] Meneses, S. R. P.; Marques, K. L.; Pires, C. K.; Santos, J. L. M.; Fernandes, E.; Lima, J. L. F. C.; Zagatto, E. A. G. Evaluation of the Total Antioxidant Capacity by Using a Multipumping Flow System with Chemiluminescent Detection. *Anal. Biochem.* 2005, *345*, 90-95.

[34] Amatatongchai, M.; Hofmann, O.; Nacapricha, D.; Chailapakul, O.; De Mello, A. J. A Microfluidic System for Evaluation of Antioxidant Capacity Based on a Peroxyoxalate Chemiluminescence Assay. *Anal. Bioanal. Chem.* 2007, *387*, 277-285.

[35] Ribeiro, J. P. N.; Magalhães, L. M.; Segundo, M. A.; Reis, S.; Lima, J. L. F. C. Hydrogen Peroxide, Antioxidant Compounds and Biological Targets: An in Vitro Approach for Determination of Scavenging Capacity Using Fluorimetric Multisyringe Flow Injection Analysis. *Talanta* 2010, *81*, 1840-1846.

[36] Li, H.; Ma, X. Y.; Dong, J.; Qian, W. P. Development of Methodology Based on the Formation Process of Gold Nanoshells for Detecting Hydrogen Peroxide Scavenging Activity. *Anal. Chem.* 2009, *81*, 8916-8922.

[37] Rao, Y. Y.; Chen, Q. F.; Dong, J. A.; Qian, W. P. Growth-Sensitive 3D Ordered Gold Nanoshells Precursor Composite Arrays as SERS Nanoprobes for Assessing Hydrogen Peroxide Scavenging Activity. *Analyst* 2011, *136*, 769-774.

In: Hydrogen Peroxide
Editors: Gilberto Aguilar and Raphael A. Guzman
ISBN: 978-1-62257-414-8

Chapter 9

HYDROGEN PEROXIDE: A VERSATILE AND ENVIRONMENTALLY BENIGN OXIDANT OF NATURAL OLEFINS ABLE WITH DIFFERENT CATALYSTS

Ligia Mendonça Maria Vieira*[1], *Milene Lopes da Silva*[1] *and Márcio José da Silva*[1]
[1]Federal University of Viçosa, Chemistry Department,
Viçosa, Minas Gerais, Brazil

ABSTRACT

This chapter summarizes our recent developments for H_2O_2 based green oxidation reactions, using two different transition metal catalysts: palladium and cobalt. Results obtained from the oxidation of the natural olefins using as oxidant the hydrogen peroxide and as catalysts palladium and cobalt chlorides are presented and discussed in mechanistic terms. Monoterpenes are abundant and renewable raw materials and when converted into oxygenates products are valuable ingredient for fragrance, agrochemicals and pharmaceutical industries. Camphene, β-pinene and 3-carene are monoterpenes of high natural occurrence, which oxygenated derivates, are highly attractive for fine chemical and fragrances industry and for these reasons were the selected substrates. The different aspects involved in the oxidation of these substrates in presence of two different catalysts were assessed. Palladium(II)-catalyzed oxidation reactions proceeds via two electrons intermediates (heterolytic oxidation); conversely, Cobalt(II)-catalyzed oxidation reactions occurs via one electrons intermediates (homolytic oxidation). In all most of the catalytic reactions, high conversion substrates were achieved. Moreover, the selectivity of the oxidation products can be controlled by molar ratio substrate:oxidant and by temperature of reaction. Hydrogen peroxide, an environmentally benign reactant, regardless catalyst employed was an efficient oxidant in the conversion of monoterpenes into valuable oxygenates derivates.

Keywords: palladium catalysts, cobalt catalysts, natural olefins oxidation, hydrogen peroxide

---

* Email: silvamj2003@ufv.br

1. Introduction

1.1. Monoterpenes: Oxidation Processes

The oxidation of natural olefins as the monoterpenes is an important method for valorization of these inexpensive, abundant, and renewable natural raw materials [1]. Oxygenated monoterpenes are largely employed in organic synthesis processes as chiral building blocks, and are valuable ingredients for the fragrance, agrochemicals, pharmaceutical and fine chemical industries [2-4]. Nevertheless, their production normally occurs via stoichiometric oxidation reactions, which results generally in low yield and poor selectivity. Moreover, the use of expensive metallic oxidants and toxic is also generally required. Thus, to develop oxidative processes based on clean oxidants like as hydrogen peroxide, which together with active catalysts can be used in the valorization of renewable raw materials becomes important from the view point economical and environmental [5].

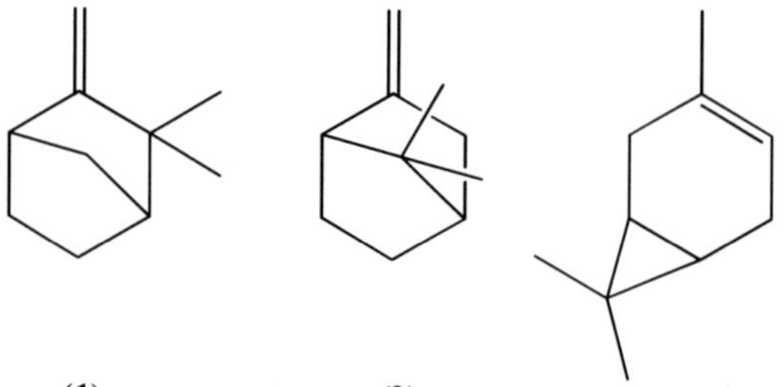

(1) (2) (3)

Figure 1. Monoterpenes: camphene (1), β-pinene (2) and 3-carene (3)

1.2. Hydrogen Peroxide as Oxidant in the Olefins Oxidation Processes

Among the stoichiometric oxidants able in metal transition-catalyzed oxidation reactions, the hydrogen peroxide can be regarded as desirable "greener oxidants", due to its high atom efficiencies, high active oxygen content, and also because it generates only water as by-product [6]. Another oxidant used in oxidation processes of olefin compounds is the *tert*-butylhydroperoxide (TBHP). Nevertheless, the best option as an oxidizing agent is hydrogen peroxide, because it presents a high amount of active oxygen by mass unit, is an inexpensive commercial product used at industrial scale, is less corrosive and easily handling than TBHP, even gives only water as a by-product [7].

The hydrogen peroxide has significant advantages comparatively to oxygen molecular, another clean oxidant commonly used in catalytic process of olefins oxidation. Hydrogen peroxide is liquid reactant and for this reason more easily handling than molecular oxygen; is also an inexpensive and available commercially reactant [8]. However, the major advantage of hydrogen peroxide is that its use on palladium-catalyzed olefin oxidation reactions prevents that a metal reoxidant is used. Indeed, on the Pd(II)-catalyzed oxidation reactions is common that Pd(0) species will be formed throughout reactions. However, they can be directly oxidized into Pd(II) by hydrogen peroxide. Contrarily, when the oxidant employed in olefins oxidation reactions is the molecular oxygen (i.e. as in the Wacker system), a reversible reoxidant (i.e. $CuCl_2$) is also requires for oxidize Pd(0) into Pd(II) [9,10]. This additional necessity, becomes the $PdCl_2/CuCl_2$ Wacker catalyst efficient only when the monoterpenes oxidized have simple structure (i.e. monocyclic or acyclic) [11]. Actually, the presence of an additional reoxidant causes concurrent reactions due to Lewis acidity of Cu(II) ions, which to favor carbon skeletal rearrangement, as well as the formation of undesired organochlorine compounds [12].

Peroxide oxidants have been employed in monoterpenes oxidation with different metal catalysts, including manganese or cobalt porphyrin complexes [13,14]. However, the laborious syntheses as well as the requirement of a co-catalyst (i.e. ammonium acetate) are the main drawbacks of porphyrin complexes catalysts.

1.3. Hydrogen Peroxide as Oxidant in Palladium-Catalyzed Olefins Oxidation Processes

Because hydrogen peroxide is able to directly oxidize the palladium catalysts, it has shown to be an attractive alternative in relation to molecular oxygen (Figure 2) [15-18]. However, there is scarce information available in literature concerning to its use in palladium-catalyzed natural olefin oxidation reactions. Mimoun and co-workers reported one of the first works employing hydrogen peroxide as the final oxidant in Pd(II)-catalyzed olefin oxidation reactions [19,20].

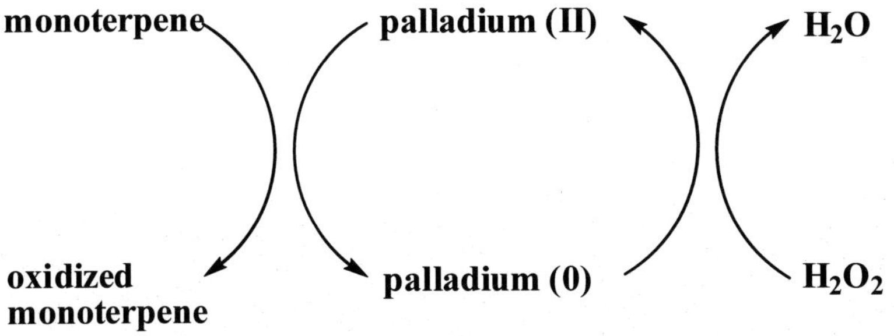

Figure 2. Redox coupling palladium catalyst/ monoterpenes/ hydrogen peroxide

The oxidation of monoterpenes catalyzed by Pd(II) salts has led to several reports relating their allylic oxidation, epoxidation and other oxidative transformations [21-23]. The oxidative processes where occur allylic oxidation are especially common in Pd-catalyzed reactions, and has been mainly developed for the synthesis of allylic esters from natural olefins [19]. Although monoterpenes oxidation into allylic products via Pd(II)-catalyzed reactions using hydrogen peroxide has been already described, the control of selectivity is still the main challenge to be overcome in those processes [22]. A remarkably instance is the monoterpenes oxidation in HOAc solutions [23]. In this case, carbon skeletal rearrangement reactions and nucleophilic addition of the solvent are most favored and compromise the selectivity of the oxidation products [23]. The use of CH_3CN prevents these problems; however, the palladium reoxidation rate is affected by solvent nature (i.e. polarity, acidity) and is faster only in protic solvents [24].

1.4. Hydrogen Peroxide as Oxidant in Cobalt-Catalyzed Olefins Oxidation Processes

The processes of catalytic oxidation have becoming most attractive when use clean and inexpensive oxidants like as hydrogen peroxide, which results in water as only by-product. However, is important to note that contrarily to processes involving molecular oxygen, where a typical two-electron transfer process occur in almost cases (heterolytic oxidation), in the oxidation reactions involving hydrogen peroxide as oxidant, a one electron transfer process is in general observed (homolytic oxidation). Several works have described the use of metal transition catalysts in radical reactions. The use of organic acids which can form stronger oxidants radical has also been described [25]. In this case, are formed radicals from organic peracids which are highly reactive troughout the oxidation reactions [26].

2. Recents Advances Obtained on the Palladium-Catalyzed Monoterpenes Oxidation Processes with Hydrogen Peroxide

Control of selectivity is one of the greatest challenges in the monoterpenes oxidation reactions. The presence of a metallic reoxidant as well as solvent acidity promotes undesirable concurrent reactions of these substrates, compromising oxidation selectivity [27]. Recently, we developed a metallic reoxidant-free system with an aprotic solvent (i.e. $PdCl_2/H_2O_2/CH_3CN$ system) which successfully promoted the oxidation of monoterpenes as camphene (**1**) [28].

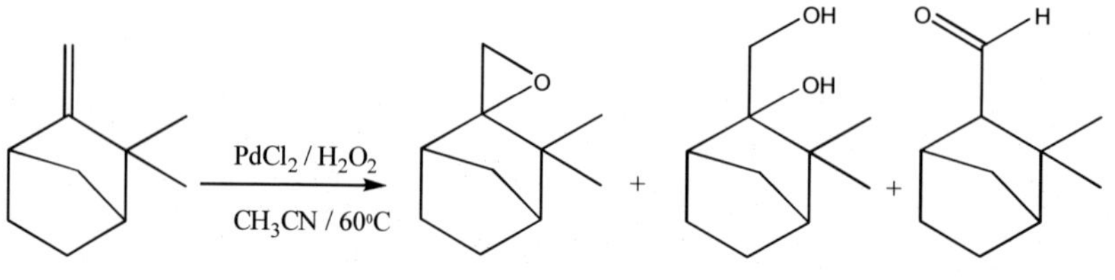

Figure 3. Oxidation products of camphene [28]

In reactions carried out in absence of catalyst containing only hydrogen peroxide (3.0 mmols) at 60 °C temperature, non detectable amounts of oxidation products was found (run 1, table 1). Contrarily, the catalyst $PdCl_2$ effectively promoted the selective conversion of camphene into oxidation products (runs 2-8, Table 1), identified by GC-MS and NMR spectroscopy as epoxy-camphene (**1a**), glycol-camphene (**1b**) and aldehyde camphene (**1c**) (Figure 3).

The effects of the $PdCl_2$ concentration and the H_2O_2/camphene molar ratio on conversion and selectivity of the reaction are presented in Table 1. An increase on $PdCl_2$ concentration increases notably the selectivity of formation of glycol camphene (**1b**), and consequently reduces the formation of aldehyde camphene (**1c**) (runs 2-5, Table 1). It should be highlighted that two isomers of camphene glycol (**1b**) were formed along of the reaction, contrarily to epoxyde (**1a**), which not presented isomers.

Table 1. Oxidation of camphene by H_2O_2 catalyzed by $PdCl_2$ in CH_3CN solutions [28][a]

Run	$PdCl_2$ (mmols)	H_2O_2 (mmols)	Conversion into oxidation products[b] (%)	Conversion into others products[b-c] (%)	Selectivity[b] (%)		
					1a	1b	1c
1	0	3.0	Traces	8	-	-	-
2	0.10	3.0	78	15	8	49	43
3	0.20	3.0	83	13	6	62	32
4[d]	0.30	3.0	86	11	2	67	31
5[d]	0.40	3.0	90	10	3	64	33
6	0.10	1.5	32	8	90	5	5
7	0.10	6.0	60	27	5	35	60
8	0.10	9.0	57	38	8	40	52

[a]Reaction conditions: camphene (2.50 mmols); H_2O_2 (1.5, 3.0, 6.0 or 9.0 mmols) were stirred at 60°C in 10.0 mL of CH_3CN; 14 h; conversion determined by GC analysis.
[b]Complexe mixture of products probably formed via hydration of olefin double bond, skeletal rearrangement allowed of H_2O addition and isomerization, such as hydroxi-camphene, borneol and trycyclene, respectively.
[c]Majority formation of a product identified as hydroxi-camphene.
[d]Conversion reached after 6 h reaction.

The presence of water in the solution can be undesirable in those reactions. It can be since beginning of reaction or can be formed throughout reaction resulting on lowering of selectivity of oxidation products (see Table 1). The hydration reaction of olefin double bond and others concurrent transformations are favored at higher amount of H_2O_2 aqueous solution. Nevertheless, a reduction on amount of hydrogen peroxide also reduces the conversion of camphene (Table 1). Indeed, the highest conversion and selectivity were reached with a H_2O_2/camphene molar ratio equal to 3:2.5 after 6h of reaction (run 5, Table 1).

In addition to oxidize the camphene, the role of the hydrogen peroxide is to reoxidize the palladium catalyst throughout reaction. Along the palladium-catalyzed reactions of camphene oxidation, was observed no metal precipitation due to formation of reduced palladium species; consequently, Pd(0) precipitation non occurred at the reaction end. Thus, it can be conclude that the hydrogen peroxide is stable in solution and efficiently maintain the palladium oxidized into medium of reaction, even during longer reaction times. The camphene can undergo a probably peroxypalladation of the double bond, which results in a π-camphene-palladium complex, that after the step of reductive elimination of HPdOOH generates finally the epoxy-camphene (**1a**) (Figure 4).

The formation of glycol camphene can be explained via two different pathways: at first, via hidroxypalladation (i) or alternatively, via a step of peroxypalladation (ii) of double bond of camphene. While the secondly option is considered, is possible that glycol-camphene (**1c**) can be formed from epoxy-camphene (**1a**).

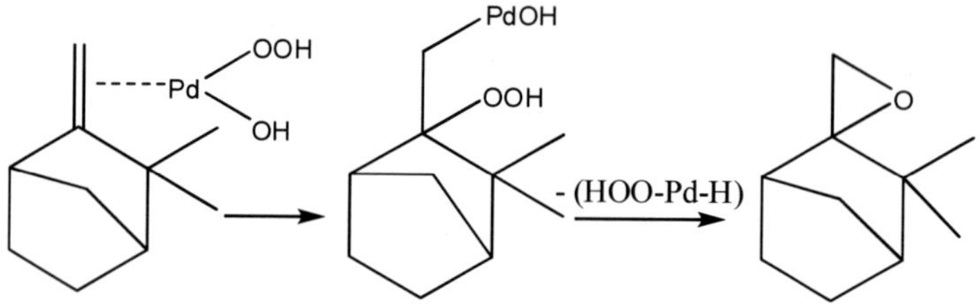

1a

Figure 4. Epoxydation of camphene via peroxypalladation step.

Although both water and hydrogen peroxide are weak nucleophile, the low selectivity for the formation of (**1a**) product indicates that possibly an epoxy ring opening could also occur in solution, which results in formation of glycol-camphene (**1b**). Since there is no hydrogen at β-carbon to palladium, a possible peroxypalladation step of olefin in a π- camphene-palladium complex followed by heterolysis of a carbon-palladium σ-bond, can then result on camphene-glycol (**1b**) [28].

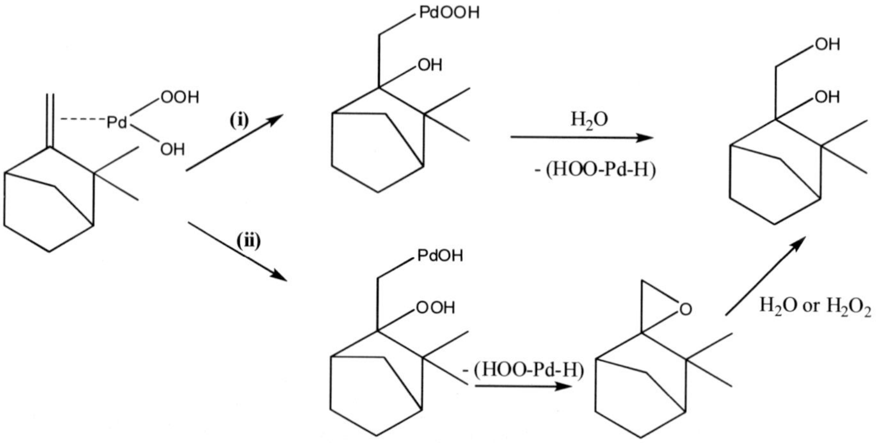

Figure 5. Oxidation of camphene into glycol-camphene via (i) hydroxypalladation or (ii) peroxypalladation steps [28]

Conversely, the formation of the aldehyde camphene (**1c**) can be explained by an uncommon anti-Markovnikov addition of the palladium to π-coordinate camphene (Figure 6).

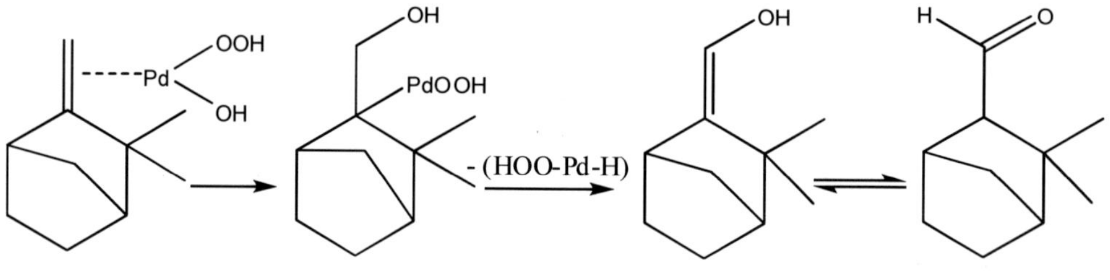

1c

Figure 6. Palladium-catalyzed camphene oxidation into aldehyde camphene (1c) [28]

Literature describes that in the palladium-catalyzed camphene deuteriation intermediates σ-vinyl-camphene-palladium are formed in $CDCl_3$ solutions [29,30]. Thus, is reasonable assumes that a probable pathway of formation of aldehyde camphene involves a similar intermediate. We suggested that probably, the β-hydride shift results in an unstable unsaturated enol, and this intermediate could then undergo a tautomerization process resulting at the end into carbonylic product (**1c**).

The reactivity of the β-pinene (**2**) was completely different from the camphene (**1**). β-pinene has two allylic hydrogens that should be easily shifted. This way, the allylic oxidation was the pathway more favourable for its conversion into oxidation products. In all catalytic runs the total conversion of β-pinene into allylic products occurred in great extension if compared to conversion of camphene. Therefore, a virtually complete conversion of (**2**) into oxidation products was observed after 6 h reaction at 60 °C (run 3, Table 2). The main products formed are pinocarveol (**2b**) and pinocarvone (**2c**) and in less extension epoxy-β-pinene (**2a**) (Figure 7).

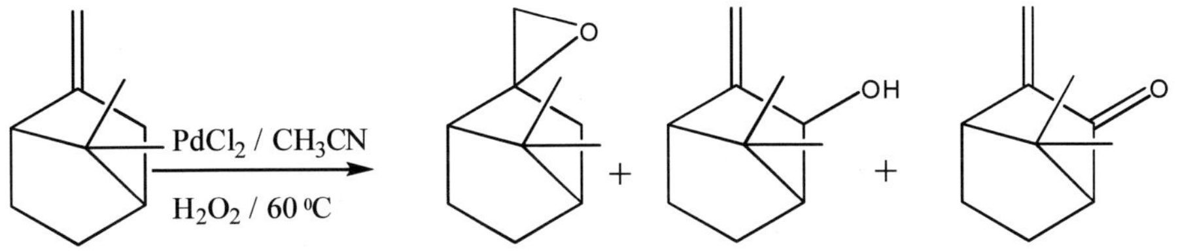

2 **2a** **2b** **2c**

Figure 7. $PdCl_2$-catalyzed oxidation products of β-pinene by H_2O_2

Under these reaction conditions, β-pinene was oxidized mainly into allylic products with a combined selectivity of approximately 60% and maximum conversion of *ca.* 58%. Although more selective than the $PdCl_2/H_2O_2/HOAc$ system, the $PdCl_2/H_2O_2/CH_3CN$ system required a longer reaction time (*ca.* 8 hours) for β-pinene oxidation. In HOAc solutions, the average reaction time was approximately 1-2 hours [31].

In general, oxidation reactions were highly selective and the conversions obtained were dependent on hydrogen peroxide concentration employed.

Table 2. $PdCl_2$-catalyzed oxidation of β-pinene by H_2O_2 in CH_3CN solutions [28][a]

Entry	H_2O_2 (mmol)	Conversion total (%)	Conversion into Oxidation products identified (%)	Selectivity oxidation products (%)		
				2a	2b	2c
1[b]	3.0	<10%	-			
2	1.5	38		-	58	32
3	3.0	58	52	19	57	24
4	6.0	54	47	5	55	45
5	9.0	55	49	4	47	49

[a]Reactions conditions: β-pinene (2.50 mmols); $PdCl_2$ (0.10 mmols); hydrogen peroxide; 8 h reaction, 60 °C.

Pinocarvone and pinocarveol were the main oxidation products formed. In addition, a small formation of numerous minoritary products of the skeletal rearrangement and hydration of double bond was detected by GC-MS analyses. When the hydrogen peroxide was employed in higher concentration, no significant increase on reaction conversion and oxidation products selectivity (see Table 2).

The $PdCl_2/CH_3CN/H_2O_2$ system was less effective on β-pinene oxidation than on the camphene oxidation, however, it was much more efficient than other catalytic systems described in literature (i.e. $PdCl_2/HOAC/H_2O_2$) [31].

As it can be observed, β-pinene oxidation proceeds via different way than camphene. Indeed, the nature of the reaction product depends strongly on the coordination mode, which, instead, is determined by the olefin structure.

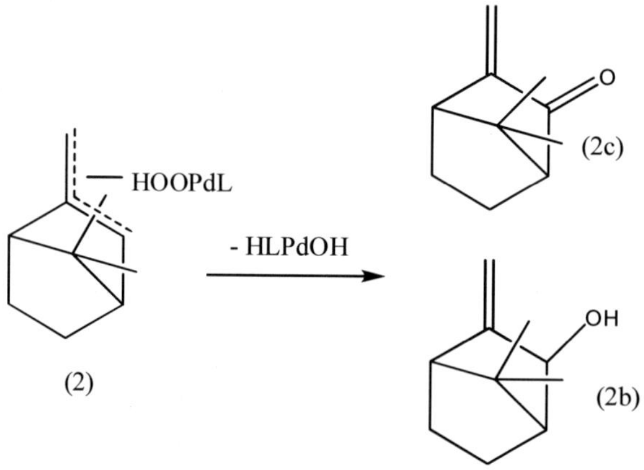

Figure 8. Palladium-catalyzed allylic oxidation of β-pinene [28]

In the camphene oxidation, an intermediate π-coordinate was suggested as probable active specie; herein, while in the β-pinene oxidation, a π-allylic intermediate seems be more reasonable. Palladium hydroperoxidic species (LPdOOH; L = H_2O, Cl), possibly obtained from the addition of H_2O_2 to $PdCl_2$, are probable active intermediate in this reaction [28].

In the 3-carene (3) oxidation, the selectivity control is main challenge to be overcomes. As 3-carene has six allylic hydrogens, which are easily shifted, a complex mixture of oxidation products can be obtained (Figure 9).

However, although numerous products can be obtained, the $PdCl_2$ catalyst was not efficient in promotes the 3-carene oxidation.

Table 3. $PdCl_2$-catalyzed oxidation of the 3-carene by H_2O_2 in CH_3CN solutions[a]

Entry	Substrate	Catalyst $PdCl_2$ (mmol)	Total conversion (%)
1	3-carene[c]	-	<10
2	3-carene	0.10	47

[a]Reactions conditions: substrates (5.0 mmol); hydrogen peroxide; (6.0 mmol); 60 °C, 10 h.

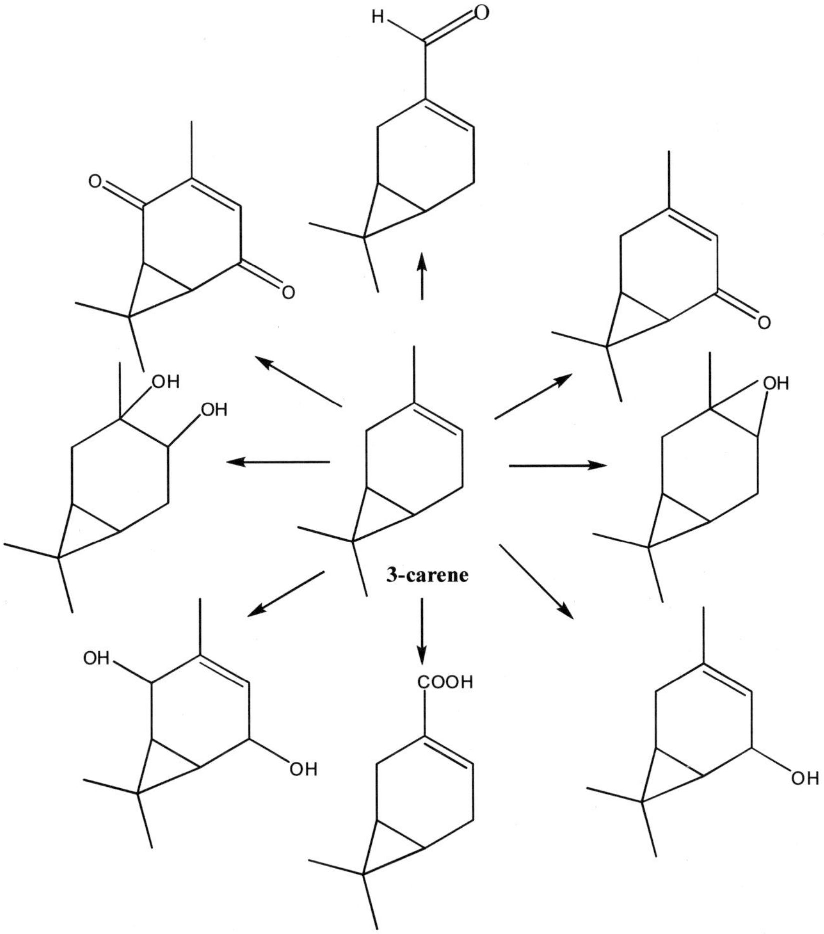

Figure 9. Possible oxidation products of 3-carene by hydrogen peroxide

Possibly, its coordination to double bond is hampered by the cicloропane ring, which becomes difficult its approximation of the substrate. Only poor conversion (*ca.* 47 %) was obtained in the $PdCl_2$-catalysed oxidation reactions (Table 3). In these reactions, no majority products were detected. However, GC-MS confirmed the presence of alcohols, carbonylic products, and epoxyde.

3.2. Recent Advances Obtained in Cobalt-Catalyzed Monoterpenes Oxidation Reactions

The developement of $CoCl_2$-catalyzed monoterpenes oxidation reactions has been performed using molecular oxygen as oxidant [32]. In this chapter, we described the advances obtained in reactions by using hydrogen peroxide as final oxidant in CH_3CN solutions.

Autoxidation of alkylbenzenes and alkanes by cobalt-catalyzed homolytic processes has been extensively studied because of their industrial importance [33]. Currently, autoxidation of olefins in the presence of cobalt complexes has attracted much less attention. In general, these reactions involve radical intermediates; a competition between the abstraction of the allylic hydrogen to give allylic oxidation products and the addition of the hydroperoxy radical to the double bond resulting in epoxide products is usually expected [33].

In this chapter, we wish to report here a comparative study on the $CoCl_2$-catalyzed oxidation reactions by hydrogen peroxide of camphene (**1**), β-pinene (**2**); 3-carene (**3**) in acetonitrile solutions.

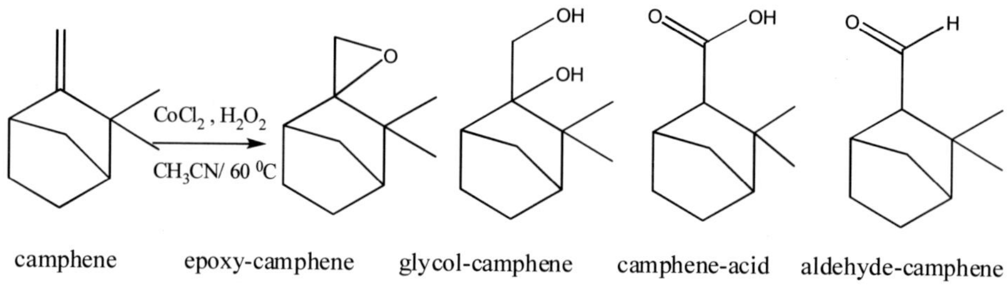

Figure 10. $CoCl_2$-catalyzed oxidation products of camphene by H_2O_2[a]

[a]Reaction conditions: camphene (5.0 mmols); $CoCl_2$ (0.2 mmols); H_2O_2 (5.0, 10.0, 15.0 and 20.0 mmols); temperature (60 °C); CH_3CN (15.0 mL).

Although high conversions (*ca.* 70-80%) of camphene have been reached by using molar ratio camphene/hydrogen peroxide equal to 1:4 and camphene/$CoCl_2$ equal to 25:1 and temperature of 60 °C, was found that camphene undergoes oligomerization throughout reaction in an extension of nearly 30-45% depending on hydrogen peroxide concentration employed. The main products identified by GC-MS analyses were probably formed via vinylic addition of radical hydroperoxide in the double bond of camphene.

Contrarily, the oxidation of β-pinene in presence of H_2O_2 and catalytic amounts of $CoCl_2$ was more selective than camphene oxidation.

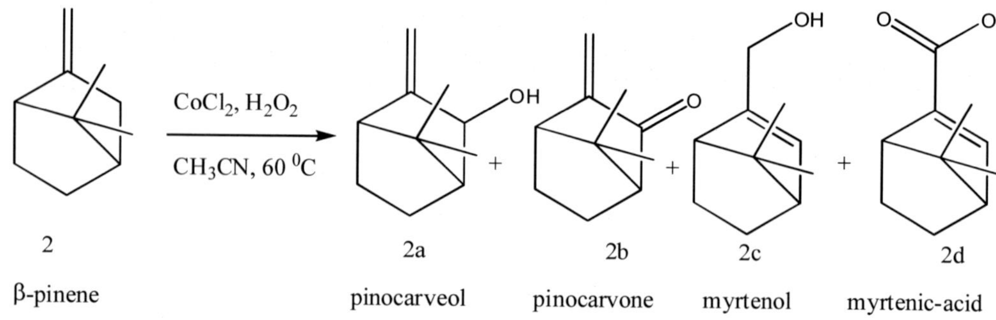

Figure 11. Oxidation products of β-pinene by H_2O_2 obtained via $CoCl_2$-catalyzed reactions.

The cobalt chloride catalyst activity in the oxidation reactions of β-pinene with hydrogen peroxide was assessed using different catalyst concentrations.

As can be observed, an increase on cobalt chloride catalyst concentration resulted in decrease on oligomers formation. Moreover, the oxidation selectivity was also increased. However, regardless catalyst concentration, pinocarveol was always the main product.

Table 4. $CoCl_2$-catalyzed oxidation of the β-pinene by H_2O_2 in CH_3CN solutions[a]

Exp.	$CoCl_2$ (mmols)	Conversion[b] (%)	Products distribution [c](%)			Oxidation selectivity [b](%)			
			PN	PI	PO	2a	2b	2c	2d
1[c]	-	5	1	2	2	-	-	-	-
2	0.1	23	9	3	11	44	16	22	18
3	0.2	49	23	4	22	49	19	19	13
4	0.4	98	31	5	62	53	20	17	10
5	0.6	99	27	2	70	53	23	14	10

[a]Reaction conditions: : β-pinene (5.0 mmols); $CoCl_2$ (0.2 mmols); H_2O_2 (20.0 mmols); temperature (60 °C); CH_3CN (15.0 mL).

[b]Determined by GC; PN = oligomers and non identified products; PI = isomers; PO = oxidation products

The $CoCl_2$-catalyzed oxidation of 3-carene by hydrogen peroxide resulted in a large amount of oxidation products (Figure 8). However, at temperature of 60 °C, 3-carene undergoes undesirable reactions of oligomerization (*ca.* 40-60% of substrate). Experiments aiming to assess this reaction under more mild conditions are in development in our Labs.

4. CONCLUSION

This work summarizes our recent developments for environmentally benign oxidation reactions based on hydrogen peroxide, catalyzed by $CoCl_2$ as well as $PdCl_2$. We have found both catalysts efficiently promoted the oxidation of renewable raw material and large occurrence natural: monoterpenes. The oxidation products of β-pinene, camphene and 3-carene are valuable ingredients for several industries, and for this reason they were the substrates selected. The activity of the $CoCl_2$ as well as $PdCl_2$ catalysts was assessed in reactions performed in acetonitrile solutions and mechanistic aspects were discussed from resulted obtained. The effects of the substrate structure and of catalyst type on conversion and reaction selectivity were investigated. It was found that the reactions of oxidation by H_2O_2 in presence of palladium catalysts were more efficient than $CoCl_2$ due to mechanism involved. The $CoCl_2$-catalyzed reactions occur via radicalar processes (homolytic processes) whereas $PdCl_2$-catalyzed reactions which occur via heterolysis of C-H bonds.

Finally, studying the oxidative processes, we have found that /H_2O_2 system was also highly effective in oxidation reactions of camphene and β-pinene comparatively to result described in the literature. Camphene and β-pinene were selectively oxidized into epoxy and allylic products with high conversion rates. Hydrogen peroxide, an environmentally benign reactant was successfully employed as oxidant in the monoterpenes oxidation reactions regardless catalyst employed.

Thus, it can be concluded that although yet non-finished, present methodology based on palladium/hydrogen peroxide combination offers several advantages such as high yields besides high selectivity for products desirable. However, improvements in cobalt-catalyzed oxidation reactions are yet necessary and deserve a continuous research work. The development of homogeneous catalyst combined with clean oxidant and inexpensive like as hydrogen peroxide is a start point for processes environmentally benign, which can be used in larger extension.

ACKNOWLEDGMENTS

The Federal University of Viçosa and Arthur Bernardes Foundation are warmly thanked for financial support. CAPES, CNPq and FAPEMIG deserves our special thanks.

REFERENCES

[1] Murphy, E. F., Mallat, T., A. Baiker, A., *Catal Today*, 2000, *vol 57*, 115-126.
[2] Torborga, C. Beller, M. *Adv Synth Catal* 2009 *vol.*351, 3027-3043.
[3] Gallezoti, P. *Catal Today* 2007 *vol.121*, 76-91.
[4] Gavrilov, K. N., Benetsky, E. B., Grishina, T. B., Tatiana, B., Zheglov, S. V, Rastorguev, E. A., Petrovskii, P. V., Davankov, V. A., Macaev, F. Z., *Tetrahedron-Asymmetr* 2007, *vol.18*, 2557-2564.
[5] Robles-Dutenhefner, P. A., Rocha, K. A. S., Sousa, E. M. B., Gusevskaya, E. V., *J Catal* 2009, *vol.265*, 72-79.
[6] Anastas, P. T., Bartlett, L. B., Kirchhoff, M. M., Williamson T. C., *Catal Today,* 2000, *vol.*55, 11-22.
[7] Campos-Martin, J. M., Blanco-Brieva, G. Fierro, J. L. G., *Angew Chem Int Ed* 2006, *vol.*45, 6962-6984.
[8] Arends, I. W. C. E., Sheldon R.A., *Topics in Catal* 2002, *vol.19*, 133-141.
[9] Collins, T. J., *Science* 2001, *vol.291*, 48-49.
[10] Cornell, C.N. Sigman, M.S. *Inorg Chem.* 2007, *vol.46*, 1903-1909.
[11] Silva, A. D., Patitucci, M. L., Bizzo, H. R., D'Elia, E., Antunes, O. A. C., *Catal Commun* 2002, *vol. 3*, 435-440.
[12] da Silva, M. J., Gusevskaya, E. V., *J Mol Catal A* 2001, *vol.176*, 23-27.
[13] Simões, M. M. Q., Silva, A. M. S., Tomé, A. C., Cavaleiro, J. A. S., Tagliatesta, P., Crestini, C., *J Mol Catal A* 2001, *vol.172*, 33-37.
[14] Santos, I. C. M. S., Simões, M. M. Q., Pereira, M. M. M. S., Martins, R. R. L., Neves, M. G. P. M. S., Cavaleiro, J. A. S., Cavaleiro, A.M.V., *J. Mol. Catal. A* 2003, *vol.195*, 253-257.
[15] Gomes, M. F. T., Antunes, O. A. C., *J Mol Catal A* 1997, *vol.121*, 145-155.
[16] Gusevskaya, E. V., Gonçalves, J. A., *J. Mol. Catal. A* 1997, *vol.121* ,131-137.
[17] Allal, B. A., El Firdoussi L., Allaoud, S., Karim, A., Castanet, Y., Mortreux, A., *J. Mol. Catal. A* 2003, *200*, 177-182.
[18] Kislenko, V. N., Berlin, A. A., *Russian Chem Rev* 1991, *vol. 60,* 949-981.
[19] Roussel, M. Mimoun, H., *J. Org. Chem.* 1980, *45*, 5387-5390.
[20] Jia, C., Miller, P., Mimoun, H., *J Mol Catal A* 1995, *vol. 101*, 127-136.
[21] Muzart, J., *J. Mol. Catal. A,* 2007, *vol. 276*, 62-72.
[22] da Silva, M. J., Gusevskaya, E. V., *J. Braz. Chem. Soc.* 2003, *vol. 14*, 83-89.

[23] Gusevskaya, E. V., Robles-Dutenhefner, P. A., Ferreira, V. M. S. *Appl Catal A* 1998, *vol. 174,* 177-186.

[24] Menini, L., da Silva, M. J., Lelis, M. F. F., Fabris, J. D., Lago, R. M., Gusevskaya, E. V., Appl Catal A, 2004, *vol. 269,* 117-121.

[25] Jin, A. H., Li, B. S., Dai, Z. J., *Petroleum Science Technol,* 2010, *vol.* 28, 700-711.

[26] Hao, L, Shen, B. X., Zhou, X. L., *Pet Sci Technol* 2005, *vol. 23,* 991-999.

[27] Carari, D. M., da Silva, M. J., *Catal Lett,* 2012, vol. 142, 251-258.

[28] de Oliveira, A. A., da Silva, M. L., da Silva, M. J., *Catal Lett* 2009, *vol.130,* 424-431.

[29] da Silva M. J., Gonçalves J. A., Alves R. B., Howarth O. W., Gusevskaya E. V. *J Organomet Chem* 2004, *vol. 689,* 302-308.

[30] da Silva, M. J., Gonçalves, J.A., Howarth O. W., Gusevskaya, E. V., Pilo-Veloso, D. *J Organomet Chem* 2005, *vol. 690,* 2996-3003.

[31] Gusevskaya, E. V., Robles-Dutenhefner, P. A., Ferreira, V.M.S. *Appl Cata A* 1998, *vol.132,* 213-217.

[32] da Silva, M. J., Robles-Dutenhefner, P. A., Menini, L., Gusevskaya, E. V., *J Mol Catal A*, 2003, vol. 201, 71-77.

[33] Sheldon, R. A., Kochi, J. K., Metal-Catalyzed Oxidations of Organic Compounds, Academic Press, London, 1981.

INDEX

B

C

D

E

F

G

H

I

J

K

L

M

N

O

P

Q

R

S

T

U

V

W

X

Y

Z